Instructor's Manual to Accompany

Chemical Principles
With Qualitative Analysis

SIXTH EDITION

Masterton/Slowinski/Stanitski

William L. Masterton
University of Connecticut
Storrs, Connecticut

Saunders Golden Sunburst Series
SAUNDERS COLLEGE PUBLISHING
Philadelphia New York Chicago
San Francisco Montreal Toronto
London Sydney Tokyo Mexico City
Rio de Janeiro Madrid

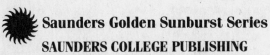

Address orders to:
383 Madison Avenue
New York, NY 10017

Address editorial correspondence to:
West Washington Square
Philadelphia, PA 19105

Instructor's Manual to accompany CHEMICAL PRINCIPLES WITH QUALITATIVE ANALYSIS, 6/e

ISBN # 0-03-008589-6

5678 066 987654321

CBS COLLEGE PUBLISHING
Saunders College Publishing
Holt, Rinehart and Winston
The Dryden Press

PREFACE

This manual is divided into two parts. The first part deals with Chapters 1 - 28. For each chapter in this part there is:

1. A list of "Basic Skills", otherwise known as learning objectives, etc., etc. These represent the skills we expect our students to acquire. You may wish to modify these, delete some, and add others.

2. "Lecture Notes" which indicate the amount of time that we devote to each chapter and the topics that we emphasize. Included are detailed lecture outlines similar to those used by one of us (WLM). These may serve as a guide for your own lectures. At a minimum, they indicate how we cover topics and how we integrate successive topics.

3. A list of "Demonstrations" illustrating topics covered in the chapter. These are taken from two sources:
-the manual "Tested Demonstrations" (6th edition) published by the Journal of Chemical Education in 1965.
-demonstrations described in the Journal of Chemical Education between 1965 and 1984.

4. "Quizzes" suitable for use in discussion sections. Each quiz is designed for 10-15 minutes. There are five quizzes for each chapter. The questions included could be used for examinations if you prefer. Note that there is also a separate testbank for the 6th edition of "Chemical Principles" which can be obtained from the publisher. The questions there are all of the multiple choice type.

5. Answers and detailed solutions to "Problems 1 - 30" in each text chapter. Note that answers to Problems 31-65 are given at the back of the text. Detailed solutions of Problems 31-65 are available in a separate solutions manual, which may be purchased by students.

The second part of this Instructor's Manual deals with Chapters 29 - 37 (Qualitative Analysis). It includes:

1. Detailed solutions to all the problems in these chapters, covering inorganic qualitative analysis.

2. Test questions for each of these chapters. Each set of test questions comprises ten multiple choice questions with five short-answer problems. Answers to all test questions are provided at the back of the manual.

CHAPTER 1

BASIC SKILLS

1. Apply the rules of significant figures to calculations based upon experimental measurements.
2. Convert between Fahrenheit, Celsius and Kelvin temperatures using the relations: $^{o}F = 1.8(^{o}C) + 32^{o}$ and $K = {}^{o}C + 273.15$
3. Use the conversion factor approach to convert lengths, volumes, masses or other quantities from one unit to another.
4. Use the equation: density = mass/volume to calculate one of these three quantities, given (or having calculated) the other two quantities.
5. Use solubility data (Figure 1.6) to determine
   a. the mass of solute that dissolves or remains in solution in a given mass of solvent at a specified temperature.
   b. the mass of solvent required to dissolve a given mass of solute at a specified temperature.
6. Use the relation $q = (\text{specific heat}) \times m \times \Delta t$ to calculate one of these four quantities, given (or having calculated) the other three quantities.
7. Differentiate between an element and a compound; between a solution and a heterogeneous mixture.
8. Describe the relationship between the wavelength of light absorbed by a species and its color.
9. Describe the principle(s) between the following separation techniques and the kinds of mixtures to which they can be applied: distillation, fractional distillation, fractional crystallization, chromatography.

LECTURE NOTES

This material ordinarily requires 2 lectures (100 min), allowing for a 10-15 min introduction to the course in the first lecture. If you're in a hurry, this can be cut to $1\frac{1}{2}$ lectures by discussing only quantitative material: significant figures, unit conversions, density, solubility, and specific heat.

A few points to keep in mind:

- nearly all of your students should be familiar with the metric system and with the Periodic Table; you should not have to spend much time on these topics. Note that the Periodic Table will be discussed later (Chapter 8) in much greater detail.

- students readily learn the rules of significant figures but typically ignore them after Chapter 1. It may help to emphasize that these are common-sense rules for estimating experimental error.

- many students (typically the weaker ones) stubbornly resist using conversion factors, preferring instead a rote method with which they became infected in high school. It may be useful to point out that conversion factors will be used throughout the course; it will help if student, instructor and textbook author speak the same language.

- density problems are solved here by a conversion factor approach (see Example 1.7b, c). You may prefer to substitute directly into the defining equation for density.

- specific heat is introduced here to lay the foundation for thermochemistry (Chapter 5). Keep the discussion elementary at this point; heat flow and calorimetry will be considered in much greater detail in Chapter 5.

LECTURE 1

I Measured Quantities

A. Length: base unit is the meter. $1\ km = 10^3\ m$; $1\ cm = 10^{-2}\ m$; $1\ mm = 10^{-3}\ m$; $1\ nm = 10^{-9}\ m$
Dimensions of very tiny particles (e.g., atoms) will be expressed in nanometers.

B. Volume Note that: $1\ L = 1000\ mL = 1000\ cm^3$
Comment briefly on use of pipet and volumetric flask.

C. Mass $1\ kg = 10^3\ g$; $1\ mg = 10^{-3}\ g$
Point out that two different kinds of balances will be used in lab. Analytical balance ($\pm$ 0.001 g) should be used only for accurate, quantitative work.

D. Temperature $^oF = 1.8(^oC) + 32^o$; $K = {}^oC + 273$
Convert 65^oF to ^oC and K:

$$^oC = (65 - 32)/1.8 = 18^oC;\quad K = 18 + 273 = 291\ K$$

II Experimental Error and Significant Figures

Suppose object is weighed on crude balance to $\pm$ 0.1 g and mass is found to be 23.6 g. This quantity contains 3 "significant figures, i.e., 3 experimentally meaningful digits. With an analytical balance, mass might be 23.582 g (5 sig. fig.).

A. Counting significant figures
1. Volume of liquid = 24.0 cm^3. Three significant figures. Zeroes at end of measured quantity, following nonzero digits, are significant.
2. Volume = 0.0240 L. Three significant figures (note that 0.0240 L = 24.0 cm^3). Zeroes at beginning of a measured quantity, preceding nonzero digits, are not significant.
3. Certain numbers are exact: 1 L = 1000 cm^3; 1 in = 2.54cm

B. Multiplication and division Keep only as many significant figures as there are in the least precise quantity. Density of piece of metal weighing 36.123 g with volume of 13.4 cm^3?

$$\text{density} = \frac{36.123\ g}{13.4\ cm^3} = 2.70\ g/cm^3$$

(Note that calculator gives 2.6957463; extra digits meaningless)

C. Addition and subtraction Keep only as many digits after the decimal point as there are in the least precise quantity. Add 1.223 g of sugar to 154.5 g of coffee:

Total mass = 1.2 g + 154.5 g = 155.7 g

Note that rule for addition and subtraction does not relate to significant figures. Number of significant figures often decreases upon subtraction:

Mass beaker + sample	=	52.169 g	; 5 sig. fig.
Mass empty beaker	=	52.120 g	; 5 sig. fig.
Mass sample		0.049 g	; 2 sig. fig.

III Conversion Factors

Will be used again and again in this course to carry out chemical conversions.

A. Simple, one-step conversions

1. A rainbow trout is measured to be 16.2 in long. Length in centimeters?

$$\text{length in cm} = 16.2\ \text{in} \times \frac{2.54\ \text{cm}}{1\ \text{in}} = 41.1\ \text{cm}$$

Note cancellation of units. To convert from centimeters to inches, would use the conversion factor 1 in/2.54 cm.

2. Barometric pressure reported on Canadian TV station as 99.6 kPa. Express in millimeters of mercury (101.3 kPa = 760 mm Hg).

$$\text{pressure (mm Hg)} = 99.6\ \text{kPa} \times \frac{760\ \text{mm Hg}}{101.3\ \text{kPa}} = 747\ \text{mm Hg}$$

Method is particularly useful with unfamiliar units.

B. Multiple conversion factors Baseball thrown at rate of 89.6 miles per hour. Speed in meters per second?

1 mile = 1.609 km = 1.609×10^3 m; 1 h = 3.6×10^3 s

$$\text{Speed} = \frac{89.6\ \text{mile}}{1\ \text{hr}} \times \frac{1.609 \times 10^3\ \text{m}}{1\ \text{mile}} \times \frac{1\ \text{h}}{3.6 \times 10^3\ \text{s}} = 40.0\ \frac{\text{m}}{\text{s}}$$

LECTURE 2

I Types of Substances

A. Elements: cannot be broken down into two or more simpler substances. Examples: nitrogen, lead, silver, sodium. Symbols: N, Pb, Ag, Na

Periodic Table: periods (horizontal rows) and groups (vertical columns of chemically similar elements). Learn to locate elements in Table.

B. Compounds: contain two or more elements with fixed mass percents.

Sodium chloride: 39.34% Na, 60.66% Cl
Glucose: 40.00% C, 6.71 % H, 53.29% O

II Properties of Substances

Used to identify a substance by comparing to properties of known substances. A colorless liquid which boils at 100°C, freezes at 0°C, has a density of 1.00 g/cm^3, and decomposes upon electrolysis to hydrogen and oxygen, must be water. Distinguish between chemical properties and physical properties, which can be observed without changing the chemical identity of the substance.

A. Density: D = m/V
Readily measured for liquid; more difficult with solid. Add 10.0 g of table salt to flask with volume of 24.2 cm^3. Find that 17.2 g of benzene (d = 0.879 g/cm^3) must be added to fill the flask. Density of salt? Draw diagram of apparatus.

$$\text{Volume of benzene} = 17.2\text{ g} \times \frac{1\text{ cm}^3}{0.879\text{ g}} = 19.6\text{ cm}^3$$

$$\text{Volume of salt} = 24.2\text{ cm}^3 - 19.6\text{ cm}^3 = 4.6\text{ cm}^3$$

$$\text{Density of salt} = \frac{10.0\text{ g}}{4.6\text{ cm}^3} = 2.2\text{ g/cm}^3$$

B. Solubility Often expressed as grams of solute per 100 g solvent. Example:

Temperature	10°C	100°C
Soly. lead nitrate (g/100 g water)	50	140

1. How much water required to dissolve 80 g lead nitrate at 100°C?

   At 100°C, 140 g lead nitrate $\hat{=}$ 100 g water

   $$\text{Mass water} = 80\text{ g lead nitrate} \times \frac{100\text{ g water}}{140\text{ g lead nitrate}}$$
   $$= 57\text{ g water}$$

2. Cool to 10°C. How much lead nitrate remains in solution?

   $$\text{Mass lead nitrate} = 57\text{ g water} \times \frac{50\text{ g lead nitrate}}{100\text{ g water}}$$
   $$= 28\text{ g lead nitrate}$$

   80 g - 28 g = 52 g lead nitrate crystallizes from soln.

C. Specific heat Amount of heat required to raise the temperature of one gram of a substance by one degree Celsius.

$$q = (\text{specific heat}) \times m \times \Delta t \;;\; \Delta t = t_{final} - t_{initial}$$

specific heat = $q/m \times \Delta t$; has units of J/g·°C
specific heat water = 4.18 J/g·°C
It is found that 6710 J of heat must be absorbed to raise the temperature of 525 g of iron from 25.00 to 51.85°C. Specific heat of iron?

$$\text{specific heat Fe} = \frac{6710\text{ J}}{525\text{ g} \times 26.85^\circ\text{C}} = 0.476\text{ J/g}\cdot^\circ\text{C}$$

D Color is due to absorption of visible light (400-700 nm). Bromine absorbs in violet and blue regions (400-500 nm); transmits red light, so has a red color.

III Separation of Mixtures

A. Distillation: separation of liquid from solid.
Fractional distillation: depends upon difference in volatility between liquids. Used in petroleum industry to separate fractions in different boiling point ranges.

B. Fractional crystallization: separation of solids (A and B)
1. Procedure. Dissolve in minimum amount of hot solvent. Cool to room temperature or below. Hope that pure A crystallizes out; all of B and some A stay in solution.
2. Requirements. A must be less soluble at low T (recall lead nitrate example). Fraction of B must be relatively small.

C. Chromatography - can be applied to gases, liquids or solids.
1. Paper chromatography - draw diagram, show separation of components.
2. VPC Mention analytical applications at parts per million level.

DEMONSTRATIONS

1. Chromatography: Test. Dem. 119, 137; J. Chem. Educ. 59 1042 (1982)

QUIZZES

Quiz 1

1. The density of a certain liquid is 69.4 lb/ft^3. Express this in grams per cubic centimeter, given:

   1 lb = 453.6 g; 1 ft = 30.48 cm

2. The solubility of potassium nitrate at 100°C is 240 g/100 g water.
   a. How much water is required to dissolve 112 g of potassium nitrate at 100°C?
   b. How much potassium nitrate can be dissolved in 56.2 g of water at 100°C?

Quiz 2

1. Assume your car has a fuel economy of 28.2 miles per gallon. Express this in kilometers per liter, given:

   1 mile = 1.609 km; 1 gal = 4 qt; 1 L = 1.057 qt

2. The density of methyl alcohol is 0.787 g/cm^3.
   a. What is the mass of 25.0 cm^3 of methyl alcohol?
   b. What volume of benzene (d = 0.879 g/cm^3) has the same mass as the methyl alcohol in (a)?

Quiz 3

1. An atom of silicon has a radius of 0.117 nm. Calculate its volume ($V = 4\pi r^3/3$) in:
   a. cubic nanometers b. cubic inches (1 in = 2.54 cm; 1 m = 10^9 nm).

2. How much heat must be absorbed to raise the temperature of 12.0 g of water (specific heat = 4.18 J/g·°C) from 16.00 to 23.24°C?

Quiz 4

1. A helium atom at room temperature is moving at an average speed of about 1.36×10^5 cm/s. Express this in miles per hour.

   1 mile = 1.609 km; 1 h = 3600 s

2. It is found that 9.00 cal is required to raise the temperature of a 36.0 g sample of metal by 2.00°C. Calculate the specific heat of the metal in J/g·°C (1 cal = 4.184 J).

Quiz 5

1. The solubility of sodium chloride at room temperature is about 35 g/100 g water. Express the solubility in pounds of sodium chloride per liter of water.

   1 lb = 453.6 g; density water = 1.00 g/cm^3

2. Using the solubility quoted in (a), calculate
   a. the mass of sodium chloride that will dissolve in one kilogram of water.
   b. the mass of water required to dissolve 1.00 g of sodium chloride.

PROBLEMS 1-30

1.1 a. length b. mass c. energy d. volume e. density f. pressure g. energy

1.2 a. 3.12 g b. $0.050\ m^3 = 5.0 \times 10^4\ cm^3$ c. 2.69 g/cm^3 d. 16.0 J

1.3 a. $°F = 1.8(-246°) + 32° = -443° + 32° = -411°F$

b. K = -246 + 273 = 27 K

1.4 $R = \frac{PV}{nT} = \frac{\text{mm Hg}\cdot\text{mL}}{\text{mol}\cdot\text{K}}$

1.5 a. 4 b. 3 c. 3 d. 3 e. 5 f. 2 or 3 (can't tell)

1.6 51.923 g/519 cm^3 = 0.100 g/cm^3 (3 sig. fig.)

1.7 a. 0.25 g/cm^3 (2 sig. fig.) b. 2.7 g/L (2 sig. fig.)

c. 53 g (± 1 g) d. 5.18 g/6.149 cm^3 = 0.842 g/cm^3 (3 s.f.)

1.8 a. $2.10\ ft^3 \times \frac{28.32\ L}{1\ ft^3} = 59.5\ L$

b. $2.10\ ft^3 \times \frac{28.32\ L}{1\ ft^3} \times \frac{1\ m^3}{10^3\ L} = 0.0595\ m^3$

c. $2.10\ ft^3 \times \frac{28.32\ L}{1\ ft^3} \times \frac{1.057\ qt}{1\ L} = 62.9\ qt$

1.9 a. $52.0\ \text{mm Hg} \times \frac{1\ atm}{760\ \text{mm Hg}} = 0.0684\ atm$

1.9 b. $52.0 \text{ mm Hg} \times \frac{1.013 \times 10^5 \text{ Pa}}{760 \text{ mm Hg}} = 6.93 \times 10^3 \text{ Pa}$

c. $52.0 \text{ mm Hg} \times \frac{14.70 \text{ lb/in}^2}{760 \text{ mm Hg}} = 1.01 \text{ lb/in}^2$

1.10 $\frac{2.0 \times 10^4}{\text{s}} \times \frac{60 \text{ s}}{1 \text{ min}} \times \frac{60 \text{ min}}{1 \text{ h}} \times \frac{24 \text{ h}}{1 \text{ d}} \times \frac{365 \text{ d}}{1 \text{ yr}} = 6.3 \times 10^{11}/\text{yr}$

1.11 $\frac{100.0 \text{ yd}}{9.08 \text{ s}} \times \frac{3 \text{ ft}}{1 \text{ yd}} \times \frac{1 \text{ mile}}{5280 \text{ ft}} \times \frac{60 \text{ s}}{1 \text{ min}} \times \frac{60 \text{ min}}{1 \text{ h}} = 22.5 \text{ mile/h}$

1.12 a. $6.51 \text{ lb} \times \frac{12 \text{ oz}}{1 \text{ lb}} \times \frac{8 \text{ drachm}}{1 \text{ oz}} = 625 \text{ drachm}$

b. $15.0 \text{ drachm} \times \frac{1 \text{ oz}}{8 \text{ drachm}} \times \frac{1 \text{ lb}}{12 \text{ oz}} = 0.156 \text{ lb}$

$15.0 \text{ drachm} \times \frac{1 \text{ oz}}{8 \text{ drachm}} = 1.88 \text{ oz}$

1.13 $12.4 \text{ g X} \times \frac{100 \text{ g mixture}}{93.2 \text{ g X}} = 13.3 \text{ g mixture}$

1.14 a. Am, Ar, As, Es, Er, He, H, In, I, Ir, Mn, Ne, Np, Ni, N, P, Pr, Pm, Pa, Ra, Rn, Re, Rh, Sm, Se, Si, Na, Sr, S, Sn, W

c. Ac, Al, As, Ca, C, Ce, Cs, Cl, Es, He, H, La, Sc, Se, S, W

1.15 a. 2 b. 8 c. 8 d. 18 e. 18

1.16 a. manganese b. sodium c. arsenic d. tungsten e. phosphorus

1.17 a. physical b. chemical c. physical d. physical

1.18 $2.62 \text{ g}/1.10 \text{ mL} = 2.38 \text{ g/mL}$

1.19 a. volume water $= 19.6 \text{ g} \times \frac{1 \text{ cm}^3}{1.00 \text{ g}} = 19.6 \text{ cm}^3$

volume metal $= 24.5 \text{ cm}^3 - 19.6 \text{ cm}^3 = 4.9 \text{ cm}^3$

b. density metal $= 16.52 \text{ g}/4.9 \text{ cm}^3 = 3.4 \text{ g/cm}^3$

1.20 $3.1 \text{ L} \times \frac{10^3 \text{ cm}^3}{1 \text{ L}} \times \frac{1.020 \text{ g}}{1 \text{ cm}^3} \times \frac{1 \text{ lb}}{453.6 \text{ g}} = 7.0 \text{ lb}$

1.21 $60.0 \text{ kg } O_2 \times \frac{10^3 \text{ g}}{1 \text{ kg}} \times \frac{1 \text{ L } O_2}{1.31 \text{ g}} \times \frac{100 \text{ L air}}{21 \text{ L } O_2} = 2.2 \times 10^5 \text{ L}$

1.22 a. $18.4 \text{ g water} \times \frac{37.0 \text{ g KCl}}{100 \text{ g water}} = 6.81 \text{ g KCl}$

b. $18.4 \text{ g KCl} \times \frac{100 \text{ g water}}{37.0 \text{ g KCl}} = 49.7 \text{ g water}$

1.23 a. approx. 150 g KNO_3/100 g water

b. $50.0 \text{ g } KNO_3 \times \frac{100 \text{ g water}}{150 \text{ g } KNO_3} = 33.3 \text{ g water}$

c. $1.00 \text{ L water} \times \frac{10^3 \text{ g water}}{1 \text{ L water}} \times \frac{150 \text{ g } KNO_3}{100 \text{ g water}} = 1.50 \times 10^3 \text{ g } KNO_3$

1.24 $q = 0.902 \frac{\text{J}}{\text{g} \cdot {}^\circ\text{C}} \times 3.50 \text{ g} \times 2.6^\circ\text{C} = 8.2 \text{ J}$

1.25 $\frac{209\ J}{44.5\ g \times 10.9^{\circ}C} = 0.431\ J/g\cdot{}^{\circ}C$

1.26 a. ultraviolet b. green c. orange d. infrared

1.27 a. filtration b. distillation c. fractional distillation
d. decantation (floats on top of water)

1.28 a. solubility KNO_3 at $10^{\circ}C$ is about 20 g KNO_3/100 g water

KNO_3 in solution = $50.0\ g\ water \times \frac{20\ g\ KNO_3}{100\ g\ water} = 10\ g\ KNO_3$

KNO_3 that crystallizes = 88 g - 10 g = 78 g

b. solubility NaCl at $10^{\circ}C$ is about 35 g NaCl/100 g water

NaCl that could be in solution = $50.0\ g\ water \times \frac{35\ g\ NaCl}{100\ g\ water}$

= 18 g NaCl

No NaCl crystallizes out

c. $\frac{10\ g}{88\ g} \times 100 = 11\%$

1.29 a. a compound contains more than one element; can be decomposed into elements
b. a mixture contains more than one substance (element or compound); can be separated into pure substances by physical means
c. solution is uniform in appearance
d. distillation is used to separate a liquid from a solid; fractional distillation is used to separate two or more volatile liquids

1.30 a, d

CHAPTER 2

BASIC SKILLS

1. State the Laws of Conservation of Mass, Constant Composition, and Multiple Proportions and indicate how they are related to the atomic theory.
2. Identify the three types of particles making up atoms; give the relative charges and masses of these particles.
3. Relate a nuclear symbol to the number of protons and neutrons in the nucleus.
4. Given the masses and abundances of the isotopes of an element, calculate its atomic mass. Given the atomic mass of an element and the masses of its two isotopes, calculate their abundances.
5. Given either the number of atoms of an element or its mass in grams, calculate the other quantity.
6. Given the formula of a species (and a table of atomic masses), calculate its molar mass.
7. Relate the number of moles of a species of known formula to the number of grams or the number of particles.
8. Given or having calculated two of the three quantities: molarity, number of moles, number of liters, determine the other quantity.

LECTURE NOTES

This chapter requires about two 50 minute lectures. If your students have a good background from high school, you may be able to skim Sections 2.1 - 2.3, saving perhaps half a lecture. Conversely, you could easily add a half lecture by discussing Section 2.7 on the experimental determination of atomic masses.

Some general observations:

- the Law of Multiple Proportions is difficult for students to grasp. It is primarily of historical importance and you may wish to skip it.

- molecules and ions are introduced here at a very elementary level. They will be considered in much greater detail in Chapter 9. Formulas of ionic compounds will be discussed in Chapter 3.

- terms which are <u>not</u> discussed in this chapter include atomic mass unit, molecular mass, and gram molecular mass. Only three "masses" are discussed: atomic mass, formula mass, and molar mass. In our opinion, that's enough.

- students have little trouble calculating atomic masses from isotopic data. The reverse process, estimating isotopic abundances from atomic masses, is much more difficult because it involves solving an algebraic equation. Many students turn white when they see "x" in an equation.

- Section 2.5 is the heart of the chapter. Mole-gram conversions will be required in many later chapters, often as the first step in a more complex problem.

- molarity is introduced here because it is likely to be used in the lab prior to the discussion of solutions in Chapter 12. At this point, we do little more than define the term; it will be covered in much greater detail in Chapter 12.

LECTURE 1

I Atomic Theory

A. Postulates. Elements consist of tiny particles called atoms, which retain their identity in reactions. In a compound, atoms of two or more elements are combined in a fixed ratio of small whole numbers (e.g., 1:1, 2:1, 3:2, - -).

B. Relation to natural laws.
1. Conservation of Mass: atoms neither created nor destroyed
2. Constant Composition: atom ratio is fixed, so mass ratio must be constant.
3. Multiple Proportions. Applies where two elements, A and B, form more than one compound (e.g., AB, AB_2). States that masses of B that combine with a fixed mass of A are in a simple, whole-number ratio such as 2:1.

II Structure of Atoms

A. Components

	relative mass	relative charge	location
proton	1	+1	nucleus
neutron	1	0	nucleus
electron	0.0005	-1	outside nucleus

B. Atomic number = number of protons in nucleus = number of electrons in neutral atom. Characteristic of a particular element: all H atoms have 1 proton, all He atoms have 2 protons, etc. Atomic number is given in Periodic Table.

C. Mass number = number of protons + number of neutrons. Atoms of same element can differ in mass number.

	protons	neutrons	at.no.	mass no.	nuclear symbol
carbon-12	6	6	6	12	${}^{12}_{6}C$
carbon-13	6	7	6	13	${}^{13}_{6}C$

${}^{12}_{6}C$, ${}^{13}_{6}C$ are referred to as isotopes

III Molecules and Ions

A. Molecule - small group of atoms held together by chemical bonds. Examples: H_2, H_2O, $C_{12}H_{22}O_{11}$. Subscripts in the molecular formula indicate the number of atoms of each type per molecule.

B. Ion = charged particle formed by gain or loss of electrons

$$F \text{ atom } (9\ p^+, 9\ e^-) + e^- \longrightarrow F^- \text{ ion } (9\ p^+, 10\ e^-)$$

$$Na \text{ atom } (11\ p^+, 11\ e^-) \longrightarrow Na^+ \text{ ion } (11\ p^+, 10\ e^-) + e^-$$

Note that number of protons is unchanged. Ionic compounds contain + ions (cations) and - ions (anions) in a ratio that maintains electrical neutrality (e.g., 1 Na^+: 1 F^-).

IV Masses of Atoms

A. Atomic mass (given in Periodic Table)

1. Meaning - gives relative mass of atom. Based on C-12 scale; most common isotope of carbon is assigned an atomic mass of exactly 12

element	B	Ca	Ni
atomic mass	10.81	40.08	58.69

A nickel atom is 58.69/40.08 = 1.464 times as heavy as a calcium atom. It is 58.69/10.81 = 5.429 times as heavy as a boron atom.

2. Calculation from isotopic masses and abundances

A.M. element = (A.M. isotope 1)(% isotope 1/100) +
(A.M. isotope 2)(% isotope 2/100) + - -

Isotope	Atomic Mass	Percent
Ne-20	20.00	90.92
Ne-21	21.00	0.26
Ne-22	22.00	8.82

$$\text{A.M. Ne} = 20.00(0.9092) + 21.00(0.0026) + 22.00(0.0882) = 20.18$$

LECTURE 2

I Masses of Atoms (cont.)

A. Masses of individual atoms Since the atomic masses of H, Cl and Ni are 1.008, 35.45, and 58.69, it follows that:

1.008 g H, 35.45 g Cl, 58.69 g of Ni

all contain the same number of atoms, N. It turns out that:

$$N = \text{Avogadro's Number} = 6.022 \times 10^{23}$$

1. Mass of H atom?

$$1 \text{ atom H} \times \frac{1.008 \text{ g H}}{6.022 \times 10^{23} \text{ H atoms}} = 1.674 \times 10^{-24} \text{ g}$$

(1.674×10^{-24} is a very small number; atoms are tiny)

2. Number of atoms in one gram of nickel?

$$1.000 \text{ g Ni} \times \frac{6.022 \times 10^{23} \text{ atoms Ni}}{58.69 \text{ g Ni}} = 1.026 \times 10^{22} \text{ atoms}$$

(1.026×10^{22} is a very large number; there are a lot of atoms in one gram of anything).

II The Mole

A. Meaning 1 mol = 6.022×10^{23} items

1 mol H = 6.022×10^{23} H atoms; mass = 1.008 g

1 mol Cl = 6.022×10^{23} Cl atoms; mass = 35.45 g

1 mol Cl_2 = 6.022×10^{23} Cl_2 molecules; mass = 70.90 g

1 mol HCl = 6.022×10^{23} HCl molecules; mass = 36.46 g

B. Molar Mass Generalizing from the above examples, the molar mass (g/mol) is numerically equal to the formula mass, obtained by summing the atomic masses of the atoms in the formula.

	formula mass	molar mass
$CaCl_2$	110.98	110.98 g/mol
$C_6H_{12}O_6$	180.18	180.18 g/mol

C. Mole-Mass conversions

1. Calculate mass in grams of 13.2 mol of $CaCl_2$.

$$1 \text{ mol } CaCl_2 = 110.98 \text{ g } CaCl_2$$

$$\text{mass} = 13.2 \text{ mol } CaCl_2 \times \frac{110.98 \text{ g } CaCl_2}{1 \text{ mol } CaCl_2} = 1.47 \times 10^3 \text{ g } CaCl_2$$

2. Calculate number of moles in 16.4 g of $C_6H_{12}O_6$.

$$1 \text{ mol } C_6H_{12}O_6 = 180.18 \text{ g } C_6H_{12}O_6$$

$$\text{no. of moles} = 16.4 \text{ g } C_6H_{12}O_6 \times \frac{1 \text{ mol } C_6H_{12}O_6}{180.18 \text{ g } C_6H_{12}O_6}$$

$$= 0.0910 \text{ mol}$$

III Molarity

The amount of solute in a solution sample depends upon two factors: the volume of solution and the concentration of solute.

$$\text{Molarity (M)} = \frac{\text{no. moles solute}}{\text{no. liters solution}}$$

1. What volume of 12 M HCl ("concentrated hydrochloric acid") must be taken to obtain 0.10 mol of HCl?

$$1 \text{ L} \triangleq 12 \text{ mol HCl}$$

$$\text{Volume} = 0.10 \text{ mol HCl} \times \frac{1 \text{ L}}{12 \text{ mol HCl}} = 0.0083 \text{ L} = 8.3 \text{ mL}$$

2. What mass of NaOH is contained in 125 mL of 6.00 M NaOH?

$$125 \text{ mL} \times \frac{1 \text{ L}}{1000 \text{ mL}} \times \frac{6.00 \text{ mol}}{1 \text{ L}} \times \frac{40.00 \text{ g}}{1 \text{ mol}} = 30.0 \text{ g NaOH}$$

DEMONSTRATIONS

1. Conservation of Mass: Test. Dem., 54

QUIZZES

(Periodic Table should be available for atomic numbers and atomic masses)

Quiz 1

1. Calculate the atomic mass of gallium, which consists of two isotopes, Ga-69 (mass = 68.92, abundance = 60.16%) and Ga-71 (mass = 70.92, abundance = 39.84%).

2. Find the number of molecules and the number of moles in 15.0 g of propane, C_3H_8.

Quiz 2

1. Determine the mass in grams and the number of atoms in 2.31 mol of Fe.

2. Dilute hydrochloric acid contains 6.00 mol of HCl per liter. What volume of this solution contains 16.0 g of HCl?

Quiz 3

1. Elements X and Y form two different compounds in which the mass percents of X are 60.0 and 69.2. Show how these data illustrate the Law of Multiple Proportions.

2. Calculate the mass in grams of 0.0138 mol of PCl_3.

Quiz 4

1. Give the nuclear symbol of an atom which contains 52 protons and 77 neutrons.

2. Calculate the approximate mass, in grams, of the atom referred to in (1).

3. What is the molarity of a solution containing 15.0 g of NaCl in 120 mL of solution?

Quiz 5

1. Chlorine (A.M. = 35.45) consists of Cl-35 (A.M. = 34.98) and Cl-37 (A.M. = 36.98). Estimate the abundances of these two isotopes.

2. Determine the number of atoms in 12.34 g of Cu.

PROBLEMS 1-30

2.1 See Table 2.1

2.2 a. Law of Conservation of Mass b. Law of Constant Composition

c. none d. Law of Multiple Proportions

2.3 a. 70.37 g F/29.63 g O = 2.375 g F/g O
54.29 g F/45.71 g O = 1.188 g F/g O

b. 2.375/1.188 = 2:1

2.4 $^{222}_{86}Rn$

2.5 Must specify either the number of neutrons or the mass number.

2.6 a. 11 b. 13 c. 11 d. 10, 11

2.7

$^{21}_{10}Ne$	0	10	11	10
$^{138}_{56}Ba$	0	56	82	56
$^{45}_{21}Sc^{3+}$	+3	21	24	18
$^{31}_{15}P^{3-}$	-3	15	16	18

2.8 a. 55.847/44.9559 = 1.2423 b. 55.847/95.94 = 0.5821
c. 55.847/127.60 = 0.43767

2.9 a. 22.98977/12.00000 = 1.915814 b. 12.011/12.000 = 1.0009
c. 20.179/12.000 = 1.6816

2.10 at. mass Cu = 62.96(0.705) + 64.96(0.295) = 63.55

2.11 79.90 = 78.92(1 - x) + 80.92 x; x = 0.49; 49% Br-81

2.12 28.086 = 27.977(0.9530 - x) + 28.977(0.0470) + 29.974 x
x = 0.031; 3.1% Si-30, 92.2% Si-28

2.13 a. 24 p^+, 21 e^- b. 24 p^+, 22 e^- c. 18 p^+, 18 e^-
d. 10 p^+, 10 e^-

2.14

$^{45}_{21}Sc^{3+}$	24	21	18
$^{32}_{16}S^{2-}$	16	16	18
$^{206}_{81}Tl^{+}$	125	81	80
$^{64}_{30}Zn^{2+}$	34	30	28

2.15 Na_2S, NaI, CaS, CaI_2

2.16 a. $1 \text{ Mo atom} \times \frac{95.94 \text{ g}}{6.022 \times 10^{23} \text{ Mo atoms}} = 1.593 \times 10^{-22} \text{ g}$

b. $1.000 \times 10^{-3} \text{ g Mo} \times \frac{6.022 \times 10^{23} \text{ atoms}}{95.94 \text{ g Mo}} = 6.277 \times 10^{18} \text{ atoms}$

2.17 a. $2 \times 10^{15} \text{ atoms} \times \frac{107.8682 \text{ g}}{6.022 \times 10^{23} \text{ Ag atoms}} = 4 \times 10^{-7} \text{ g}$

b. $1.000 \text{ lb Ag} \times \frac{453.6 \text{ g}}{1 \text{ lb}} \times \frac{6.022 \times 10^{23} \text{ atoms}}{107.8682 \text{ g Ag}}$

$= 2.532 \times 10^{24} \text{ atoms}$

2.18 $0.063 \text{ g} \times 6.022 \times 10^{23} \times \frac{1 \text{ metric ton}}{10^6 \text{ g}} = 3.8 \times 10^{16} \text{ met. ton}$

2.19 a. 35 b. $35 \times 6.022 \times 10^{23} = 2.108 \times 10^{25}$

c. $0.0187 \text{ mol Br} \times \frac{2.108 \times 10^{25} e^-}{1 \text{ mol Br}} = 3.94 \times 10^{23} e^-$

d. $0.0187 \text{ g Br} \times \frac{1 \text{ mol Br}}{79.90 \text{ g Br}} \times \frac{2.108 \times 10^{25} e^-}{1 \text{ mol Br}}$

$= 4.93 \times 10^{21} e^-$

2.20 a. 28.0855 g/mol

b. 28.0855 g/mol + 4(35.453)g/mol = 169.898 g/mol

c. 12(12.011)g/mol + 22(1.00794)g/mol + 11(15.9994)g/mol = 342.299 g/mol

2.21 a. $1.34 \text{ g } H_2O \times \frac{1 \text{ mol } H_2O}{18.02 \text{ g } H_2O} = 0.0744 \text{ mol } H_2O$

b. $1.34 \text{ g Cu} \times \frac{1 \text{ mol Cu}}{63.55 \text{ g Cu}} = 0.0211 \text{ mol Cu}$

c. $1.34 \text{ g } N_2O \times \frac{1 \text{ mol } N_2O}{44.01 \text{ g } N_2O} = 0.0304 \text{ mol } N_2O$

2.22 a. $2.42 \text{ mol H} \times \frac{1.008 \text{ g H}}{1 \text{ mol H}} = 2.44 \text{ g H}$

b. $2.42 \text{ mol } H_2 \times \frac{2.016 \text{ g } H_2}{1 \text{ mol } H_2} = 4.88 \text{ g } H_2$

c. $2.42 \text{ mol } H_2O \times \frac{18.02 \text{ g } H_2O}{1 \text{ mol } H_2O} = 43.6 \text{ g } H_2O$

d. $2.42 \text{ mol } H_2O_2 \times \frac{34.02 \text{ g } H_2O_2}{1 \text{ mol } H_2O_2} = 82.3 \text{ g } H_2O_2$

2.23 molar mass C_3H_6O = 3(12.01)g/mol + 6(1.008)g/mol + 16.00 g/mol = 58.08 g/mol

no. grams	no. moles	no. molecules	no. C atoms
0.0880	0.00152	9.12×10^{20}	2.74×10^{21}
0.290	0.00500	3.01×10^{21}	9.03×10^{21}
9.6×10^{-14}	1.7×10^{-15}	1.0×10^{9}	3.0×10^{9}
3.2×10^{-3}	5.5×10^{-5}	3.3×10^{19}	1.0×10^{20}

2.24 a. MM = 2(12.01)g/mol + 6(1.008)g/mol + 16.00 g/mol = 46.07 g/mol

b. $252 \text{ mL} \times \frac{0.785 \text{ g}}{1 \text{ mL}} \times \frac{1 \text{ mol}}{46.07 \text{ g}} = 4.29 \text{ mol}$

c. $1.62 \text{ mol} \times \frac{46.07 \text{ g}}{1 \text{ mol}} = 74.6 \text{ g}$

2.25 a. In 0.240 L of 3.00 M NaOH, there is:

0.240 x 3.00 x 40.00 g NaOH = 28.8 g NaOH. Weigh out 28.8 g NaOH, dissolve in water to form 240 mL of solution

2.25 b. In 0.100 L of 6.00 M KNO_3, there is:

$$0.100 \text{ L} \times \frac{6.00 \text{ mol}}{1 \text{ L}} \times \frac{101.11 \text{ g } KNO_3}{1 \text{ mol}} = 60.7 \text{ g } KNO_3$$

Weigh out 60.7 g KNO_3; dissolve in water to form 100 mL solution.

2.26 a. $10.2 \text{ mL} \times \frac{1 \text{ L}}{10^3 \text{ mL}} \times \frac{6.00 \text{ mol HCl}}{1 \text{ L}} = 6.12 \times 10^{-2} \text{ mol HCl}$

b. $0.100 \text{ mol HCl} \times \frac{1 \text{ L}}{6.00 \text{ mol HCl}} = 0.0167 \text{ L} = 16.7 \text{ mL}$

2.27 molarity $NaHCO_3 = \frac{(2.52/84.01) \text{ mol}}{0.125 \text{ L}} = 0.240 \text{ mol/L}$

mass $C_3H_8O_3 = 0.800 \text{ L} \times \frac{3.50 \text{ mol}}{1 \text{ L}} \times \frac{92.08 \text{ g}}{1 \text{ mol}} = 258 \text{ g}$

V $SrCl_2 = 2.30 \text{ g} \times \frac{1 \text{ mol}}{158.53 \text{ g}} \times \frac{1 \text{ L}}{1.45 \text{ mol}} = 1.00 \times 10^{-2} \text{ L}$

mass $Fe(NO_3)_3 = 0.300 \text{ L} \times \frac{0.275 \text{ mol}}{1 \text{ L}} \times \frac{241.87 \text{ g}}{1 \text{ mol}} = 20.0 \text{ g}$

2.28 a. false (same number of protons)
b. false (about ½ as heavy)
c. true (added electrons increase mass slightly)
d. false (6×10^{23} times as great)

2.29a. 17 + 19 + 1 + 3 = 40

b.

C	17 x 12.01 g/mol =	204.17 g/mol	
H	19 x 1.008 g/mol =	19.15 g/mol	
N	=	14.01 g/mol	N contributes least
O	3 x 16.00 g/mol =	48.00 g/mol	
		285.33 g/mol	

c. $10.0 \times 10^{-3} \text{g} \times \frac{1 \text{ mol}}{285.33 \text{ g}} \times \frac{6.022 \times 10^{23} \text{ molecules}}{1 \text{ mol}} \times \frac{17 \text{ C atoms}}{1 \text{ molecule}}$

$= 3.59 \times 10^{20}$ molecules

2.30 Compare numbers of atoms

Cl_2 molecule: 2 Cl atoms

$1.0 \times 10^{-23} \text{ mol Cl} \times \frac{6.022 \times 10^{23} \text{ Cl atoms}}{1 \text{ mol Cl}} = 6.0 \text{ Cl atoms}$

$1.0 \times 10^{-23} \text{ g Cl} \times \frac{1 \text{ mol Cl}}{35.45 \text{ g Cl}} \times \frac{6.022 \times 10^{23} \text{ Cl atoms}}{1 \text{ mol}}$

= 0.17 Cl atom

b > a > d > c

CHAPTER 3

BASIC SKILLS

1. Given the formula of a compound, calculate the mass percents of the elements.
2. Determine the simplest formula of a compound, given the mass percents of the elements or data from which the masses of the elements in a sample of the compound can be determined.
3. Given the simplest formula of a compound and its approximate molar mass, obtain its molecular formula.
4. Given a Periodic Table, write formulas for ionic compounds, including ones containing transition metal ions and/or polyatomic ions.
5. Name ionic compounds and binary molecular compounds.
6. Write and balance simple chemical equations.
7. Use a balanced equation to relate the numbers of moles or grams of reactants and products.
8. Given the numbers of moles or grams of each reactant, determine the limiting reactant and calculate the theoretical yield of product.
9. Relate the actual yield of product to the theoretical yield and percent yield.

LECTURE NOTES

This chapter contains some very basic material which students must master. Depending on the backrounds of your students, you may need 2 - 3 lectures to cover this chapter. We find that about $2\frac{1}{2}$ fifty minute lectures are required. Points to keep in mind include the following:

1. It is important to emphasize early on that the subscripts in a formula give not only the atom ratio but also the mole ratio. It is necessary that students realize this if they are to follow the logic of obtaining simplest formulas from mass percents.

2. Students ordinarily have little trouble calculating formulas from mass percents, perhaps because they memorize the procedure followed. They are much less adept at obtaining formulas from analytical data such as that given in Example 3.3.

3. Students should be told to learn the charges of transition

metal ions (Figure 3.3) and the charges and formulas of polyatomic ions (Table 3.1). They will need this information throughout the rest of the course; now is the time to learn it.

4. In our opinion, little is gained at this point by spending a lot of time balancing equations. The equations referred to in the next several chapters are relatively simple ones which can be balanced by the simple approach described here. It is important to get across the point that an equation describes what happens in a reaction carried out in the laboratory. Including in the equation the physical states of reactants and products helps to emphasize this point.

5. When you discuss mass relations in reactions, some students will want to revert to the "ratio-and-proportion" method they claim to have learned in high school. The comments made in Chapter 1 about conversion factors apply here as well.

6. There are many different ways to find the limiting reactant and calculate the theoretical yield. We've tried most of them and recommend the approach described in Section 3.8.

LECTURE 1

I Formulas

A. Types of formulas

1. Simplest formula: gives simplest atom ratio = mole ratio

K_2CrO_4 2 atom K:1 atom Cr: 4 atom O
2 mol K: 1 mol Cr: 4 mol O

CH_2O 1 atom C : 2 atom H : 1 atom O
1 mol C : 2 mol H : 1 mol O

2. Molecular formula gives composition of molecule. May be identical with the simplest formula or an integral multiple of it.

formaldehyde: molecular formula CH_2O
glucose: molecular formula $C_6H_{12}O_6$

B. Mass percent from formula

Percentage composition of K_2CrO_4?

molar mass = (78.20 + 52.00 + 64.00) g/mol = 194.20 g/mol

$$\% \text{ K} = \frac{78.20}{194.20} \times 100 = 40.27; \quad \% \text{ Cr} = \frac{52.00}{194.20} \times 100 = 26.78$$

$$\% \text{ O} = \frac{64.00}{194.20} \times 100 = 32.96$$

Note that percents must add to 100; could have obtained % O by difference from 100.

C. Simplest formula from percent composition

Find mass of each element in sample of compound. Then find numbers of moles of each element and finally the mole ratio.

1. Simplest formula of compound containing 26.6% K, 35.4% Cr and 38.0% O?

Work with 100 g sample: 26.6 g K, 35.4 g Cr, 38.0 g O

no. moles K = 26.6 g x $\frac{1 \text{ mol}}{39.10 \text{ g}}$ = 0.680 mol K

no. moles Cr = 35.4 g x $\frac{1 \text{ mol}}{52.00 \text{ g}}$ = 0.681 mol Cr

no. moles O = 38.0 g x $\frac{1 \text{ mol}}{16.00 \text{ g}}$ = 2.38 mol O

Note that 2.38/0.680 = 3.50 = 7/2 (<u>not</u> 3 or 4).

Simplest formula: $K_2Cr_2O_7$

2. A sample of acetic acid (C, H and O atoms) weighing 1.000 g burns to give 1.466 g of CO_2 and 0.6001 g H_2O. Simplest formula?

Find mass of C in sample (from mass of CO_2), then mass of H (from H_2O), and finally mass of oxygen by difference.

mass C = 1.466 g CO_2 x $\frac{12.01 \text{ g C}}{44.01 \text{ g } CO_2}$ = 0.4001 g C

mass H = 0.6001 g H_2O x $\frac{2.02 \text{ g H}}{18.02 \text{ g } H_2O}$ = 0.0673 g H

mass O = 1.000 g - 0.400 g - 0.067 g = 0.533 g O

no. moles C = 0.4001 g C x $\frac{1 \text{ mol C}}{12.01 \text{ g C}}$ = 0.0333 mol C

no. moles H = 0.0673 g H x $\frac{1 \text{ mol H}}{1.01 \text{ g H}}$ = 0.0666 mol H

no. moles O = 0.533 g O x $\frac{1 \text{ mol O}}{16.00 \text{ g O}}$ = 0.0333 mol O

Simplest formula is CH_2O, as with formaldehyde and glucose

D. <u>Molecular formula from simplest formula</u>
Must know molar mass. For acetic acid, MM = 60 g/mol
Formula mass CH_2O = 30; 60/30 = 2; molecular formula $C_2H_4O_2$

LECTURE 2

I <u>Formulas</u> (cont.)

A. <u>Prediction of formulas of ionic compounds</u>
Must know charges of ions

1. Charges of monatomic ions of maingroup elements can be predicted from position in Periodic Table. Group 1 = +1, Group 2 = +2, Group 6 = -2, Group 7 = -1
2. Transition metal cations have charges indicated in Figure 3.3 (e.g., Cr^{3+}, Ni^{2+}, Ag^{+}).
3. Polyatomic ions: see Table 3.1. Examples: NH_4^{+}, NO_3^{-}, SO_4^{2-}, PO_4^{3-}

Predict the formulas of barium nitrate, potassium phosphate, silver oxide.

cation	anion	formula
Ba^{2+}	NO_3^-	$Ba(NO_3)_2$
K^+	PO_4^{3-}	K_3PO_4
Ag^+	O^{2-}	Ag_2O

II Names of Compounds

A. Ionic compounds Give name of cation followed by that of anion. If metal forms more than one cation, as with many transition metals, indicate charge by Roman numeral.

Na_2SO_4	$FeSO_4$	$Fe_2(SO_4)_3$
sodium sulfate	iron(II) sulfate	iron(III) sulfate

B. Binary molecular compounds (two nonmetals) Indicate number of atoms of each element using Greek prefixes.

PCl_3 phosphorus trichloride (ordinarily omit prefix mono)

N_2O_5 dinitrogen pentoxide

N_2H_4 dinitrogen tetrahydride

III Chemical Equations

A. Balancing Must have same number of atoms of each type on both sides. Achieve this by adjusting coefficients in front of formulas. Example: combustion of propane in air to give carbon dioxide and water.

$$C_3H_8(g) + O_2(g) \longrightarrow CO_2(g) + H_2O(l)$$

Balance C: $C_3H_8(g) + O_2(g) \longrightarrow 3\ CO_2(g) + H_2O(l)$

Balance H: $C_3H_8(g) + O_2(g) \longrightarrow 3\ CO_2(g) + 4\ H_2O(l)$

Balance O: $C_3H_8(g) + 5\ O_2(g) \longrightarrow 3\ CO_2(g) + 4\ H_2O(l)$

Meaning: 1 mol C_3H_8 reacts with 5 mol O_2 to form 3 mol CO_2 and 4 mol H_2O

B. Mass relations in reactions

1. Moles of CO_2 produced when 1.65 mol of C_3H_8 burns?

$$1\text{ mol } C_3H_8 \simeq 3\text{ mol } CO_2$$

$$1.65\text{ mol } C_3H_8 \times \frac{3\text{ mol } CO_2}{1\text{ mol } C_3H_8} = 4.95\text{ mol } CO_2$$

2. Moles H_2O from combustion of 20.0 g C_3H_8?

$$1\text{ mol } C_3H_8 = 44.09\text{ g } C_3H_8;\quad 1\text{ mol } C_3H_8 \simeq 4\text{ mol } H_2O$$

$$20.0\text{ g } C_3H_8 \times \frac{1\text{ mol } C_3H_8}{44.09\text{ g } C_3H_8} \times \frac{4\text{ mol } H_2O}{1\text{ mol } C_3H_8} = 1.81\text{ mol } H_2O$$

3. Mass of O_2 required to react with 12.0 g of C_3H_8?

$$12.0\text{ g } C_3H_8 \times \frac{1\text{ mol } C_3H_8}{44.09\text{ g } C_3H_8} \times \frac{5\text{ mol } O_2}{1\text{ mol } C_3H_8} \times \frac{32.00\text{ g } O_2}{1\text{ mol } O_2}$$

$$= 43.6\text{ g } O_2$$

LECTURE $2\frac{1}{2}$

I Yield of Product in Reaction

A. Limiting reactant, theoretical yield Ordinarily, reactants are not present in the exact ratio required for reaction. Instead, one reactant is in excess; some of it is left when the reaction is over. The other, limiting reactant, is completely converted to give the theoretical yield of product.

To calculate the theoretical yield and identify the limiting reactant:

1. Calculate the yield to be expected if the first reactant is limiting.
2. Repeat this calculation for the second reactant.
3. The theoretical yield is the smaller of these two quantities. The reactant that gives the smaller calculated yield is the limiting reactant.

Example: $2\ Ag(s) + I_2(s) \rightarrow 2\ AgI(s)$

Calculate the theoretical yield of AgI and determine the limiting reactant starting with

a. 1.00 mol Ag, 1.00 mol I_2

theor. yield AgI if Ag is limiting:

$$1.00 \text{ mol Ag} \times \frac{2 \text{ mol AgI}}{2 \text{ mol Ag}} = 1.00 \text{ mol AgI}$$

theor. yield AgI if I_2 is limiting:

$$1.00 \text{ mol } I_2 \times \frac{2 \text{ mol AgI}}{1 \text{ mol } I_2} = 2.00 \text{ mol AgI}$$

Theoretical yield = 1.00 mol AgI; Ag is limiting reactant

b. 1.00 g Ag, 1.00 g I_2

theor. yield AgI if Ag is limiting:

$$1.00 \text{ g Ag} \times \frac{469.54 \text{ g AgI}}{215.74 \text{ g Ag}} = 2.18 \text{ g AgI}$$

theor. yield if I_2 is limiting:

$$1.00 \text{ g } I_2 \times \frac{469.54 \text{ g AgI}}{253.80 \text{ g } I_2} = 1.85 \text{ g AgI}$$

Theoretical yield = 1.85 g AgI; I_2 is limiting reactant

B. Actual yield, percent yield

$$\% \text{ yield} = \frac{\text{Actual Yield}}{\text{Theoretical Yield}} \times 100$$

Suppose actual yield of AgI in this reaction is 1.50 g.

$$\text{Percent yield} = \frac{1.50}{1.85} \times 100 = 81.1$$

DEMONSTRATIONS

1. Reaction of Al with I_2: Test. Dem. 31, 88

QUIZZES

(A Periodic Table should be available)

Quiz 1

1. Give the formulas of the following ionic compounds:
   a. lead nitrate   b. chromium(III) sulfate
   c. magnesium phosphate

2. Determine the simplest formula of a compound of carbon, hydrogen and chlorine in which the mass percent of C is 49.02 and that of H is 2.74.

Quiz 2

1. What is the molecular formula of a hydrocarbon whose simplest formula is CH if its molar mass is about 90 g/mol?

2. Consider the reaction: $2\ Cr(s) + 3\ Cl_2(g) \longrightarrow 2\ CrCl_3(s)$
   a. What mass of $CrCl_3$ can be formed from 62.5 g of Cl_2?
   b. What is the theoretical yield of $CrCl_3$, starting with 1.00 g of Cr and 2.00 g of Cl_2?

Quiz 3

1. Write balanced equations for the formation of ionic compounds when:
   a. aluminum reacts with fluorine   b. nickel reacts with oxygen
   c. potassium reacts with nitrogen

2. Combustion of a 1.000 g sample of a compound containing the three elements C, H and Cl gives 2.325 g of CO_2 and 0.397 g of H_2O. What are the mass percents of C, H and Cl in the compound?

Quiz 4

1. Name:
   a. Cu_2SO_4 b. P_4O_6 c. $MnCO_3$ d. $Fe(ClO_4)_3$

2. Consider the reaction between silver and sulfur.
   a. Write a balanced equation for the reaction, assuming the product is ionic.
   b. Calculate the number of moles of silver required to react with 15.0 g of sulfur.
   c. How many grams of sulfur are required to react with 1.00 g of silver?

Quiz 5

1. Give the formula of the hydroxide and carbonate of:
   a. rubidium   b. strontium   c. aluminum

2. Consider the reaction: $Fe^{3+}(aq) + 3\ OH^-(aq) \longrightarrow Fe(OH)_3(s)$

   What is the theoretical yield of $Fe(OH)_3$ starting with

   a. 1.20 mol Fe^{3+}, 2.62 mol OH^-?

   b. 1.00 g Fe^{3+}, 1.00 g OH^-?

PROBLEMS 1-30

3.1 C: 15 x 12.01 g = 180.15 g
H: 11 x 1.008 g = 11.09 g
N = 14.01 g
O: 4 x 16.00 g = 64.00 g
I: 4 x 126.90 g = 507.60 g
776.85 g

mass % C = 100 x 180.15/776.85 = 23.190
mass % O = 100 x 64.00/776.85 = 8.238
mass % H = 100 x 11.09/776.85 = 1.428
mass % N = 100 x 14.01/776.85 = 1.803
mass % I = 100 x 507.60/776.85 = 65.341

3.2 a. $1.15\text{ g Cl} \times \dfrac{58.44\text{ g NaCl}}{35.45\text{ g Cl}} = 1.90\text{ g NaCl}$

b. $\dfrac{1.90}{3.50} \times 100 = 54.3$

3.3 mass C $= 3.383\text{ g } CO_2 \times \dfrac{12.01\text{ g C}}{44.01\text{ g } CO_2} = 0.9232\text{ g C}$

mass H = 1.000 g - 0.9232 g = 0.077 g H

mass % C $= \dfrac{0.9232}{1.000} \times 100 = 92.32$; mass % H $= \dfrac{0.077}{1.000} \times 100 = 7.7$

3.4 no. moles Fe $= 2.561\text{ g} \times \dfrac{1\text{ mol}}{55.85\text{ g}} = 0.04585\text{ mol}$

no. moles S $= 2.206\text{ g} \times \dfrac{1\text{ mol}}{32.06\text{ g}} = 0.06881\text{ mol}$

0.06881/0.04585 = 1.50; Fe_2S_3

3.5 a. no. moles K $= 43.2\text{ g K} \times \dfrac{1\text{ mol K}}{39.10\text{ g K}} = 1.10\text{ mol K}$

no. moles Cl $= 39.1\text{ g Cl} \times \dfrac{1\text{ mol Cl}}{35.45\text{ g Cl}} = 1.10\text{ mol Cl}$ KClO

no. moles O $= 17.7\text{ g O} \times \dfrac{1\text{ mol O}}{16.00\text{ g O}} = 1.11\text{ mol O}$

b. no. moles C $= 62.1\text{ g C} \times \dfrac{1\text{ mol C}}{12.01\text{ g C}} = 5.17\text{ mol C}$

no. moles H $= 5.21\text{ g H} \times \dfrac{1\text{ mol H}}{1.008\text{ g H}} = 5.17\text{ mol H}$

no. moles N $= 12.1\text{ g N} \times \dfrac{1\text{ mol N}}{14.01\text{ g N}} = 0.864\text{ mol N}$

no. moles O $= 20.7\text{ g O} \times \dfrac{1\text{ mol O}}{16.00\text{ g O}} = 1.29\text{ mol O}$

5.17/0.864 = 5.98; 1.29/0.864 = 1.49; $C_{12}H_{12}N_2O_3$

3.5 c. no. moles Co = 32.3 g Co x $\frac{1 \text{ mol Co}}{58.93 \text{ g Co}}$ = 0.548 mol Co

no. moles H = 1.66 g H x $\frac{1 \text{ mol H}}{1.008 \text{ g H}}$ = 1.65 mol H

no. moles N = 7.68 g N x $\frac{1 \text{ mol N}}{14.01 \text{ g N}}$ = 0.548 mol N

no. moles Cl = 58.3 g Cl x $\frac{1 \text{ mol Cl}}{35.45 \text{ g Cl}}$ = 1.64 mol Cl

CoH_3NCl_3

3.6 a. mass C = 2.257 g CO_2 x $\frac{12.01 \text{ g C}}{44.01 \text{ g } CO_2}$ = 0.6159 g C

mass H = 0.9241 g H_2O x $\frac{2.016 \text{ g H}}{18.02 \text{ g } H_2O}$ = 0.1034 g H

% C = 100 x 0.6159/1.540 = 39.99
% H = 100 x 0.1034/1.540 = 6.714
% O = 100.00 - 39.99 - 6.71 = 53.30

b. in 100 g:

no. moles C = 39.99 g C x $\frac{1 \text{ mol C}}{12.01 \text{ g C}}$ = 3.330 mol C

no. moles H = 6.714 g H x $\frac{1 \text{ mol H}}{1.008 \text{ g H}}$ = 6.661 mol H

no. moles O = 53.30 g O x $\frac{1 \text{ mol O}}{16.00 \text{ g O}}$ = 3.331 mol O

simplest formula = CH_2O

c. formula mass = 12 + 2 + 16 = 30; 60/30 = 2 ; $C_2H_4O_2$

3.7 mass C = 6.162 g CO_2 x $\frac{12.01 \text{ g C}}{44.01 \text{ g } CO_2}$ = 1.682 g C

mass H = 0.9008 g H_2O x $\frac{2.016 \text{ g H}}{18.02 \text{ g } H_2O}$ = 0.1008 g H

mass Cl = 3.200 g - 1.682 g - 0.101 g = 1.417 g Cl

no. moles C = 1.682 g C x $\frac{1 \text{ mol C}}{12.01 \text{ g C}}$ = 0.1400 mol C

no. moles H = 0.1008 g H x $\frac{1 \text{ mol H}}{1.008 \text{ g H}}$ = 0.1000 mol H

no. moles Cl = 1.417 g Cl x $\frac{1 \text{ mol Cl}}{35.45 \text{ g Cl}}$ = 0.03997 mol Cl

0.1400/0.03997 = 3.50; 0.1000/0.03997 = 2.50; $C_7H_5Cl_2$

3.8 mass O = 0.592 g H_2O x $\frac{16.00 \text{ g O}}{18.02 \text{ g } H_2O}$ = 0.526 g O

mass Cu = 2.612 g - 0.526 g = 2.086 g Cu

no. moles O = 0.526 g O x $\frac{1 \text{ mol O}}{16.00 \text{ g O}}$ = 0.0329 mol O

CuO

no. moles Cu = 2.086 g Cu x $\frac{1 \text{ mol Cu}}{63.55 \text{ g Cu}}$ = 0.0328 mol Cu

3.9 no. moles $H_2O = 0.0556 \text{ g } H_2O \times \frac{1 \text{ mol } H_2O}{18.02 \text{ g } H_2O} = 0.0309 \text{ mol } H_2O$

moles $MgI_2 = 1.072 \text{ g } MgI_2 \times \frac{1 \text{ mol } MgI_2}{278.12 \text{ g } MgI_2} = 0.003854 \text{ mol } MgI_2$

0.0309/0.003854 = 8.02; x = 8

3.10 formula mass = 2(12) + 4 + 16 = 44; 90/44 = 2; $C_4H_8O_2$

3.11 a. K^+, $Cr_2O_7^{2-}$ ions; $K_2Cr_2O_7$

b. Sn^{2+}, PO_4^{3-} ions; $Sn_3(PO_4)_2$

c. Au^+, S^{2-} ions; Au_2S d. Al^{3+}, O^{2-} ions; Al_2O_3

3.12 a. lithium hydroxide   b. $KMnO_4$ c. NH_4Cl

d. calcium fluoride   e. barium hydroxide   f. $Ca(NO_3)_2$

g. iron(II) carbonate

3.13 a. boron trichloride   b. XeF_6 c. NF_3

d. tetrasulfur tetranitride   e. SiI_4 f. ammonia

3.14 a. Li_2O; 0.600 mol ions   b. $Ba(OH)_2$; 1.05 mol ions

c. $Fe_3(PO_4)_2$: $1.67 \text{ g} \times \frac{1 \text{ mol}}{357.49 \text{ g}} \times 5 = 0.0234$ mol ions

3.15 a. $2\ C_6H_{14}(l) + 19\ O_2(g) \longrightarrow 12\ CO_2(g) + 14\ H_2O(l)$

b. $2\ Al(s) + 6\ HCl(aq) \longrightarrow 2\ AlCl_3(s) + 3\ H_2(g)$

c. $4\ KClO_3(s) \longrightarrow 3\ KClO_4(s) + KCl(s)$

3.16 a. $2\ Al(s) + 3\ Br_2(l) \longrightarrow 2\ AlBr_3(s)$; Al^{3+}, Br^- ions

b. $Ba(s) + Br_2(l) \longrightarrow BaBr_2(s)$; Ba^{2+}, Br^- ions

c. $2\ K(s) + Br_2(l) \longrightarrow 2\ KBr(s)$; K^+, Br^- ions

d. $Ni(s) + Br_2(l) \longrightarrow NiBr_2(s)$; Ni^{2+}, Br^- ions

e. $2\ Ag(s) + Br_2(l) \longrightarrow 2\ AgBr(s)$; Ag^+, Br^- ions

3.17 a. $3\ Mg(s) + N_2(g) \longrightarrow Mg_3N_2(s)$

b. $2\ Cu_2O(s) + O_2(g) \longrightarrow 4\ CuO(s)$

c. $2\ CH_3OH(l) + 3\ O_2(g) \longrightarrow 2\ CO_2(g) + 4\ H_2O(l)$

d. $2\ NaN_3(s) \longrightarrow 2\ Na(s) + 3\ N_2(g)$

e. $2\ Al(s) + 3\ F_2(g) \longrightarrow 2\ AlF_3(s)$

3.18 a. $8.60 \text{ mol } Ca_{10}F_2(PO_4)_6 \times \frac{7 \text{ mol } CaSO_4}{1 \text{ mol } Ca_{10}F_2(PO_4)_6} = 60.2 \text{ mol } CaSO_4$

b. $7.25 \text{ mol } Ca(H_2PO_4)_2 \times \frac{7 \text{ mol } H_2SO_4}{3 \text{ mol } Ca(H_2PO_4)_2} = 16.9 \text{ mol } H_2SO_4$

3.18 c. $0.990 \text{ mol } H_2SO_4 \times \frac{1 \text{ mol } Ca_{10}F_2(PO_4)_6}{7 \text{ mol } H_2SO_4} = 0.141 \text{ mol}$

d. $3.330 \text{ mol } Ca_{10}F_2(PO_4)_6 \times \frac{3 \text{ mol } Ca(H_2PO_4)_2}{1 \text{ mol } Ca_{10}F_2(PO_4)_6} = 9.990 \text{ mol}$

3.19 a. $0.660 \text{ mol } Ca_{10}F_2(PO_4)_6 \times \frac{7 \text{ mol } CaSO_4}{1 \text{ mol } Ca_{10}F_2(PO_4)_6} \times \frac{136.14 \text{ g } CaSO_4}{1 \text{ mol } CaSO_4}$

$= 629 \text{ g } CaSO_4$

b. $100.0 \text{ g HF} \times \frac{1 \text{ mol HF}}{20.006 \text{ g HF}} \times \frac{1 \text{ mol } Ca_{10}F_2(PO_4)_6}{2 \text{ mol HF}}$

$= 2.499 \text{ mol } Ca_{10}F_2(PO_4)_6$

c. $1.06 \text{ g } CaSO_4 \times \frac{1 \text{ mol } CaSO_4}{136.14 \text{ g } CaSO_4} \times \frac{1 \text{ mol } Ca_{10}F_2(PO_4)_6}{7 \text{ mol } CaSO_4}$

$\times \frac{1008.62 \text{ g } Ca_{10}F_2(PO_4)_6}{1 \text{ mol } Ca_{10}F_2(PO_4)_6} = 1.12 \text{ g } Ca_{10}F_2(PO_4)_6$

d. $16.25 \text{ g } H_2SO_4 \times \frac{1 \text{ mol } H_2SO_4}{98.08 \text{ g } H_2SO_4} \times \frac{2 \text{ mol HF}}{7 \text{ mol } H_2SO_4} \times \frac{20.006 \text{ g HF}}{1 \text{ mol HF}}$

$= 0.9470 \text{ g HF}$

3.20 a. $2\ C_2H_2(g) + 5\ O_2(g) \longrightarrow 4\ CO_2(g) + 2\ H_2O(l)$

b. $0.524 \text{ mol } C_2H_2 \times \frac{4 \text{ mol } CO_2}{2 \text{ mol } C_2H_2} = 1.05 \text{ mol } CO_2$

c. $2.46 \text{ mol } C_2H_2 \times \frac{5 \text{ mol } O_2}{2 \text{ mol } C_2H_2} \times \frac{32.00 \text{ g } O_2}{1 \text{ mol } O_2} = 197 \text{ g } O_2$

3.21 a. $454 \text{ g } C_6H_{12}O_6 \times \frac{1 \text{ mol } C_6H_{12}O_6}{180.16 \text{ g } C_6H_{12}O_6} \times \frac{2 \text{ mol } C_2H_5OH}{1 \text{ mol } C_6H_{12}O_6}$

$\times \frac{46.07 \text{ g } C_2H_5OH}{1 \text{ mol } C_2H_5OH} = 232 \text{ g } C_2H_5OH$

b. $1.00 \text{ gal. gasohol} \times \frac{4 \text{ qt}}{1 \text{ gal.}} \times \frac{10^3 \text{ cm}^3}{1.057 \text{ qt}} \times \frac{10}{100} \times \frac{0.79 \text{ g } C_2H_5OH}{1 \text{ cm}^3}$

$\times \frac{180.16 \text{ g } C_6H_{12}O_6}{92.14 \text{ g } C_2H_5OH} = 580 \text{ g glucose}$

3.22 $0.702 \text{ g } CO_2 \times 5 \times \frac{1 \text{ mol } CO_2}{44.01 \text{ g } CO_2} \times \frac{4 \text{ mol } KO_2}{4 \text{ mol } CO_2} \times \frac{71.10 \text{ g } KO_2}{1 \text{ mol } KO_2}$

$= 5.67 \text{ g } KO_2$

3.23 1500 sets of figures → 1500 chapters
500 sets of tables → 500 chapters
1000 sets of photos → 1000 chapters

limit is 500 chapters

3.24 a. $H_2(g) + Cl_2(g) \longrightarrow 2\ HCl(g)$

3.24 b. 10.0 mol H_2 ⟶ 20.0 mol HCl
12.0 mol Cl_2 ⟶ 24.0 mol HCl H_2 is limiting

c. 20.0 mol HCl d. 12.0 mol - 10.0 mol = 2.0 mol Cl_2

3.25 a. $2\ Na(s) + H_2(g) \longrightarrow 2\ NaH(s)$

b. $6.75\ g\ Na \times \frac{1\ mol\ Na}{22.99\ g\ Na} \times \frac{2\ mol\ NaH}{2\ mol\ Na} \times \frac{24.00\ g\ NaH}{1\ mol\ NaH}$

= 7.05 g NaH

$3.03\ g\ H_2 \times \frac{1\ mol\ H_2}{2.016\ g\ H_2} \times \frac{2\ mol\ NaH}{1\ mol\ H_2} \times \frac{24.00\ g\ NaH}{1\ mol\ NaH}$

= 72.1 g NaH; Na is limiting

c. 7.05 g NaH

d. 100 x 4.00/7.05 = 56.7

3.26 $1.00 \times 10^6\ g\ Na_2SO_4 \times \frac{1\ mol\ Na_2SO_4}{142.04\ g\ Na_2SO_4} \times \frac{2\ mol\ NaHCO_3}{1\ mol\ Na_2SO_4}$

$\times \frac{84.01\ g\ NaHCO_3}{1\ mol\ NaHCO_3} \times (0.85)^3 \times \frac{1\ kg}{10^3\ g} = 7.3 \times 10^2\ kg\ NaHCO_3$

3.27 a. $50.0\ g\ HBr \times \frac{1\ mol\ HBr}{80.91\ g\ HBr} \times \frac{1\ mol\ NaBr}{1\ mol\ HBr} \times \frac{102.89\ g\ NaBr}{1\ mol\ NaBr}$

= 63.6 g NaBr

$50.0\ g\ HBr \times \frac{1\ mol\ HBr}{80.91\ g\ HBr} \times \frac{1\ mol\ H_3PO_4}{3\ mol\ HBr} \times \frac{97.99\ g\ H_3PO_4}{1\ mol\ H_3PO_4}$

x 1.40 = 28.3 g H_3PO_4

b. 63.6 g/0.80 = 80 g; 28.3 g/0.80 = 35 g

3.28 a. false; simplest formula CH_2O

b. true: $Br^- \longrightarrow Br + e^-$

c. may or may not be true; must express in terms of amount required for reaction

d. false; % C greater in CO

e. false; acetylene vs benzene

3.29 a. oxidation b. oxidation c. reduction

3.30 In 100 g of compound:

no. moles H = $18.3\ g\ H \times \frac{1\ mol\ H}{1.008\ g\ H} = 18.2\ mol\ H$

no. moles C = $81.7\ g\ C \times \frac{1\ mol\ C}{12.01\ g\ C} = 6.80\ mol\ C$

18.2/6.80 = 2.67 = 8/3; simplest formula = C_3H_8

$C_3H_8(g) + 5\ O_2(g) \longrightarrow 3\ CO_2(g) + 4\ H_2O(l)$

CHAPTER 4

BASIC SKILLS

1. Describe how the following nonmetals are extracted from the atmosphere or the earth: N_2, O_2, Ar, S.
2. Describe, using chemical equations, the processes used to extract the following nonmetals from natural sources: H_2, F_2, Cl_2, Br_2, I_2.
3. Describe, using chemical equations, how the following metals are extracted from natural sources: Na, Mg, Al, Fe, Cu.
4. Distinguish between oxidation and reduction; explain what is meant by electrolysis.

LECTURE NOTES

This is the first of several optional descriptive chapters. Its purpose is to show how the basic principles introduced in Chapters 1-3 are applied in the "real world". This is achieved by discussing methods used commercially to extract elements from the air, the oceans, or the earth's crust. The mole concept (Chapter 2), chemical formulas (Chapter 3), and mass relations in balanced equations (Chapter 3) are reviewed in exercises within Chapter 4 and problems at the end of the chapter.

Depending upon your preferences and the needs of your students, you may choose to follow any of three paths with Chapter 4:

1. Skip it completely. The material in this chapter is of an applied nature; no new principles are introduced. Future chapters do not build upon this one, so omitting Chapter 4 will not cause any organizational problems.

2. Make it a reading assignment, not covered in lecture. If you choose this alternative, you will probably want to assign several of the problems at the end of the chapter. Note that Problems 4.1-4.12 and 4.31-4.42 are descriptive; they refer to the chemistry involved in the extraction of elements. The remaining problems are of a review nature and are classified according to topic (e.g., Problems 4.27-4.30 and 4.57-4.60 cover limiting reactant and theoretical yield).

3. Cover it in lecture in the usual manner. Here, you may wish to spend much of your time working examples that review principles covered in Chapters 1-3. It may also be useful to show demonstrations illustrating some of the chemistry discussed in this

chapter (see p. 31). It has been our experience that a long, unbroken monologue devoted entirely to descriptive chemistry can be deadly dull.

Assuming you follow the third alternative, about two lectures will be required.

LECTURE 1

I Sources of the Nonmetals

A. N_2, O_2 and Ar. Occur in air (mole % N_2 = 78, O_2 = 21, Ar = 1). Recovered by fractional distillation. Liquid air at -200°C allowed to warm up; N_2 (bp = -196°C) boils off first, followed by Ar (bp = -186°C) and O_2 (bp = -183°C). Other noble gases (Ne, Kr, Xe) obtained in trace amounts.

B. Sulfur Obtained by Frasch process from underground deposits. Sulfur (mp = 119°C) melted by superheated water, brought to surface by compressed air.

C. Hydrogen

1. Cheapest source is methane, which is heated with steam:

$$CH_4(g) + 2\ H_2O(g) \longrightarrow CO_2(g) + 4\ H_2(g)$$

CO_2 is removed by bubbling through water. Mass percent of H_2 in CO_2-H_2 mixture?

$$\text{mass } CO_2 = 1 \text{ mol } CO_2 \times \frac{44.01 \text{ g } CO_2}{1 \text{ mol } CO_2} = 44.01 \text{ g } CO_2$$

$$\text{mass } H_2 = 4 \text{ mol } H_2 \times \frac{2.016 \text{ g } H_2}{1 \text{ mol } H_2} = 8.06 \text{ g } H_2$$

$$\%\ H_2 = \frac{8.06 \text{ g}}{52.07 \text{ g}} \times 100 = 15.5$$

2. Electrolysis of water: $2\ H_2O(l) \longrightarrow 2\ H_2(g) + O_2(g)$

3. Reaction of zinc with acid:

$Zn(s) \longrightarrow Zn^{2+}(aq) + 2\ e^-$ oxidation (loss of electrons)

$2\ H^+(aq) + 2\ e^- \longrightarrow H_2(g)$ reduction (gain of electrons)

$Zn(s) + 2\ H^+(aq) \longrightarrow Zn^{2+}(aq) + H_2(g)$

Volume of 12.0 M HCl (H^+, Cl^- ions) required to produce 1.00 g H_2?

$$1.00 \text{ g } H_2 \times \frac{1 \text{ mol } H_2}{2.016 \text{ g } H_2} \times \frac{2 \text{ mol } H^+}{1 \text{ mol } H_2} \times \frac{1 \text{ L HCl}}{12.0 \text{ mol } H^+} = 0.0827 \text{ L}$$

D. Halogens. Produced by oxidation of halide ions:

$$2\ X^- \longrightarrow X_2 + 2\ e^-$$

1. Br_2 and I_2 produced by oxidation of Br^-, I^- with Cl_2:

$Cl_2(g) + 2\ Br^-(aq) \longrightarrow 2\ Cl^-(aq) + Br_2(l)$

$Cl_2(g) + 2\ I^-(aq) \longrightarrow 2\ Cl^-(aq) + I_2(s)$

Theoretical yield I_2, starting with 1.00 g Cl_2, 1.00 g I^-?

If Cl_2 is limiting:

$$1.00\text{ g } Cl_2 \times \frac{1\text{ mol } Cl_2}{70.90\text{ g } Cl_2} \times \frac{1\text{ mol } I_2}{1\text{ mol } Cl_2} \times \frac{253.8\text{ g } I_2}{1\text{ mol } I_2} = 3.58\text{ g } I_2$$

If I^- is limiting: mass I_2 = 1.00 g (all $I^- \longrightarrow I_2$)

Hence, I^- is limiting, Cl_2 is in excess, theor. yield = 1.00 g

2. Cl_2 and F_2 produced by electrolysis.

a. $2\ HF(l) \longrightarrow H_2(g) + F_2(g)$

b. $2\ Cl^-(aq) + 2\ H_2O \longrightarrow Cl_2(g) + H_2(g) + 2\ OH^-(aq)$

What mass of Cl_2 can be produced by the electrolysis of 10.0 L of 2.50 M NaCl?

$$10.0\text{ L} \times \frac{2.50\text{ M } Cl^-}{1\text{ L}} \times \frac{1\text{ mol } Cl_2}{2\text{ mol } Cl^-} \times \frac{70.90\text{ g } Cl_2}{1\text{ mol } Cl_2} = 886\text{ g } Cl_2$$

LECTURE 2

I Recovery of Metals From Ores

A. May occur as chlorides (NaCl), carbonates ($CaCO_3$), phosphates ($LaPO_4$), sulfides (Cu_2S), oxides (Al_2O_3, Fe_2O_3).
Simplest formula dolomite (13.18% Mg, 21.74% Ca, 13.03% C, 52.06% O)?

moles Mg = $\frac{13.18}{24.30}$ = 0.5424; moles Ca = $\frac{21.74}{40.08}$ = 0.5424

moles C = $\frac{13.03}{12.01}$ = 1.085; moles O = $\frac{52.06}{16.00}$ = 3.254

simplest formula: $MgCaC_2O_6$; $MgCO_3 \cdot CaCO_3$

B. Na from NaCl

Electrolysis: $2\ NaCl(l) \longrightarrow 2\ Na(l) + Cl_2(g)$

Na^+ ions reduced to Na atoms, Cl^- ions oxidized to Cl_2

C. Mg from seawater

1. $Mg^{2+}(aq) + 2\ OH^-(aq) \longrightarrow Mg(OH)_2(s)$

What mass of $Ca(OH)_2$ is requured to precipitate the Mg^{2+} in 0.500 L of seawater (M of Mg^{2+} = 0.052 mol/L)?

$$0.500\text{ L} \times \frac{0.052\text{ mol } Mg^{2+}}{1\text{ L}} \times \frac{1\text{ mol } Ca(OH)_2}{1\text{ mol } Mg^{2+}} \times \frac{74.10\text{ g } Ca(OH)_2}{1\text{ mol } Ca(OH)_2}$$

= 1.93 g $Ca(OH)_2$

2. $Mg(OH)_2(s) + 2\ H^+(aq) \longrightarrow Mg^{2+}(aq) + 2\ H_2O$

evaporation gives solid $MgCl_2$

3. $MgCl_2(l) \longrightarrow Mg(s) + Cl_2(g)$; electrolysis molten $MgCl_2$

D. Al from Al_2O_3

Electrolysis: $2\ Al_2O_3(l) \longrightarrow 4\ Al(s) + 3\ O_2(g)$

Cryolite, Na_3AlF_6, added to lower melting point.

E. Fe from Fe_2O_3

1. Reduction of Fe^{3+} ions to Fe:

   $Fe_2O_3(s) + 3\ CO(g) \longrightarrow 2\ Fe(l) + 3\ CO_2(g)$

2. Slag formation with SiO_2 impurity:

   $CaCO_3(s) + SiO_2(s) \longrightarrow CaSiO_3(s) + CO_2(g)$

3. Conversion of pig iron to steel. Burn C to CO_2, Si to SiO_2, Mn to MnO, P to P_4O_{10}.

F. Cu from Cu_2S

Ore is concentrated by flotation, then "roasted" by heating in air or O_2:

$Cu_2S(s) + O_2(g) \longrightarrow 2\ Cu(s) + SO_2(g)$

Impure ("blister") copper formed is purified by electrolysis. Suppose 1.00 kg of an ore containing 12.2% Cu_2S produces 95.8 g of blister copper (89.0% Cu). Percent yield?

$$\text{theor. yield} = 122\text{ g } Cu_2S \times \frac{1\text{ mol } Cu_2S}{159.16\text{ g } Cu_2S} \times \frac{2\text{ mol Cu}}{1\text{ mol } Cu_2S}$$

$$\times \frac{63.55\text{ g Cu}}{1\text{ mol Cu}} = 97.4\text{ g Cu}$$

actual yield = 95.8 g x 0.890 = 85.3 g Cu

$$\%\text{ yield} = \frac{85.3}{97.4} \times 100 = 87.6$$

DEMONSTRATIONS

1. Electrolysis of water: Test Dem. 5; J. Chem. Educ. 58 1017 (1981)
2. Preparation of O_2 from H_2O_2: Test. Dem. 7, 55, 127, 152
3. Reaction of O_2 with H_2: Test. Dem. 9, 59
4. Flotation of ores: Test. Dem. 25, 93
5. Photochemical reaction of H_2 with Cl_2: Test. Dem. 43, 154

QUIZZES

Quiz 1

1. Write a balanced equation for the reaction that occurs when
   a. Chlorine gas is bubbled through a solution of sodium iodide.
   b. Iron(III) oxide is reduced by CO.
   c. Aluminum oxide is electrolyzed.
2. For the reaction in (1b), calculate:
   a. the mass of iron formed from 2.54 kg of iron(III) oxide.
   b. the number of moles of CO required to react with 2.54 kg of iron(III) oxide.

Quiz 2

1. Give the formula of the principal ore of
   a. iron b. aluminum c. sulfur d. fluorine

2. Explain, in 25 words or less, how steel differs from pig iron.

3. The principal ore of boron is borax, which is a hydrate of general composition $Na_2B_4O_7 \cdot xH_2O$. Heating a 1.000 g sample of borax drives off all the water, forming 0.4724 g of $H_2O(g)$. What is the value of x in the formula of borax?

Quiz 3

1. Write a balanced equation for the reaction that occurs when
   a. Methane reacts with steam to form hydrogen gas.
   b. Magnesium chloride is electrolyzed.
   c. $CaCO_3$ reacts with silicon dioxide.

2. Consider the reaction that occurs in the blast furnace:

$$Fe_2O_3(s) + 3\ CO(g) \longrightarrow 2\ Fe(l) + 3\ CO_2(g)$$

   What is the theoretical yield of iron, starting with 1.00 g of Fe_2O_3 and 0.500 g of CO?

Quiz 4

1. Describe, with the aid of balanced equations, how magnesium metal is obtained from seawater (three steps).

2. Consider the roasting of zinc sulfide ore:

$$2\ ZnS(s) + 3\ O_2(g) \longrightarrow 2\ ZnO(s) + 2\ SO_2(g)$$

   a. What mass of ZnO can be formed from 1.62×10^3 kg of ZnS?
   b. What volume of SO_2 (d = 2.61 g/L) is produced when 1.29 mol of ZnS reacts?

Quiz 5

1. Write a balanced equation for the reaction that occurs when
   a. Copper(I) sulfide is roasted in air.
   b. Sodium chloride is electrolyzed.
   c. Chlorine gas is passed through a solution of sodium bromide.

2. How many moles of metal can be obtained from 1.00 kg of each of the following ores?

   a. Al_2O_3 b. Cu_2S c. $MgCO_3 \cdot CaCO_3$

PROBLEMS 1-30

4.1 a. earth b. air c. earth d. earth e. oceans f. air g. earth

4.2 a. oxide b. sulfide c. oxide d. neither e. sulfide f. neither

4.3 a. fractional distillation liquid air
b. treatment of salt brines with chlorine
c, d, e. See discussion Section 4.3

4.4 a. Boiling point Ar closer to O_2 than N_2.
b. Natural gas is a cheaper source.

4.4 c. Argon is much less reactive than sodium or chlorine.
d. cryolite lowers the melting point.

4.5 a. Steel contains small amounts of carbon.
b. Copper metal is formed.
c. Copper is formed by reduction of Cu^{2+} ions.
d. Chlorine molecules are involved, not Cl^- ions.

4.6 a. $CH_4(g) + 2\ H_2O(g) \longrightarrow CO_2(g) + 4\ H_2(g)$
b. $2\ H_2O(l) \longrightarrow 2\ H_2(g) + O_2(g)$
c. $Cl_2(g) + 2\ Br^-(aq) \longrightarrow 2\ Cl^-(aq) + Br_2(l)$
d. $2\ NaCl(l) \longrightarrow 2\ Na(l) + Cl_2(g)$

4.7 a. $Cu_2S(s) + O_2(g) \longrightarrow 2\ Cu(s) + SO_2(g)$
b. $Fe_2O_3(s) + 3\ CO(g) \longrightarrow 2\ Fe(l) + 3\ CO_2(g)$
c. $CaCO_3(s) \longrightarrow CaO(s) + CO_2(g)$
d. $2\ Al_2O_3(l) \longrightarrow 4\ Al(l) + 3\ O_2(g)$

4.8 a. $Br_2(l) + 2\ I^-(aq) \longrightarrow 2\ Br^-(aq) + I_2(s)$
b. $6\ CaO(s) + P_4O_{10}(s) \longrightarrow 2\ Ca_3(PO_4)_2(s)$

4.9 a. $Cl_2(g) + 2\ e^- \longrightarrow 2\ Cl^-(aq)$ reduction
$2\ Br^-(aq) \longrightarrow Br_2(l) + 2\ e^-$ oxidation
b. $2\ Na^+ + 2\ e^- \longrightarrow 2\ Na(s)$ reduction
$2\ Cl^- \longrightarrow Cl_2(g) + 2\ e^-$ oxidation
c. $Mg(s) \longrightarrow Mg^{2+}(aq) + 2\ e^-$ oxidation
$Cl_2(g) + 2\ e^- \longrightarrow 2\ Cl^-(aq)$ reduction

4.10 a. loss of electrons
b. passage of electric current through liquid, causing a reaction to take place.
c. natural mineral from which metal can be profitably extracted.
d. metal toward the center of Periods 4, 5, 6.

4.11 a. synthesis of NH_3 b. production of H_2SO_4
c. preparation of lead tetraethyl d. structural metal
e. coinage, alloys

4.12 a. $6.7 \times 10^7/5.0 \times 10^8 = 0.13$ b. $4.5 \times 10^6/1.6 \times 10^7 = 0.28$
c. $1.2 \times 10^6/7.8 \times 10^6 = 0.15$ d. $1.1 \times 10^6/5.4 \times 10^6 = 0.20$
e. $4.0 \times 10^4/7.6 \times 10^5 = 0.053$

4.13 a. no. moles $= 100.0\ g \times \frac{1\ mol}{159.70\ g} = 0.6262$ mol Fe_2O_3
b. no. moles $= 100.0\ g \times \frac{1\ mol}{101.96\ g} = 0.9808$ mol Al_2O_3
c. no. moles $= 100.0\ g \times \frac{1\ mol}{159.16\ g} = 0.6283$ mol Cu_2S
d. no. moles $= 100.0\ g \times \frac{1\ mol}{58.44\ g} = 1.711$ mol NaCl

4.14 a. $1.65 \text{ mol} \times \frac{159.81 \text{ g}}{1 \text{ mol}} = 264 \text{ g } Br_2$

b. $1.65 \text{ mol} \times \frac{63.55 \text{ g}}{1 \text{ mol}} = 105 \text{ g Cu}$

c. $1.65 \text{ mol} \times \frac{65.38 \text{ g}}{1 \text{ mol}} = 108 \text{ g Zn}$

d. $1.65 \text{ mol} \times \frac{32.06 \text{ g}}{1 \text{ mol}} = 52.9 \text{ g S}$

e. $1.65 \text{ mol} \times \frac{123.90 \text{g}}{1 \text{ mol}} = 204 \text{ g } P_4$

4.15 formula mass $CuFeS_2 = 63.55 + 55.85 + 2(32.06) = 183.52$

mass % Cu = 100 x 63.55/183.52 = 34.63

$0.1250 \text{ mol} \times \frac{183.52 \text{ g}}{1 \text{ mol}} = 22.94 \text{ g}$; $22.94 \text{ g} \times 0.3463 = 7.944 \text{ g Cu}$

$6.36 \text{ g} \times \frac{1 \text{ mol}}{183.52 \text{ g}} = 0.0347 \text{ mol}$; $6.36 \text{ g} \times 0.3463 = 2.20 \text{ g Cu}$

$\frac{14.82 \text{ g}}{0.3463} = 42.80 \text{ g } CuFeS_2$; $42.80 \text{ g} \times \frac{1 \text{ mol}}{183.52 \text{ g}} = 0.2332 \text{ mol}$

4.16 a. $\frac{(34.0 \times 10^{-3})/24.30 \text{ mol}}{1 \text{ L}} = 1.40 \times 10^{-3} \text{ mol/L}$

b. $2.50 \text{ mol} \times \frac{1 \text{ L}}{1.40 \times 10^{-3} \text{ mol}} = 1.79 \times 10^3 \text{ L}$

4.17 a. $0.750 \text{ L} \times \frac{18.0 \text{ mol}}{1 \text{ L}} = 13.5 \text{ mol}$

b. $13.5 \text{ mol} \times \frac{98.08 \text{ g}}{1 \text{ mol}} = 1.32 \times 10^3 \text{ g}$

c. total mass = $750 \text{ cm}^3 \times 1.84 \text{ g/cm}^3 = 1380 \text{ g}$

mass water = 1380 g - 1320 g = 60 g

4.18 a. formula mass = 58.93 + 74.92 + 32.06 = 165.91

mass % Co = 100 x 58.93/165.91 = 35.52

b. $75.0 \text{ g Co} \times \frac{165.91 \text{ g CoAsS}}{58.93 \text{ g Co}} = 211 \text{ g CoAsS}$

4.19 a. no. moles Cu = $57.5 \text{ g Cu} \times \frac{1 \text{ mol Cu}}{63.55 \text{ g Cu}} = 0.905 \text{ mol Cu}$

no. moles C = $5.43 \text{ g C} \times \frac{1 \text{ mol C}}{12.01 \text{ g C}} = 0.452 \text{ mol C}$

no. moles O = $36.2 \text{ g O} \times \frac{1 \text{ mol O}}{16.00 \text{ g O}} = 2.26 \text{ mol O}$

no. moles H = $0.914 \text{ g H} \times \frac{1 \text{ mol H}}{1.008 \text{ g H}} = 0.907 \text{ mol H}$

simplest formula: $Cu_2CO_5H_2$

b. $CuCO_3 \cdot Cu(OH)_2$

4.20 no. moles Mn = $63.2 \text{ g Mn} \times \frac{1 \text{ mol Mn}}{54.94 \text{ g Mn}} = 1.15 \text{ mol Mn}$

no. moles O = $36.8 \text{ g O} \times \frac{1 \text{ mol O}}{16.00 \text{ g O}} = 2.30 \text{ mol O}$

simplest formula MnO_2

4.21 a. formula mass = 2(55.85) + 3(16.00) = 159.70
% Fe = 100 x 111.70/159.70 = 69.94

b. formula mass = 40.08 + 32.06 + 64.00 + 2(18.02) = 172.18
% Ca = 100 x 40.08/172.18 = 23.28

c. formula mass = 2(26.98) + 3(16.00) + 2(18.02) = 138.00
% Al = 100 x 53.96/138.00 = 39.10

d. formula mass = 2(65.38) + 28.09 + 4(16.00) = 222.85
% Zn = 100 x 130.76/222.85 = 58.68

4.22 a. $2.00 \times 10^3 \text{ g } Al_2O_3 \times \frac{53.96 \text{ g Al}}{101.96 \text{ g } Al_2O_3} = 1.06 \times 10^3 \text{ g Al}$

b. $2.00 \times 10^3 \text{ g } Al_2O_3 \times \frac{48.00 \text{ g O}}{101.96 \text{ g } Al_2O_3} = 9.42 \times 10^2 \text{ g O}$

c. $9.42 \times 10^2 \text{ g } O_2 \times 0.15 \times \frac{1 \text{ mol } O_2}{32.00 \text{ g } O_2} \times \frac{1 \text{ mol } CO_2}{1 \text{ mol } O_2}$
$= 4.42 \text{ mol } CO_2$

4.23 a. $3.50 \text{ mol Fe} \times \frac{3 \text{ mol CO}}{2 \text{ mol Fe}} = 5.25 \text{ mol CO}$

b. $0.500 \text{ mol CO} \times \frac{1 \text{ mol } Fe_2O_3}{3 \text{ mol CO}} \times \frac{159.70 \text{ g } Fe_2O_3}{1 \text{ mol } Fe_2O_3} = 26.6 \text{ g } Fe_2O_3$

c. $1.00 \times 10^6 \text{ g } Fe_2O_3 \times \frac{1 \text{ mol } Fe_2O_3}{159.70 \text{ g } Fe_2O_3} \times \frac{3 \text{ mol C}}{1 \text{ mol } Fe_2O_3}$
$\times \frac{12.01 \times 10^{-3} \text{ kg C}}{1 \text{ mol C}} = 2.26 \times 10^2 \text{ kg C}$

4.24 a. $CaSiO_3(s) + 6\ HF(l) \longrightarrow CaF_2(s) + 3\ H_2O(l) + SiF_4(g)$

b. $1.40 \text{ mol HF} \times \frac{3 \text{ mol } H_2O}{6 \text{ mol HF}} = 0.700 \text{ mol } H_2O$

c. $10.0 \text{ g } CaSiO_3 \times \frac{1 \text{ mol } CaSiO_3}{116.17 \text{ g } CaSiO_3} \times \frac{6 \text{ mol HF}}{1 \text{ mol } CaSiO_3} \times \frac{20.01 \text{ g HF}}{1 \text{ mol HF}}$
$= 10.3 \text{ g HF}$

d. $10.0 \text{ g } CaSiO_3 \times \frac{1 \text{ mol } CaSiO_3}{116.17 \text{ g } CaSiO_3} \times \frac{1 \text{ mol } CaF_2}{1 \text{ mol } CaSiO_3}$
$\times \frac{78.08 \text{ g } CaF_2}{1 \text{ mol } CaF_2} = 6.72 \text{ g } CaF_2$

4.25 a. $1.00 \text{ g } Mg(OH)_2 \times \frac{1 \text{ mol } Mg(OH)_2}{58.32 \text{ g } Mg(OH)_2} \times \frac{2 \text{ mol } H^+}{1 \text{ mol } Mg(OH)_2}$
$= 0.0343 \text{ mol } H^+$

b. $0.0343 \text{ mol HCl} \times \frac{1 \text{ L}}{6.00 \text{ mol HCl}} = 0.00572 \text{ L} = 5.72 \text{ cm}^3$

4.26 mass O_2 = 0.550 L x 1.31 g/L = 0.720 g
mass Cu = $0.720 \text{ g } O_2 \times \frac{1 \text{ mol } O_2}{32.00 \text{ g } O_2} \times \frac{2 \text{ mol Cu}}{1 \text{ mol } O_2} \times \frac{63.55 \text{ g Cu}}{1 \text{ mol Cu}}$
= 2.86 g Cu
mass % Cu = 100 x 2.86/10.0 = 28.6

4.27 a. mass CO = 1.00×10^3 L x 1.14 g/L = 1.14×10^3 g

mass Fe_2O_3 = 1.20×10^3 g

$$1.14 \times 10^3 \text{ g CO} \times \frac{1 \text{ mol CO}}{28.01 \text{ g CO}} \times \frac{2 \text{ mol Fe}}{3 \text{ mol CO}} \times \frac{55.85 \text{ g Fe}}{1 \text{ mol Fe}}$$

= 1.52×10^3 g Fe

$$1.20 \times 10^3 \text{ g } Fe_2O_3 \times \frac{1 \text{ mol } Fe_2O_3}{159.70 \text{ g } Fe_2O_3} \times \frac{2 \text{ mol Fe}}{1 \text{ mol } Fe_2O_3} \times \frac{55.85 \text{gFe}}{1 \text{ molFe}}$$

= 8.39×10^2 g Fe

Fe_2O_3 is limiting

b. 839 g c. 100 x 612/839 = 72.9

4.28 a. $1.25 \text{ mol } NH_4^+ \times \frac{1 \text{ mol } N_2}{1 \text{ mol } NH_4^+} = 1.25 \text{ mol } N_2$

$1.30 \text{ mol } NO_2^- \times \frac{1 \text{ mol } N_2}{1 \text{ mol } NO_2^-} = 1.30 \text{ mol } N_2$

NH_4^+ is limiting

b. $1.00 \text{ g } NH_4^+ \times \frac{1 \text{ mol } NH_4^+}{18.04 \text{ g } NH_4^+} \times \frac{1 \text{ mol } N_2}{1 \text{ mol } NH_4^+} = 0.0554 \text{ mol } N_2$

$1.00 \text{ g } NO_2^- \times \frac{1 \text{ mol } NO_2^-}{46.01 \text{ g } NO_2^-} \times \frac{1 \text{ mol } N_2}{1 \text{ mol } NO_2^-} = 0.0217 \text{ mol } N_2$

NO_2^- is limiting

c. $15.6 \text{ g } NH_4^+ \times \frac{1 \text{ mol } NH_4^+}{18.04 \text{ g } NH_4^+} \times \frac{1 \text{ mol } N_2}{1 \text{ mol } NH_4^+} = 0.865 \text{ mol } N_2$

$0.864 \text{ mol } NO_2^- \times \frac{1 \text{ mol } N_2}{1 \text{ mol } NO_2^-} = 0.864 \text{ mol } NO_2^-$

Within experimental error, there are equivalent amounts of the two reactants.

4.29 a. $2.40 \text{ mol } Cl_2 \times \frac{1 \text{ mol } Br_2}{1 \text{ mol } Cl_2} = 2.40 \text{ mol } Br_2$

$5.00 \text{ L} \times \frac{0.500 \text{ mol } Br^-}{1 \text{ L}} \times \frac{1 \text{ mol } Br_2}{2 \text{ mol } Br^-} = 1.25 \text{ mol } Br_2$

Br^- is limiting

b. 1.25 mol Br_2 c. 100 x 1.24/1.25 = 99.2%

4.30 a. $5.00 \times 10^2 \text{ g } MnO_2 \times \frac{1 \text{ mol } MnO_2}{86.94 \text{ g } MnO_2} \times \frac{3 \text{ mol Mn}}{3 \text{ mol } MnO_2} \times \frac{54.94 \text{ g Mn}}{1 \text{ mol Mn}}$

= 316 g Mn

$1.00 \times 10^2 \text{ g Al} \times \frac{1 \text{ mol Al}}{26.98 \text{ g Al}} \times \frac{3 \text{ mol Mn}}{4 \text{ mol Al}} \times \frac{54.94 \text{ g Mn}}{1 \text{ mol Mn}}$

= 153 g Mn

Al is limiting

4.30 b. 153 g

c. $5.00 \times 10^2 \text{ g } MnO_2 \times \frac{1 \text{ mol } MnO_2}{86.94 \text{ g } MnO_2} \times \frac{4 \text{ mol Al}}{3 \text{ mol } MnO_2}$

$\times \frac{26.98 \text{ g Al}}{1 \text{ mol Al}} = 207 \text{ g Al}$

207 g Al - 100 g Al = 107 g Al

CHAPTER 5

BASIC SKILLS

1. Relate the direction of heat flow in a reaction (exothermic or endothermic) to the sign of ΔH.

2. Given a thermochemical equation, calculate:
   a. ΔH for a specified amount of reactant or product.
   b. the amount of reactant or product required to produce a specified ΔH.

3. Use Hess' Law to calculate ΔH for
   a. a stepwise reaction, knowing the value of ΔH for each step.
   b. a step in a reaction, knowing ΔH for every other step and for the overall reaction.

4. Use the general relation: $\Delta H = \sum \Delta H_f$ products $- \sum \Delta H_f$ reactants to calculate:
   a. ΔH for a reaction, knowing the heats of formation of all species.
   b. ΔH_f for one species, knowing the heats of formation of all other species as well as ΔH for the reaction.

5. Use calorimetric data (temperature change, mass of water, specific heat of water, calorimeter constant) to determine the heat flow for a reaction.

6. Use the First Law of Thermodynamics to calculate one of the three quantities ΔE, q or w, knowing the values of the other two quantities.

7. Knowing the value of $\Delta(PV)$ for a reaction, relate ΔH to ΔE.

LECTURE NOTES

Typically, students find this chapter more difficult than any covered previously. In part this reflects the fact that this is new material, seldom covered in high school. Moreover, thermochemistry is an abstract subject; students have trouble relating it to the real world. For this reason, Section 5.1 includes a brief introduction to calorimetry with the hope that it will help students to understand where ΔH values come from.

Thermochemical equations (Section 5.2) extend the conversion factor approach to heat flow. It is important to emphasize that ΔH is directly proportional to amount of reactant or product. Hess' Law is not stressed heavily in this chapter; instead we emphasize Equation 5.10, which relates ΔH to heats of formation.

In discussing calorimetry, it is important to point out that ΔH for the reaction is equal in magnitude but opposite in sign to q water + q bomb. Students find the idea of the "calorimeter constant", C, hard to grasp. It may help to point out that it is really the product of mass times specific heat.

Many instructors delete the discussion of the First Law (Section 5.5); it is not required as background in any subsequent chapter. If you cover it, keep in mind that college freshmen seldom appreciate the elegant logic of thermodynamics. Try to relate ΔE and ΔH to real processes. Section 5.6, on sources of energy, can also be deleted or made a reading assignment. However, this is an area that students are curious about; covering it in lecture adds interest to a rather abstract chapter.

This chapter requires a minimum of two lectures and could expand to three if you spend a lot of time on Sections 5.5 and 5.6. In the outlines that follow, we strike an average and include material for $2\frac{1}{2}$ lectures.

LECTURE 1

I Thermochemistry

A. General concepts

1. Exothermic reaction: heat flows into surroundings, raising their temperature. Examples include the combustion of natural gas and the condensation of steam. Endothermic reaction: heat absorbed from surroundings, lowering their temperature. Examples: melting of ice, vaporization of water.
2. Heat flow is accounted for by a change in "heat flow" or enthalpy of the reaction system.
   a. Endothermic reaction: $\Delta H > 0$ (i.e., H products > H reactants). Heat absorbed goes to increase the enthalpy of the reaction system.
   b. Exothermic reaction: $\Delta H < 0$ (i.e., H products < H reactants). Heat is evolved at the expense of the reaction system.
3. Thermochemical equation: specify ΔH in kilojoules

   $CH_4(g) + 2O_2(g) \longrightarrow CO_2(g) + 2H_2O(l)$; $\Delta H = -890.3$ kJ

   890.3 kJ of heat is evolved when one mole of CH_4 burns

   $H_2O(s) \longrightarrow H_2O(l)$; $\Delta H = +6.00$ kJ

   6.00 kJ of heat must be absorbed to melt one mole of ice.

B. Laws of Thermochemistry

1. ΔH is directly proportional to amount of reactants or products. When one mole of ice melts, 6.00 kJ of heat is absorbed, $\Delta H = +6.00$ kJ. If one gram of ice melts, $\Delta H =$ 6.00 kJ/18.0 = +0.333 kJ.

   In general, ΔH can be related to amount by the conversion factor approach:

   $CH_4(g) + 2O_2(g) \longrightarrow CO_2(g) + 2H_2O(l)$; $\Delta H = -890.3$ kJ

Value of ΔH when 2.16 g CH_4 burns:

$$1 \text{ mol } CH_4 \simeq -890.3 \text{ kJ}$$

$$\Delta H = 2.16 \text{ g } CH_4 \times \frac{1 \text{ mol } CH_4}{16.04 \text{ g } CH_4} \times \frac{-890.3 \text{ kJ}}{1 \text{ mol } CH_4} = -120 \text{ kJ}$$

2. ΔH for a reaction is equal in magnitude but opposite in sign to ΔH for the reverse reaction.

$H_2O(s) \rightarrow H_2O(l)$; $\Delta H = +6.00$ kJ; 6.00 kJ absorbed

$H_2O(l) \rightarrow H_2O(s)$; $\Delta H = -6.00$ kJ; 6.00 kJ evolved

3. Hess' Law. If Equation 3 = Equation 1 + Equation 2, then

$$\Delta H_3 = \Delta H_1 + \Delta H_2$$

Often used to calculate ΔH for one step, knowing ΔH for all other steps and for the overall reaction.

$C(s) + \frac{1}{2} O_2(g) \rightarrow CO(g)$; $\Delta H_1 = ?$

$CO(g) + \frac{1}{2} O_2(g) \rightarrow CO_2(g)$; $\Delta H_2 = -283.0$ kJ

$C(s) + O_2(g) \rightarrow CO_2(g)$; $\Delta H_3 = -393.5$ kJ

$\Delta H_1 - 283.0$ kJ $= -393.5$ kJ; $\Delta H_1 = -110.5$ kJ

C. Heats of Formation

1. Meaning. ΔH_f of compound = ΔH when one mole of compound is formed from the elements in their stable states.

$2Ag(s) + Cl_2(g) \rightarrow 2AgCl(s)$; $\Delta H = -254.0$ kJ
ΔH_f AgCl = -127.0 kJ/mol

$HgO(s) \rightarrow Hg(l) + \frac{1}{2}O_2(g)$; $\Delta H = +90.7$ kJ
ΔH_f HgO = -90.7 kJ/mol

Heats of formation are usually negative; heat is evolved when a compound is formed.

2. Usefulness For any thermochemical equation:

$$\Delta H = \sum \Delta H_f \text{ products} - \sum \Delta H_f \text{ reactants}$$

Take ΔH_f of element in its stable state to be zero.

$CH_4(g) + 2\,O_2(g) \rightarrow CO_2(g) + 2\,H_2O(l)$

$\Delta H = \Delta H_f\, CO_2(g) + 2\,\Delta H_f\, H_2O(l) - \Delta H_f\, CH_4(g)$
$= -393.5$ kJ $+2(-285.8$ kJ$) -(-74.8$ kJ$) = -890.3$ kJ

LECTURE 2

I Thermochemistry

A. Heats of Formation (cont.) $\Delta H = \sum \Delta H_f$ products $- \sum \Delta H_f$ react.

1. Can apply to ions, setting $\Delta H_f\ H^+ = 0$

$Zn(s) + 2H^+(aq) \rightarrow Zn^{2+}(aq) + H_2(g)$

$\Delta H = \Delta H_f\ Zn^{2+} = -152.4$ kJ

2. If ΔH and all but one ΔH_f are known, can calculate that ΔH_f

$2C_2H_6(g) + 7\ O_2(g) \rightarrow 4CO_2(g) + 6H_2O(l)$; $\Delta H = -3204.1$ kJ

$\Delta H_f\ CO_2(g) = -393.5$ kJ/mol, $\Delta H_f\ H_2O(l) = -285.8$ kJ/mol

$-3204.1 \text{ kJ} = 4(-393.5 \text{ kJ}) + 6(-285.8 \text{ kJ}) - 2 \text{ mol}(\Delta H_f\ C_2H_6)$

Solving, $\Delta H_f\ C_2H_6 = -84.7$ kJ/mol

B. Calorimetry

1. Coffee-cup calorimeter

ΔH reaction = - q water (heat given off absorbed by water)

$q \text{ water} = 4.18 \frac{J}{g\cdot{}^{o}C} \times m \text{ water} \times \Delta t$

Suppose heat given off by reaction is absorbed by 412 g of water, increasing the temperature from 20.12 to 29.86oC. ΔH reaction?

$q \text{ water} = 4.18 \frac{J}{g\cdot{}^{o}C} \times 412 \text{ g} \times 9.74^{o}C = 1.68 \times 10^3 \text{ J}$

$\Delta H = -1.68 \times 10^3$ J (reaction evolves heat absorbed by water)

2. Bomb calorimeter

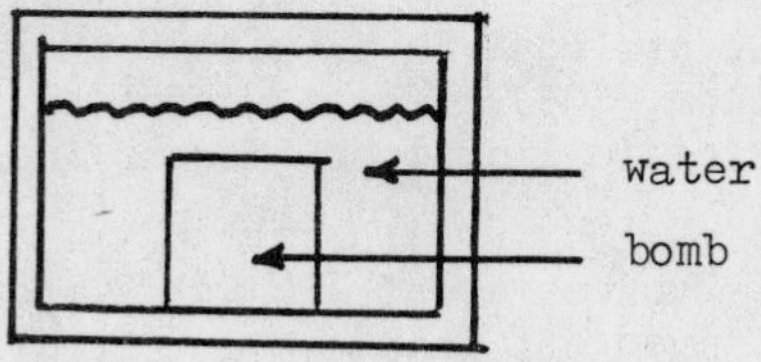

some heat is absorbed by metal in the bomb

$q \text{ reaction} = -(q \text{ water} + q \text{ bomb})$

$= -(4.18 \frac{J}{g\cdot{}^{o}C} \times m \text{ water} \times \Delta t + C\Delta t)$

where C is a constant (mass bomb times its specific heat)

Suppose combustion of 1.60 g CH_4 in a bomb calorimeter raises t by 5.16oC. Calorimeter contains 3970 g of water, C = 562 J/oC. Calculate q reaction for:

$CH_4(g) + 2\ O_2(g) \rightarrow CO_2(g) + 2\ H_2O(l)$;

For 1.60 g CH_4:

$q \text{ reaction} = -(4.18 \frac{J}{g\cdot{}^{o}C} \times 3970 \text{ g} \times 5.16^{o}C + 562 \frac{J}{{}^{o}C} \times 5.16^{o}C)$

$= -88500 \text{ J} = -88.5 \text{ kJ}$

For one mole of methane:

$q \text{ reaction} = 1.00 \text{ mol } CH_4 \times \frac{16.0 \text{ g } CH_4}{1 \text{ mol } CH_4} \times \frac{-88.5 \text{ kJ}}{1.60 \text{ g } CH_4} = -885 \text{ kJ}$

C. First Law: ΔH and ΔE

1. $\Delta E = q + w$, where ΔE = change in energy of system
q = heat flowing into system
w = work done on system

Suppose gas absorbs 50 J of heat, expands and does 20 J of work upon surroundings:

$\Delta E = 50 \text{ J} - 20 \text{ J} = +30 \text{ J}$

2. $\Delta H = \Delta E + \Delta(PV)$ P = pressure, V = volume

Ordinarily, $\Delta(PV)$ in a reaction is small, much smaller than ΔH or ΔE. For a thermochemical equation:

$$\Delta H \text{ (in kJ)} = \Delta E \text{ (in kJ)} + 2.5\,\Delta n_{gas}$$

where Δn_{gas} is the net change in the number of moles of gas

In a bomb calorimeter, q reaction = ΔE reaction, because w = 0.

$$CH_4(g) + 2\ O_2(g) \longrightarrow CO_2(g) + 2\ H_2O(l)$$

Since q reaction = -885 kJ, ΔE = -885 kJ

ΔH = -885 kJ - 2(2.5)kJ = -890 kJ

LECTURE 2½

I <u>Sources of Energy</u>

About 1/3 of our petroleum is imported; supply limited and uncertain. Possible alternative sources of energy:

A. <u>Coal</u> Main use is in generation of electricity. U.S. has large reserves. However, coal is a "dirty" fuel, a major source of air pollution (SO_2). To be useful in transportation, must be converted to liquid or gaseous fuel.

B. <u>Nuclear energy</u>

1. Fission of U-235 Evolves about 10^8 kJ/g, about 2 million times the energy given off when a gram of petroleum burns. No problem with air pollution. However, intensely radioactive isotopes produced by fission create a safety hazard and a long-term storage problem.
2. Fusion of heavy hydrogen (^2_1H, ^3_1H). Evolves even more energy than fission. Heavy hydrogen is relatively plentiful, products are not radioactive. However, nuclei must be accelerated to very high velocities for reaction to occur. Appears to be impractical as an energy source in this century.

C. <u>Solar energy</u>

1. Solar cells Convert sunlight into electrical energy. Too expensive; very pure Si required.
2. Solar heating In an "active" system, sun's heat is absorbed by water or other material. Circulating system carries heat through house. Must have backup system in cold, cloudy weather. Not economical in cold climates. In passive solar heating, house is designed to absorb as much of sun's heat in winter as feasible.

D. Conservation - simplest, cheapest way to reduce oil imports
More efficient automobiles
Thermostatic control of heating, air conditioning
Insulation of buildings

DEMONSTRATIONS

1. Thermite reaction: Test Dem. 17, 80, 168; J. Chem. Educ. <u>42</u> A607 (1965), <u>56</u> 675 (1979)
2. Endothermic reaction (dissolving NH_4NO_3 in water) Test. Dem. 17
3. Exothermic reactions: Test. Dem. 80

QUIZZES

<u>Quiz 1</u>

1. Given the thermochemical equation:

   $SnO_2(s) + 2\ CO(g) \longrightarrow Sn(s) + 2\ CO_2(g)\ ;\ \Delta H = +14.7$ kJ

   calculate:

   a. the amount of heat absorbed when 2.06 g of tin are formed.

   b. ΔH for the reaction:

   $2\ Sn(s) + 4\ CO_2(g) \longrightarrow 2\ SnO_2(s) + 4\ CO(g)$

   c. the heat of formation of SnO_2, given that the heats of formation of CO_2 and CO are -393.5 kJ/mol and -110.5 kJ/mol, in that order.

<u>Quiz 2</u>

1. When 2.50 g of $KClO_3$ decomposes to KCl and O_2, 908 J of heat is evolved. Calculate ΔH for the thermochemical equation:

   $2\ KClO_3(s) \longrightarrow 2\ KCl(s) + 3\ O_2(g)$

2. Suppose the 908 J of heat evolved in (1) is absorbed by 45.0 g of water (specific heat = 4.18 J/g·°C). What is the increase in temperature?

<u>Quiz 3</u>

1. Using the table of heats of formation in your text, calculate ΔH for the reaction:

   $C_2H_4(g) + 3\ O_2(g) \longrightarrow 2\ CO_2(g) + 2\ H_2O(l)$

2. When 1.00 g of AgCl is formed by the reaction

   $Ag^+(aq) + Cl^-(aq) \longrightarrow AgCl(s)$

   in a coffee-cup calorimeter, the temperature of 20.0 g of water increases from 15.00 to 20.47°C. Taking the specific heat of water to be 4.18 J/g·°C, calculate:

   a. the amount of heat absorbed by the water.

   b. ΔH when one mole of AgCl is formed by this reaction.

<u>Quiz 4</u>

1. The heat of fusion of benzene, C_6H_6, is 9.84 kJ/mol. Calculate ΔH when 1.00 g of liquid benzene freezes.

2. Given the thermochemical equation:

   $SiO_2(s) + 4\ HF(g) \longrightarrow SiF_4(g) + 2\ H_2O(l);\ \Delta H = -185.8$ kJ

   and using the table of heats of formation in your text, calculate ΔH_f $SiF_4(g)$

Quiz 5

1. Given the thermochemical equation:

$CO_3^{2-}(aq) + 2\ H^+(aq) \longrightarrow CO_2(g) + H_2O(l)$; $\Delta H = -3.0$ kJ

calculate:

a. ΔH when 1.40 g of CO_3^{2-} reacts with excess acid.

b. ΔH for the reaction:

$HCO_3^-(aq) + H^+(aq) \longrightarrow CO_2(g) + H_2O(l)$

given that:

$CO_3^{2-}(aq) + H^+(aq) \longrightarrow HCO_3^-(aq)$; $\Delta H = -14.8$ kJ

c. ΔH for the reaction:

$CO_2(g) + H_2O(l) \longrightarrow CO_3^{2-}(aq) + 2\ H^+(aq)$

PROBLEMS 1-30

5.1 a. true b. true c. false d. false

5.2 a. exothermic b. negative

5.3 a. exothermic

b. the product, $FeBr_2$, lies 249.8 kJ below the reactants, Fe and Br_2, on the enthalpy diagram

c. $10.0\text{ g } FeBr_2 \times \frac{1\text{ mol } FeBr_2}{215.66\text{ g } FeBr_2} \times \frac{-249.8\text{ kJ}}{1\text{ mol } FeBr_2} = -11.6\text{ kJ}$

d. $1.00\text{ kJ} \times \frac{1\text{ mol Fe}}{249.8\text{ kJ}} \times \frac{55.85\text{ g Fe}}{1\text{ mol Fe}} = 0.224\text{ g Fe}$

5.4 a. +65.5 kJ

b. $1.00\text{ g AgCl} \times \frac{1\text{ mol AgCl}}{143.32\text{ g AgCl}} \times \frac{65.5\text{ kJ}}{1\text{ mol AgCl}} = 0.457\text{ kJ}$

5.5 a. $4\ KClO_3(s) \longrightarrow KCl(s) + 3\ KClO_4(s)$

b. $1.00\text{ mol } KClO_3 \times \frac{122.55\text{ g } KClO_3}{1\text{ mol } KClO_3} \times \frac{-0.350\text{ kJ}}{1.000\text{ g } KClO_3}$

$= -42.9\text{ kJ}$

c. $0.250\text{ mol } KClO_4 \times \frac{4\text{ mol } KClO_3}{3\text{ mol } KClO_4} \times \frac{-42.9\text{ kJ}}{1\text{ mol } KClO_3} = -14.3\text{ kJ}$

5.6 $q_{water} = 4.18\ \frac{J}{g \cdot {}^\circ C} \times 1.00 \times 10^3\text{ g} \times 5.00^\circ C = 2.09 \times 10^4\text{ J}$

$= 20.9\text{ kJ}$

$\Delta H = -20.9\text{ kJ}$

mass glucose $= -20.9\text{ kJ} \times \frac{1\text{ mol } C_6H_{12}O_6}{-2820\text{ kJ}} \times \frac{180.16\text{ g } C_6H_{12}O_6}{1\text{ mol } C_6H_{12}O_6}$

$= 1.34\text{ g } C_6H_{12}O_6$

5.7 mercury: $\Delta H = 100.0 \text{ g} \times \frac{1 \text{ mol}}{200.59 \text{ g}} \times \frac{2.33 \text{ kJ}}{1 \text{ mol}} = 1.16 \text{ kJ}$

benzene: $\Delta H = 25.0 \text{ g} \times \frac{1 \text{ mol}}{78.11 \text{ g}} \times \frac{30.8 \text{ kJ}}{1 \text{ mol}} = 9.86 \text{ kJ}$

More heat absorbed with benzene

5.8
$$\begin{array}{ll} Na^{+}(g) + Cl^{-}(g) \longrightarrow Na(g) + Cl(g) & \Delta H = -147 \text{ kJ} \\ Na(g) + Cl(g) \longrightarrow Na(s) + \tfrac{1}{2} Cl_2(g) & \Delta H = -230 \text{ kJ} \\ Na(s) + \tfrac{1}{2} Cl_2(g) \longrightarrow NaCl(s) & \Delta H = -411 \text{ kJ} \\ \hline Na^{+}(g) + Cl^{-}(g) \longrightarrow NaCl(s) & \Delta H = -788 \text{ kJ} \end{array}$$

5.9 $+18.4 \text{ kJ} = \Delta H_f \, FeBr_2 - \Delta H_f \, FeBr_3$

$\Delta H_f \, FeBr_3 = \Delta H_f \, FeBr_2 - 18.4 \text{ kJ} = -249.8 \text{ kJ} - 18.4 \text{ kJ}$
$= -268.2 \text{ kJ}$

5.10 a. $\Delta H_f \, Al_2O_3 = \frac{-3339.6 \text{ kJ}}{2} = -1669.8 \text{ kJ}$

b. $10.0 \text{ g } Al_2O_3 \times \frac{1 \text{ mol } Al_2O_3}{101.96 \text{ g } Al_2O_3} \times \frac{-1669.8 \text{ kJ}}{1 \text{ mol } Al_2O_3} = -164 \text{ kJ}$

5.11 For one mole $CaCl_2$:

$\Delta H = \Delta H_f \, Ca^{2+}(aq) + 2 \, \Delta H_f \, Cl^{-}(aq) - \Delta H_f \, CaCl_2(s)$
$= -543.0 \text{ kJ} + 2(-167.4 \text{ kJ}) + 795.0 \text{ kJ} = -82.8 \text{ kJ}$

For one gram $CaCl_2$:

$\Delta H = 1.00 \text{ g } CaCl_2 \times \frac{1 \text{ mol } CaCl_2}{110.99 \text{ g } CaCl_2} \times \frac{-82.8 \text{ kJ}}{1 \text{ mol } CaCl_2} = -0.746 \text{ kJ}$

5.12 a. $\Delta H = 2 \, \Delta H_f \, I^{-} - 2 \, \Delta H_f \, Cl^{-} = 2(-55.9 \text{ kJ}) - 2(-167.4 \text{ kJ})$
$= +223.0 \text{ kJ}$

b. $\Delta H = 2 \, \Delta H_f \, H_2O(l) + \Delta H_f \, Mn^{2+}(aq) - 2 \, \Delta H_f \, Cl^{-}(aq)$
$- \Delta H_f \, MnO_2(s)$
$= 2(-285.8 \text{ kJ}) - 218.8 \text{ kJ} - 2(-167.4 \text{ kJ}) + 519.7 \text{ kJ}$
$= +64.1 \text{ kJ}$

c. $\Delta H = \Delta H_f \, SO_4^{2-}(aq) - \Delta H_f \, H_2SO_4(l) = -907.5 \text{ kJ} + 811.3 \text{ kJ}$
$= -96.2 \text{ kJ}$

5.13 a. $C_2H_2(g) + \frac{5}{2} O_2(g) \longrightarrow 2 \, CO_2(g) + H_2O(l)$

$\Delta H = 2 \, \Delta H_f \, CO_2(g) + \Delta H_f \, H_2O(l) - \Delta H_f \, C_2H_2(g)$
$= 2(-393.5 \text{ kJ}) - 285.8 \text{ kJ} - 226.7 \text{ kJ} = -1299.5 \text{ kJ}$

b. $C_2H_2(g) + 5/2 \, O_2(g) \longrightarrow 2 \, CO_2(g) + H_2O(g)$

$\Delta H = 2 \, \Delta H_f \, CO_2(g) + \Delta H_f \, H_2O(g) - \Delta H_f \, C_2H_2(g)$
$= 2(-393.5 \text{ kJ}) - 241.8 \text{ kJ} - 226.7 \text{ kJ} = -1255.5 \text{ kJ}$

5.14 $-89.5\ kJ = \Delta H_f\ CaSiO_3(s) - \Delta H_f\ SiO_2(s) - \Delta H_f\ CaO(s)$

$\Delta H_f\ CaSiO_3(s) = -89.5\ kJ + \Delta H_f\ SiO_2(s) + \Delta H_f\ CaO(s)$

$= -89.5\ kJ - 859.4\ kJ - 635.5\ kJ = -1584.4\ kJ$

5.15 $-88.0\ kJ = 2\ \Delta H_f\ NO_2(g) + \Delta H_f\ H_2O(l) + \Delta H_f\ SO_4^{2-}(aq)$

$- 2\ \Delta H_f\ NO_3^-(aq) - \Delta H_f\ SO_3^{2-}(aq)$

$\Delta H_f\ SO_3^{2-}(aq) = 88.0\ kJ + 2\ \Delta H_f\ NO_2(g) + \Delta H_f\ H_2O(l)$

$+ \Delta H_f\ SO_4^{2-}(aq) - 2\ \Delta H_f\ NO_3^-(aq)$

$= 88.0\ kJ + 2(33.9\ kJ) - 285.8\ kJ - 907.5\ kJ$

$-2(-206.6\ kJ) = -624.3\ kJ$

5.16 a. $C_8H_{18}(l) + 25/2\ O_2(g) \longrightarrow 8\ CO_2(g) + 9\ H_2O(l)$

$C_2H_5OH(l) + 3\ O_2(g) \longrightarrow 2\ CO_2(g) + 3\ H_2O(l)$

b. C_2H_5OH:

$\Delta H = 2\ \Delta H_f\ CO_2(g) + 3\ \Delta H_f\ H_2O(l) - \Delta H_f\ C_2H_5OH(l)$

$= 2(-393.5\ kJ) + 3(-285.8\ kJ) + 277.6\ kJ = -1366.8\ kJ$

C_8H_{18}:

$\Delta H = 8\ \Delta H_f\ CO_2(g) + 9\ \Delta H_f\ H_2O(l) - \Delta H_f\ C_8H_{18}(l)$

$= 8(-393.5\ kJ) + 9(-285.8\ kJ) + 269.7\ kJ = -5450.5\ kJ$

c. C_2H_5OH:

$$\Delta H = 1.000\ g\ C_2H_5OH \times \frac{1\ mol\ C_2H_5OH}{46.06\ g\ C_2H_5OH} \times \frac{-1366.8\ kJ}{1\ mol\ C_2H_5OH}$$

$= -29.67\ kJ$

C_8H_{18}:

$$\Delta H = 1.000\ g\ C_8H_{18} \times \frac{1\ mol\ C_8H_{18}}{114.22\ g\ C_8H_{18}} \times \frac{-5450.5\ kJ}{1\ mol\ C_8H_{18}}$$

$= -47.72\ kJ$

Octane gives off more heat per gram

5.17 a. $q_{water} = 4.18\ \frac{J}{g \cdot {}^\circ C} \times 100.0\ g \times 10.7^\circ C = 4.47 \times 10^3\ J$

$= 4.47\ kJ$

$\Delta H = -4.47\ kJ$

b. $$\Delta H = 1.00\ mol\ CaCl_2 \times \frac{110.99\ g\ CaCl_2}{1\ mol\ CaCl_2} \times \frac{-4.47\ kJ}{6.00\ g\ CaCl_2}$$

$= -82.7\ kJ$

5.18 $$\Delta H = 1.00\ g\ Cu \times \frac{1\ mol\ Cu}{63.55\ g\ Cu} \times \frac{-128.4\ kJ}{1\ mol\ Cu} = -2.02\ kJ$$

$q_{water} = 2.02\ kJ = 2020\ J = 4.18\ \frac{J}{g \cdot {}^\circ C} \times 125\ g \times \Delta t$

$$\Delta t = \frac{2020}{4.18 \times 125} = +3.87^\circ C$$

5.19 For 3.85 g of sucrose:

$q_{water} = 4.18 \dfrac{J}{g\cdot{}^oC} \times 6000\ g \times 2.24^oC = 5.62 \times 10^4\ J = 56.2 kJ$

$q_{bomb} = 3180 \dfrac{J}{^oC} \times 2.24^oC = 7.12 \times 10^3\ J = 7.1\ kJ$

$q_{reaction} = -(q_{water} + q_{bomb}) = -63.3\ kJ$

For one mole of sucrose:

$q_{reaction} = 1.00\ mol\ C_{12}H_{22}O_{11} \times \dfrac{342.30\ g}{1\ mol} \times \dfrac{-63.3\ kJ}{3.85\ g}$

$= -5.63 \times 10^3\ kJ$

5.20 $5.290 \times 10^3\ J = 4.184 \dfrac{J}{g\cdot{}^oC} \times 526.0\ g \times 2.020^oC + C \times 2.020^oC$

$C = \dfrac{5.290 \times 10^3\ J - 4.446 \times 10^3\ J}{2.020^oC} = 418\ J/^oC$

5.21 $45.4 \times 10^3\ J = 4.18 \dfrac{J}{g\cdot{}^oC} \times 4800\ g \times \Delta t + 2540 \dfrac{J}{^oC} \Delta t$

$\Delta t = \dfrac{45400}{20060 + 2540} = 2.01^oC;\ t_f = 25.00^oC + 2.01^oC = 27.01^oC$

5.22 a. $\Delta E = 60\ J + 20\ J = 80\ J$

b. $q = \Delta E - w = 89\ J + 85\ J = 174\ J$

5.23 $\Delta H = \Delta E + \Delta(PV) = -1296\ kJ - 3.8\ kJ = -1300\ kJ$

5.24 heat from natural gas $= 8.0 \times 10^{16}\ kJ \times 0.26 = 2.1 \times 10^{16}\ kJ$

$2.1 \times 10^{16}\ kJ \times \dfrac{1\ g}{43\ kJ} \times \dfrac{1\ L}{0.73\ g} \times \dfrac{1\ m^3}{10^3\ L} = 6.7 \times 10^{11}\ m^3$

5.25 $\Delta H = 1.00\ g\ hydrate \times \dfrac{-78.7\ kJ}{322.20\ g\ hydrate} = -0.244\ kJ$

$q_{water} = 244\ J = 4.18 \dfrac{J}{g\cdot{}^oC} \times m_{water} \times 40.0^oC$

$m_{water} = 1.46\ g$

5.26 a. $4.18 \dfrac{J}{g\cdot{}^oC} \times 1.50 \times 10^5\ g \times 15^oC = 9.4 \times 10^6\ J$

b. $9.4 \times 10^6\ J \times \dfrac{1\ kWh}{3.60 \times 10^6\ J} \times \dfrac{6.0\ ¢}{1\ kWh} = 16\ ¢$

5.27 a. matter of opinion; a discussion question

5.28 a. ΔH_f Na(s) is 0; ΔH_f Na(l) is a positive quantity

b. $\Delta H = \Delta E + \Delta(PV)$; $\Delta(PV)$ is small, but not ordinarily 0

c. exothermic

d. energy is not "consumed"; used, maybe

5.29 $$S.H. = \frac{1034\ J}{27.1\ mL \times 0.785\ \frac{g}{mL} \times 20.0^{\circ}C} = 2.43\ J/g{\cdot}^{\circ}C$$

5.30 $q_{metal} = -q_{water}$

$$(S.H.)(129\ g)(-55.0^{\circ}C) = -4.18\ \frac{J}{g{\cdot}^{\circ}C}(44.6\ g)(15.0^{\circ}C)$$

$$S.H. = \frac{4.18(44.6)(15.0)}{(129)(55.0)} = 0.394\ J/g{\cdot}^{\circ}C$$

could well be brass

CHAPTER 6

BASIC SKILLS

1. Use the Ideal Gas Law to:
   a. determine the effect of a change in one or more variables (e.g., n, T, P) on the value of another variable (e.g., V).
   b. solve for one variable (P, V, n or T), knowing the values of the other three variables.
   c. calculate the density or molar mass of a gas, given the necessary data (molar mass, pressure and temperature for density; temperature, pressure and density for molar mass).
   d. relate the volume of a gas involved in a reaction to the amount of another reactant or product.

2. Relate the volumes of gases (measured at the same T and P) in a reaction.

3. Use Dalton's Law to find the partial pressure of a gas in a mixture; relate partial pressure to mole fraction.

4. Use Graham's Law to relate the molar masses of two gases to their rates or times of effusion.

5. Calculate the average speed of a gas molecule at a given temperature.

LECTURE NOTES

As indicated by the basic skills listed above, the Ideal Gas Law is at the heart of this chapter. It is unnecessary to spend much time on relationships between variables such as Boyle's Law or Charles' Law. For one thing, students almost certainly got a heavy dose of problems of that type in high school. Moreover, such relationships have very limited applicability to chemistry.

Dalton's Law is another one that students have little trouble with. The relation between mole fraction and partial pressure, though, is important; it comes up later in the guise of Raoult's Law or Henry's Law. We ordinarily spend relatively little time on deviations from the Ideal Gas Law.

The basic equation of kinetic theory is: $\frac{1}{2}mu^2$ = constant x T, where the constant is the same for all gases (constant = 3R/2N). This leads to Graham's Law and to the equation for the "average" (root mean square) velocity. The concept of a distribution of molecular velocities is introduced here and serves as background for chemical kinetics (Chapter 16).

This chapter is readily covered in two lectures; students find the material much easier than that in Chapter 5.

LECTURE 1

I The Ideal Gas Law

A. Variables: V = volume (liters, milliliters, cubic centimeters)
n = amount in moles
T = temperature (K)
P = pressure (atmospheres, millimeters of mercury: 1 atm = 760 mm Hg)

Relation between variables:

$$PV = nRT$$

where R is a true constant, equal to 0.0821 L·atm/mol·K

B. Initial and final state problems Use Ideal Gas Law to find necessary relations.

An open flask holds 0.100 mol of air at 25°C. How many moles of air are contained in the flask at 50°C? Note that P and V are constant (P is the atmospheric pressure, V is the volume of the flask). Need relation between n and T:

$$n_2T_2 = n_1T_1 = PV/R$$

$$n_2 = n_1 \times \frac{T_1}{T_2} = 0.100 \text{ mol} \times \frac{298 \text{ K}}{323 \text{ K}} = 0.0923 \text{ mol}$$

(some air escapes from flask when it is heated)

C. Calculation of P, V, n or T

What is the pressure exerted by 15.0 mol of O_2 in a 50.0 L tank at 25°C?

$$P = \frac{nRT}{V} = \frac{(15.0 \text{ mol})(0.0821 \text{ L·atm/mol·K})(298 \text{ K})}{50.0 \text{ L}} = 7.34 \text{ atm}$$

Note that since R = 0.0821 L·atm/mol·K, V must be in liters, P in atmospheres, T in K, n in moles)

D. Calculation of density or molar mass

$$PV = \frac{gRT}{MM} \; ; \; \frac{g}{V} = d = \frac{P(MM)}{RT}$$

Density of $O_2(g)$ at 27°C, 735 mm Hg?

$$d = \frac{(735/760 \text{ atm})(32.0 \text{ g/mol})}{(0.0821 \text{ L·atm/mol·K})(300 \text{ K})} = 1.26 \text{ g/L}$$

Can also use this relation to determine molar mass experimentally; measure gas density at known P, T.

E Volumes of gases involved in reactions

1. $Zn(s) + 2\,H^+(aq) \longrightarrow Zn^{2+}(aq) + H_2(g)$

Mass of Zn required to form 16.0 L of $H_2(g)$ at 20°C, 735 mm Hg?

Path to follow: $V\ H_2 \rightarrow n\ H_2 \rightarrow$ moles Zn $\leftrightarrow$ mass Zn

$$n\ H_2 = \frac{PV}{RT} = \frac{(735/760 \text{ mm Hg})\ (16.0 \text{ L})}{(0.0821 \text{ L.atm/mol.K})(293 \text{ K})} = 0.643 \text{ mol } H_2$$

$$\text{mass Zn} = 0.643 \text{ mol } H_2 \times \frac{1 \text{ mol Zn}}{1 \text{ mol } H_2} \times \frac{65.38 \text{ g Zn}}{1 \text{ mol Zn}} = 42.0 \text{ g Zn}$$

2. Gay Lussac's Law: Volumes of gases involved in reaction (same P, T) are in same ratio as number of moles.

$$2\ H_2O(l) \rightarrow 2\ H_2(g) + O_2(g)\ ; \quad \frac{V\ H_2}{V\ O_2} = \frac{n\ H_2}{n\ O_2} = 2$$

LECTURE 2

I Dalton's Law (gas mixtures)

A. $P_{tot} = P_1 + P_2 + - - -$ where P_1 = partial pressure of gas 1, etc.. Most often used in collection of gases over water:

$P_1 = P_{tot} - P\ H_2O$

where P_{tot} is the measured pressure and $P\ H_2O$ is the vapor pressure of water, which is a constant at a given temperature (e.g., 5 mm Hg at 0°C, 24 mm Hg at 25°C).

b. $P_1 = X_1 P_{tot}$

where X_1 = mole fraction of gas in mixture. Partial pressure of O_2 in air ($X\ O_2 = 0.2095$), when barometric pressure = 734 mm Hg?

$P\ O_2 = 0.2095(734 \text{ mm Hg}) = 154 \text{ mm Hg}$

II Real Gases

Deviate at least slightly from Ideal Gas Law because of two factors:

1. gas molecules attract one another
2. gas molecules occupy a finite volume

Both of these factors are neglected in the Ideal Gas Law. Both increase in importance when molecules are close together (high P, low T).

Van der Waals equation: $\left(P + \frac{n^2a}{V^2}\right)\left(V - nb\right) = nRT$

n^2a/V^2 corrects for attraction between molecules; nb $\approx$ volume of gas molecules

III Kinetic Theory of Gases

$E_{trans} = \frac{1}{2}mu^2 = C \times T$

where m = mass of molecule, u = average velocity, T = temperature in K, and C is a constant which has the same value for all gases.

A. Graham's Law: relates m and u for two different gases, same T

$m_2u_2^2 = m_1u_1^2$

$$\frac{u_2^2}{u_1^2} = \frac{m_1}{m_2} \quad ; \quad \frac{u_2}{u_1} = \left(\frac{MM_1}{MM_2}\right)^{\frac{1}{2}}$$

$$\frac{\text{rate 2}}{\text{rate 1}} = \left(\frac{MM_1}{MM_2}\right)^{\frac{1}{2}} = \frac{\text{time 1}}{\text{time 2}}$$

Certain gas takes 2.42 times as long to effuse as O_2 at same T, P. MM of gas?

$$\left(\frac{MM}{32.0\text{ g/mol}}\right)^{\frac{1}{2}} = 2.42; \quad MM = (2.42)^2 \times 32.0\text{ g/mol} = 187\text{ g/mol}$$

B. Calculation of average velocity, u

$$C = \frac{3R}{2N}; \quad \frac{mu^2}{2} = \frac{3RT}{2N} \; ; \; u^2 = \frac{3RT}{mN} = \frac{3RT}{MM}$$

To find u in cm/s, need R in $cm^2 \cdot g/s^2 \cdot K \cdot mol$

$R = 8.31 \times 10^7\ g \cdot cm^2/s^2 \cdot mol \cdot K$

average velocity H_2 molecule at 0^oC?

$$u^2 = \left(\frac{3 \times 8.31 \times 10^7 \times 273}{2.016}\right)\frac{cm^2}{s^2} = 3.38 \times 10^{10}\ cm^2/s^2$$

$u = 1.84 \times 10^5$ cm/s

DEMONSTRATIONS

1. Effusion of gases: Test Dem. 9, 64, 128, 160, 204
2. Principle of Barometer: Test. Dem. 13
3. Dalton's Law: Test. Dem. 156, 195
4. Charles' Law: J. Chem. Educ. 56 823 (1979)
5. Boyle's Law: J. Chem. Educ. 56 322 (1979)
6. Properties of Gases: J. Chem. Educ. 60 67 (1983)

QUIZZES

Quiz 1

1. Consider the reaction of aluminum with acid:

   $$2\ Al(s) + 6\ H^+(aq) \longrightarrow 2\ Al^{3+}(aq) + 3\ H_2(g)$$

   What mass of Al (at. mass = 26.98) is required to produce 1.00 L of hydrogen, measured at 740 mm Hg and 15^oC?

2. A gas mixture at a total pressure of 842 mm Hg contains 0.100 mol N_2 and 0.800 mol He. What are the partial pressures of N_2 and He?

Quiz 2

1. Calculate the density of $SO_3(g)$ at 22°C and 734 mm Hg.
2. At a certain temperature and pressure, it takes 52 s for 0.100 mol of H_2 to effuse through a pinhole. How long will it take for the same amount of Cl_2 to effuse, at the same temperature and pressure?

Quiz 3

1. How many moles of N_2 are there in a 20.0 L cylinder at 25°C if the pressure is 2190 lb/in^2? (14.7 lb/in^2 = 1 atm)
2. Calculate the average velocity of a CH_4 molecule at 16°C. ($R = 8.31 \times 10^7$ g·cm^2/s^2·mol·K)

Quiz 4

1. What volume of O_2, at 751 mm Hg and 23°C, is formed by the decomposition of 6.48 g of $KClO_3$:

   $2\ KClO_3(s) \longrightarrow 2\ KCl(s) + 3\ O_2(g)$
2. How many moles of H_2 are there in a sample collected over water if the sample volume is 225 mL at 15°C and a total pressure of 740 mm Hg? (vapor pressure water at 15°C = 13 mm Hg)

Quiz 5

1. What is the molar mass of a gas which has a density of 0.00249 g/mL at 20°C and 744 mm Hg?
2. Sketch a graph of, for an ideal gas:

   a. P vs T   b. u vs P (constant T)

PROBLEMS 1-30

6.1 $P = 748 \text{ mm Hg} + 150 \text{ mm Hg} = 898 \text{ mm Hg} \times \dfrac{1 \text{ atm}}{760 \text{ mm Hg}} = 1.18 \text{ atm}$

6.2 $745 \text{ mm Hg} \times \dfrac{1 \text{ atm}}{760 \text{ mm Hg}} = 0.980 \text{ atm}$

$745 \text{ mm Hg} \times \dfrac{101.3 \text{ kPa}}{760 \text{ mm Hg}} = 99.3 \text{ kPa}$

$1.40 \text{ atm} \times \dfrac{760 \text{ mm Hg}}{1 \text{ atm}} = 1.06 \times 10^3 \text{ mm Hg}$

$1.40 \text{ atm} \times \dfrac{101.3 \text{ kPa}}{1 \text{ atm}} = 142 \text{ kPa}$

$97.3 \text{ kPa} \times \dfrac{760 \text{ mm Hg}}{101.3 \text{ kPa}} = 730 \text{ mm Hg}$

$97.3 \text{ kPa} \times \dfrac{1 \text{ atm}}{101.3 \text{ kPa}} = 0.961 \text{ atm}$

6.3 $P_2V_2 = P_1V_1$; $V_2 = V_1 \times \dfrac{P_1}{P_2}$

a. $V_2 = 8.50 \text{ L} \times \dfrac{0.980 \text{ atm}}{1.30 \text{ atm}} = 6.41 \text{ L}$

b. $V_2 = 8.50 \text{ L} \times \dfrac{0.980 \text{ atm}}{0.490 \text{ atm}} = 17.0 \text{ L}$

6.4 $\frac{V_2}{T_2} = \frac{V_1}{T_1}$; $V_2 = V_1 \times \frac{T_2}{T_1}$

a. $V_2 = 1.82\ L \times \frac{323\ K}{298\ K} = 1.97\ L$

b. $V_2 = 1.82\ L \times \frac{285\ K}{298\ K} = 1.74\ L$

6.5 $\frac{P_2V_2}{T_2} = \frac{P_1V_1}{T_1}$; $V_2 = V_1 \times \frac{P_1}{P_2} \times \frac{T_2}{T_1}$

$V_2 = 3.20\ cm^3 \times \frac{2.45\ atm}{1.12\ atm} \times \frac{292\ K}{281\ K} = 7.27\ cm^3$

6.6 $\frac{n_2T_2}{P_2} = \frac{n_1T_1}{P_1}$; $n_2 = n_1 \times \frac{P_2}{P_1} \times \frac{T_1}{T_2}$

$P_2 = 1.02\ atm \times 760\ mm\ Hg/1\ atm = 775\ mm\ Hg$

$T_1 = \frac{62 - 32}{1.8}\ ^{\circ}C = 17^{\circ}C = 290\ K$

$n_2 = 0.100\ mol \times \frac{775\ mm\ Hg}{742\ mm\ Hg} \times \frac{290\ K}{304\ K} = 0.0996\ mol$

6.7 $P_2V_2 = P_1V_1$; $V_2 = V_1 \times \frac{P_1}{P_2}$

$V_1 = 0.25 \times 1.50\ m^3 = 0.38\ m^3$

$P_1 = 65\ lb/in^2$; $P_2 = 35\ lb/in^2$

$V_2 = 0.38\ m^3 \times \frac{65\ lb/in^2}{35\ lb/in^2} = 0.70\ m^3$; $\Delta V = 0.32\ m^3$

6.8 $V = \frac{nRT}{P} = \frac{(44.3/4.003\ mol)(0.0821\ L\cdot atm/mol\cdot K)(310\ K)}{2.50\ atm} = 113\ L$

6.9 $n = \frac{PV}{RT} = \frac{(739/760\ atm)(0.125\ L)}{(0.0821\ L\cdot atm/mol\cdot K)(291\ K)} = 5.09 \times 10^{-3}\ mol$

6.10 moles He = $10.0\ g \times \frac{1\ mol}{4.003\ g} = 2.50\ mol$

$V = \frac{(2.50\ mol)(0.0821\ L\cdot atm/mol\cdot K)(273\ K)}{4.00\ atm} = 14.0\ L$

moles He = $0.800\ g \times \frac{1\ mol}{4.003\ g} = 0.200\ mol$

$P = \frac{(0.200\ mol)(0.0821\ L.atm/mol\cdot K)(298\ K)}{0.105\ L} = 46.6\ atm$

n He = $\frac{(751/760\ atm)(2.50\ L)}{(0.0821\ L\cdot atm/mol\cdot K)(373\ K)} = 0.0807\ mol$

g He = $0.0807\ mol \times \frac{4.003\ g}{1\ mol} = 0.323\ g$

6.10 (cont.)

$$\text{moles He} = 20.0 \text{ g} \times \frac{1 \text{ mol}}{4.003 \text{ g}} = 5.00 \text{ mol}$$

$$T = \frac{(202/101.3 \text{ atm})(61.5 \text{ L})}{(5.00 \text{ mol})(0.0821 \text{ L·atm/mol·K})} = 299 \text{ K} = 26^{\circ}\text{C}$$

6.11 $$d = \frac{P(MM)}{RT} = \frac{(1.00 \text{ atm})(58.12 \text{ g/mol})}{(0.0821 \text{ L·atm/mol·K})(298 \text{ K})} = 2.38 \text{ g/L}$$

6.12 a. $$MM = \frac{gRT}{PV} = \frac{(0.564 \text{ g})(0.0821 \text{ L·atm/mol·K})(372 \text{ K})}{(757/760 \text{ atm})(0.240 \text{ L})} = 72.1 \frac{\text{g}}{\text{mol}}$$

b. mass C in 1 mol = 72.1 g x 0.666 = 48.0 g = 4 mol C
mass H in 1 mol = 72.1 g x 0.112 = 8.08 g = 8 mol H
mass O in 1 mol = 72.1 g x 0.222 = 16.0 g = 1 mol O

C_4H_8O

6.13 a. $$MM = 0.745 \times 28.01 \frac{\text{g}}{\text{mol}} + 0.157 \times 32.00 \frac{\text{g}}{\text{mol}} + 0.036 \times 44.01 \frac{\text{g}}{\text{mol}} + 0.062 \times 18.02 \frac{\text{g}}{\text{mol}} = 28.6 \frac{\text{g}}{\text{mol}}$$

b. $$d = \frac{P(MM)}{RT} = \frac{(1.00 \text{ atm})(28.6 \text{ g/mol})}{(0.0821 \text{ L·atm/mol·K})(300 \text{ K})} = 1.16 \text{ g/L}$$ vs 1.18 g/L

6.14 a. $$MM = \frac{gRT}{PV} = \frac{(1.95 \text{ g})(0.0821 \text{ L·atm/mol·K})(288 \text{ K})}{(1.00 \text{ atm})(0.740 \text{ L})} = 62.3 \frac{\text{g}}{\text{mol}}$$

b. $$\frac{62.3 - 6}{2} = 28.1 = \text{at. mass Si}$$

6.15 $$3.0 \text{ L } O_2 \times \frac{2 \text{ L } SO_3}{1 \text{ L } O_2} = 6.0 \text{ L } SO_3$$

6.16 a. $$n_{H_2} = \frac{(1.00 \text{ atm})(1.00 \text{ L})}{(0.0821 \text{ L·atm/mol·K})(273 \text{ K})} = 0.0446 \text{ mol } H_2$$

$$0.0446 \text{ mol } H_2 \times \frac{1 \text{ mol Zn}}{1 \text{ mol } H_2} \times \frac{65.38 \text{ g Zn}}{1 \text{ mol Zn}} = 2.92 \text{ g Zn}$$

b. $$n_{H_2} = 1.00 \text{ g Zn} \times \frac{1 \text{ mol Zn}}{65.38 \text{ g Zn}} \times \frac{1 \text{ mol } H_2}{1 \text{ mol Zn}} = 0.0153 \text{ mol } H_2$$

$$V = \frac{(0.0153 \text{ mol})(0.0821 \text{ L·atm/mol·K})(295 \text{ K})}{(729/760 \text{ mm Hg})} = 0.386 \text{ L}$$

6.17 $$n\ B_2H_6 = 26.0 \text{ g } NaBH_4 \times \frac{1 \text{ mol } NaBH_4}{37.83 \text{ g } NaBH_4} \times \frac{2 \text{ mol } B_2H_6}{3 \text{ mol } NaBH_4} = 0.458 \text{ mol } B_2H_6$$

$$V = \frac{(0.458 \text{ mol})(0.0821 \text{ L·atm/mol·K})(295 \text{ K})}{(758/760 \text{ atm})} = 11.1 \text{ L}$$

6.18 a. $$2\ C_8H_{18}(l) + 25\ O_2(g) \longrightarrow 16\ CO_2(g) + 18\ H_2O(l)$$

b. $$n\ O_2 = 2.00 \text{ g } C_8H_{18} \times \frac{1 \text{ mol } C_8H_{18}}{114.22 \text{ g } C_8H_{18}} \times \frac{25 \text{ mol } O_2}{2 \text{ mol } C_8H_{18}} = 0.219 \text{ mol } O_2$$

$$V = \frac{(0.219 \text{ mol})(0.0821 \text{ L·atm/mol·K})(323 \text{ K})}{1.00 \text{ atm}} = 5.81 \text{ L}$$

6.18 c. $5.81 \text{ L } O_2 \times \frac{100 \text{ L air}}{21 \text{ L } O_2} = 28 \text{ L air}$

6.19 $P\ N_2 = 751 \text{ mm Hg} \times 0.745 = 559 \text{ mm Hg}$

$P\ O_2 = 751 \text{ mm Hg} \times 0.157 = 118 \text{ mm Hg}$

$P\ CO_2 = 751 \text{ mm Hg} \times 0.036 = 27 \text{ mm Hg}$

$P\ H_2O = 751 \text{ mm Hg} \times 0.062 = 47 \text{ mm Hg}$

6.20 a. 768 mm Hg - 21.0 mm Hg = 747 mm Hg

b. $n = \frac{(747/760 \text{ atm})(0.275 \text{ L})}{(0.0821 \text{ L·atm/mol·K})(296 \text{ K})} = 0.0111 \text{ mol}$

mass $O_2 = 0.0111 \text{ mol} \times \frac{32.00 \text{ g}}{1 \text{ mol}} = 0.356 \text{ g}$

6.21 $P\ O_2 = 749 \text{ mm Hg} - 24 \text{ mm Hg} = 725 \text{ mm Hg}$

$V = \frac{(1.25/32.00 \text{ mol})(0.0821 \text{ L·atm/mol.K})(298 \text{ K})}{(725/760 \text{ atm})} = 1.00 \text{ L}$

6.22 a. more ideally b. less ideally

6.23 a. $P = \frac{nRT}{V} = \frac{(1.00 \text{ mol})(0.0821 \text{ L·atm/mol·K})(273 \text{ K})}{0.225 \text{ L}} = 99.6 \text{ atm}$

b. $P = \frac{nRT}{V - nb} - \frac{n^2a}{V^2}$

$= \frac{(1.00 \text{ mol})(0.0821 \text{ L·atm/mol·K})(273 \text{ K})}{0.198 \text{ L}} - \frac{0.244}{(0.225)^2} \text{ atm}$

$= 113 \text{ atm} - 4.82 \text{ atm} = 108 \text{ atm}$

6.24 $\frac{\text{rate } UF_6}{\text{rate } H_2} = \left(\frac{2.016}{352.0}\right)^{\frac{1}{2}} = 0.07568$

6.25 $\frac{t \text{ unknown}}{t} = \frac{44.0}{22.0} = \left(\frac{\text{MM unknown}}{16.04}\right)^{\frac{1}{2}} = 2.00$

$\frac{\text{MM unknown}}{16.04 \text{ g/mol}} = 4.00$; MM = 64.2 g/mol; SO_2

6.26 $u = (3RT/MM)^{\frac{1}{2}}$

a. $u = \left(\frac{3 \times 8.31 \times 10^7 \text{ g·cm}^2/\text{s}^2\text{·mol·K} \times 323 \text{ K}}{32.00 \text{ g/mol}}\right)^{\frac{1}{2}}$

$= 5.02 \times 10^4 \text{ cm/s} = 502 \text{ m/s}$

b. $u = \left(\frac{3 \times 8.31 \times 10^7 \text{ g·cm}^2/\text{s}^2\text{·mol·K} \times 223 \text{ K}}{28.01 \text{ g/mol}}\right)^{\frac{1}{2}}$

$= 4.46 \times 10^4 \text{ cm/s} = 446 \text{ m/s}$

6.27 a. K. E. = constant x T; $T_2 = \frac{T_1}{2} = \frac{298 \text{ K}}{2} = 149 \text{ K} = -124^\circ\text{C}$

b. u^2 = constant x T; $T_2 = 4T_1 = 4(298 \text{ K}) = 1192 \text{ K} = 919^\circ\text{C}$

6.28 P = constant x d x h; dh is same for both liquids

h water $= 760 \text{ mm} \times \frac{13.6 \text{ g/cm}^3}{1.00 \text{ g/cm}^3} = 1.03 \times 10^4 \text{ mm}$ ($\approx$ 10 m)

6.29 a. 1.00 b. 20.18/2.016 = 10.0 c. 1.00 d. 0.500

6.30

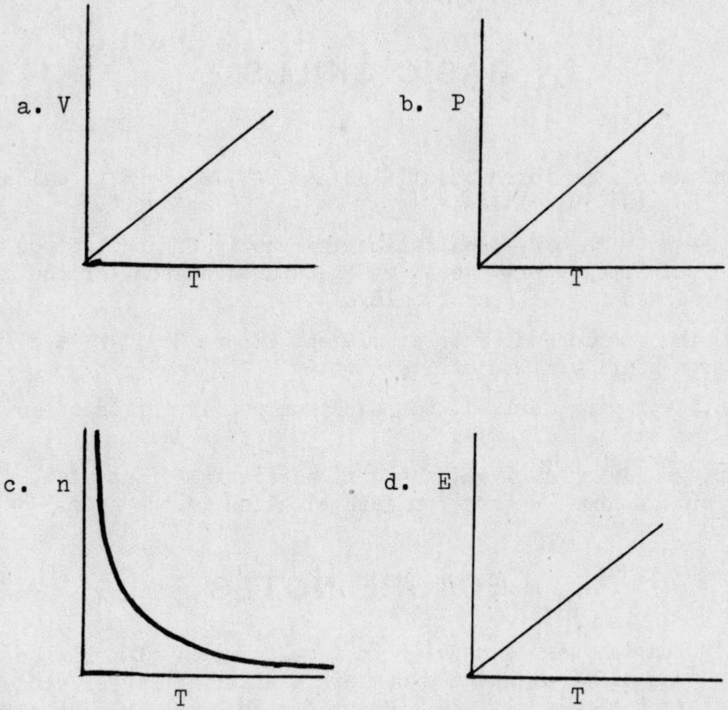

CHAPTER 7

BASIC SKILLS

1. Given one of the three quantities: λ , ν , $E_{hi} - E_{lo}$, calculate the other two quantities.
2. Use the Bohr theory to calculate the energy of an electron in a given principal energy level of the hydrogen atom, or the difference in energy between two levels.
3. Given the atomic number of an element (up to 36), write the electron configuration of its atom.
4. Given, or having derived, the electron configuration of an atom, draw its orbital diagram.
5. Given, or having derived, the orbital diagram of an atom, write the four quantum numbers for each electron in the atom.

LECTURE NOTES

This chapter can be covered in two lectures. Experience suggests that all four quantum numbers are difficult for students to digest at a single sitting. We prefer to break them up between two lectures.

Note that the text discussion of electronic structure stops with atomic number 36. Beyond that, it is perhaps best to deduce the "aufbau" order from the Periodic Table, to be discussed in Chapter 8.

LECTURE 1

I Electronic Structure

A. Quantum Theory Electrons in atoms can have only certain discrete energies, refered to as energy states or energy levels. Normally, the electron is in the state of lowest energy, called the ground state. By absorbing a certain definite amount of energy, the electron can move to a higher level, called an excited state. When electrons return to lower energy levels, energy may be given off as light. The difference in energy between the levels can be deduced from the wavelength or frequency of the light:

$$E_{hi} - E_{lo} = \frac{hc}{\lambda} = \frac{1.196 \times 10^5}{\lambda} \frac{\text{kJ}\cdot\text{nm}}{\text{mol}} \; ; \; \nu = \frac{c}{\lambda} = \frac{2.998 \times 10^8}{\lambda} \text{ m/s}$$

Suppose λ = 452 nm (in the blue region); $E_{hi} - E_{lo}$ = ? ν ?

$$E_{hi} - E_{lo} = \frac{1.196 \times 10^5}{452} \frac{kJ}{mol} = 265 \text{ kJ/mol}$$

(i.e., the two energy states differ by 265 kJ/mol, comparable to the ΔE in reactions)

$$\nu = \frac{2.998 \times 10^8 \text{ m/s}}{452 \times 10^{-9} \text{ m}} = 6.63 \times 10^{14}/\text{s}$$

B. Bohr Theory Hydrogen Atom Bohr postulated that electron moves about nucleus in circular orbit of fixed radius. By absorbing energy, it moves to higher orbit of larger radius; energy given off as light when electron returns.

$$E = \frac{-1312}{n^2} \text{ kJ/mol} \quad ; \quad n = 1, 2, 3, - -$$

in ground state, n = 1; in excited states, n = 2, 3, - -

Suppose electron moves from n = 3 to n = 2; λ = ?

$$E_3 = \frac{-1312}{9} = -145.8 \text{ kJ} \quad ; \quad E_2 = \frac{-1312}{4} = -328.0 \frac{kJ}{mol}$$

$$E_{hi} - E_{lo} = 182.2 \text{ kJ/mol}; \quad \lambda = \frac{1.196 \times 10^5}{182.2} \text{ nm} = 656.4 \text{ nm}$$

this is one of the lines in the Balmer series ($n_{lo} = 2$)

C. Quantum Mechanical Atom
Bohr model did not agree with experiment for any atom with more than one electron. Idea of electron moving about nucleus in circular orbit had to be abandoned. In the quantum-mechanical atom:

1. Can only refer to the probability of finding an electron in a region; cannot specify path.
2. Four quantum numbers needed to describe energy of electron completely.

D. Principal Energy Levels
n = 1, 2, 3, - - . Value of n is the main factor (but not the only one) that determines the energy of an electron and its distance from the nucleus. Maximum capacity = $2n^2$

n	1	2	3
maximum no. e^-	2	8	18

E. Sublevels
1. Quantum number ℓ = 0, 1, 2, - - (n - 1)
   n = 1 ℓ = 0 (one sublevel)
   n = 2 ℓ = 0, 1 (two sublevels)
   n = 3 ℓ = 0, 1, 2 (three sublevels)
   (in general, no. sublevels = n)
2. Sublevel designations: s, p, d, f

Value of ℓ	0	1	2	3
Letter designation	s	p	d	f
Capacity for e^-	2	6	10	14

1st prin. level: 1s (2 e^-)
2nd prin. level: 2s (2 e^-), 2p (6 e^-)
3rd prin. level: 3s (2 e^-), 3p (6 e^-), 3d (10 e^-)

LECTURE 2

I Electronic Structure

A. Electron Configuration Indicate by a superscript the number of electrons in each sublevel.

H $1s^1$	Li $1s^2 2s^1$	Na $1s^2 2s^2 2p^6 3s^1$	K [Ar] $4s^1$
He $1s^2$	Be $1s^2 2s^2$	Mg $1s^2 2s^2 2p^6 3s^2$	Ca [Ar] $4s^2$
	B $1s^2 2s^2 2p^1$	Al $1s^2 2s^2 2p^6 3s^2 3p^1$	Sc [Ar] $4s^2 3d^1$
	⋮	⋮	⋮
	Ne $1s^2 2s^2 2p^6$	Ar $1s^2 2s^2 2p^6 3s^2 3p^6$	
			Zn [Ar] $4s^2 3d^{10}$
			⋮
			Kr [Ar] $4s^2 3d^{10} 4p^6$

B. 3rd and 4th Quantum Numbers

1. Orbital designated by m_ℓ = ℓ, - - -, 0, - - -, $-\ell$

   ℓ = 0 (s sublevel); m_ℓ = 0 (one s orbital)

   ℓ = 1 (p sublevel); m_ℓ = 1, 0, -1 (three p orbitals)

   ℓ = 2 (d sublevel); m_ℓ = 2, 1, 0, -1, -2 (five d orbitals)

   Each orbital has a capacity of two electrons. s orbitals are spherical about nucleus; p orbitals are dumbell shaped and at right angles to each other.

2. Electron in an orbital can have either of two spins: $m_s = +\frac{1}{2}, -\frac{1}{2}$

C. Orbital Diagram Show number of electrons in each orbital and spin of each electron (↑ or ↓)

	1s	2s	2p	
H	(↑)			2 e^- in same orbital have opposed spins
He	(↑↓)			
Li	(↑↓)	(↑)		
Be	(↑↓)	(↑↓)		
B	(↑↓)	(↑↓)	(↑)( )( )	
C	(↑↓)	(↑↓)	(↑)(↑)( )	When several orbitals of same sublevel are available, electrons enter singly with parallel spins.
N	(↑↓)	(↑↓)	(↑)(↑)(↑)	
O	(↑↓)	(↑↓)	(↑↓)(↑)(↑)	

Abbreviated electron configuration, orbital diagram of Fe?

[Ar] $4s^2 3d^6$

4s	3d
(↑↓)	(↑↓)(↑)(↑)(↑)(↑)

D. Summary of Quantum Numbers

Give quantum numbers of each electron in the carbon atom

n	1	1	2	2	2	2
ℓ	0	0	0	0	1	1
m_ℓ	0	0	0	0	1	0
m_s	$+\frac{1}{2}$	$-\frac{1}{2}$	$+\frac{1}{2}$	$-\frac{1}{2}$	$+\frac{1}{2}$	$+\frac{1}{2}$

Note that no two electrons in an atom have the same set of four quantum numbers.

QUIZZES

Quiz 1

1. Consider the germanium atom (at. no. = 32)
   a. Write the electron configuration of Ge.
   b. Write the abbreviated orbital diagram of Ge.
   c. Give a set of four quantum numbers for each electron beyond the Ar core in Ge.

Quiz 2

1. Calculate the wavelength of the line in the Lyman series of hydrogen resulting from the transition: n = 3 to n = 1

   $E = -1312/n^2$; $E_{hi} - E_{lo} = \frac{1.196 \times 10^5}{\lambda} \frac{\text{kJ·nm}}{\text{mol}}$

2. Write the electron configuration of the Mn atom (at. no. = 25)

Quiz 3

1. A certain spectral line originates from an electron transition between two levels that differ in energy by 282 kJ/mol. Calculate the wavelength and frequency of the line.

   $E_{hi} - E_{lo} = \frac{1.196 \times 10^5}{\lambda} \frac{\text{kJ·nm}}{\text{mol}}$; $\nu = \frac{2.998 \times 10^8}{\lambda}$ m/s

2. Write the abbreviated electron configuration of the Ni atom (at. no. = 28). How many unpaired electrons are there in this atom?

Quiz 4

1. Sketch the geometry of an s orbital; of the three p orbitals.
2. Give a set of four quantum numbers for each electron in the F atom (at. no. = 9).

Quiz 5

1. Calculate the difference in energy for each of the first four lines in the Lyman series of hydrogen (n lo = 1).

   $E = -1312/n^2$

2. Give the electron configuration and orbital diagram of an atom of atomic number 22.

PROBLEMS 1-30

7.1 Higher energy in excited state

7.2 a. $\nu = \dfrac{2.998 \times 10^{8} \text{ m/s}}{540.0 \times 10^{-9} \text{ m}} = 5.552 \times 10^{14}/\text{s}$

b. $E_{hi} - E_{lo} = \dfrac{1.986 \times 10^{-25}}{540.0 \times 10^{-9}} \text{ J/particle} = 3.678 \times 10^{-19} \dfrac{\text{J}}{\text{particle}}$

c. $E_{hi} - E_{lo} = \dfrac{1.196 \times 10^{5}}{540.0} \text{ kJ/mol} = 221.5 \text{ kJ/mol}$

7.3 a. $\lambda = \dfrac{1.196 \times 10^{5}}{E_{hi} - E_{lo}} \text{ nm} = \dfrac{1.196 \times 10^{5}}{305} \text{ nm} = 392 \text{ nm}$

b. $\nu = \dfrac{2.998 \times 10^{8} \text{ m/s}}{392 \times 10^{-9} \text{ m}} = 7.65 \times 10^{14}/\text{s}$

7.4 a. $\lambda = \dfrac{2.998 \times 10^{8} \text{ m/s}}{4.51 \times 10^{14}/\text{s}} = 6.65 \times 10^{-7} \text{ m} = 665 \text{ nm}$

b. $E_{hi} - E_{lo} = \dfrac{1.986 \times 10^{-25}}{665 \times 10^{-9}} \text{ J/part.} = 2.99 \times 10^{-19} \text{ J/part.}$

c. $E_{hi} - E_{lo} = \dfrac{1.196 \times 10^{5}}{665} \text{ kJ/mol} = 180 \text{ kJ/mol}$

7.5 a, c absorbed; b, d emitted

7.6 $E = \dfrac{-1312}{n^2} \dfrac{\text{kJ}}{\text{mol}}$;

n	1	2	3	4
E	-1312	-328.0	-145.8	-82.00

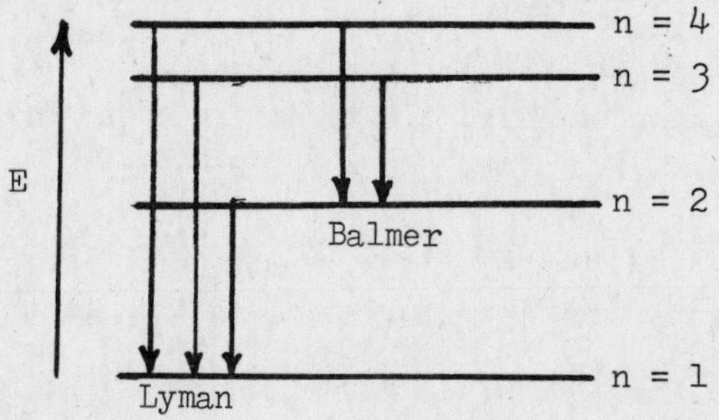

7.7 $E_3 = \dfrac{-1312}{9} = -145.8 \dfrac{\text{kJ}}{\text{mol}}$; $E_5 = \dfrac{-1312}{25} = -52.48 \dfrac{\text{kJ}}{\text{mol}}$

$E_{hi} - E_{lo} = -52.48 \dfrac{\text{kJ}}{\text{mol}} + 145.8 \dfrac{\text{kJ}}{\text{mol}} = +93.3 \dfrac{\text{kJ}}{\text{mol}}$

$\lambda = \dfrac{1.196 \times 10^{5}}{93.3} \text{ nm} = 1.28 \times 10^{3} \text{ nm}$

7.8 longest wavelength = smallest ΔE = transition from n = 6 to 5

$E_6 = -\frac{1312}{36} = -36.44\ \frac{kJ}{mol}$; $E_5 = -\frac{1312}{25} = -52.48\ \frac{kJ}{mol}$

$E_6 - E_5 = 16.04$ kJ/mol; $\lambda = \frac{1.196 \times 10^5}{16.04} = 7456$ nm

7.9 $E_2 = -\frac{1312}{4} = -328.0\ \frac{kJ}{mol}$

$E_{ion} = 0 - (-328.0\ kJ/mol) = 328.0\ kJ/mol$

7.10 $\lambda = h/mv = \frac{6.626 \times 10^{-34}\ J \cdot s}{(9.11 \times 10^{-31}\ kg)(3.0 \times 10^6\ m/s)} = 2.4 \times 10^{-10}$ m

7.11 a. -1, 0, +1 b. -2, -1, 0, +1, +2 c. 2s: 0, 2p: -1, 0, +1

7.12 ℓ = 0 (s sublevel); ℓ = 1 (p sublevel); ℓ = 2 (d sublevel)
ℓ = 3 (f sublevel)

7.13 a. 18 b. 10 c. 2

7.14 a. n = 4 b. f c. 9 d. 4

7.15 a. $1s^2 2s^2 2p^6 3s^2 3p^3$
b. $1s^2 2s^2 2p^6 3s^2 3p^6 4s^2 3d^2$
c. $1s^2 2s^2 2p^6 3s^2 3p^6 4s^2 3d^{10} 4p^3$
d. $1s^2 2s^2 2p^6 3s^2 3p^1$

7.16 a. $[{}_{10}Ne]\ 3s^2 3p^5$
b. $[{}_{18}Ar]\ 4s^2 3d^8$
c. $[{}_{18}Ar]\ 4s^2 3d^{10}$
d. $[{}_{18}Ar]\ 4s^2 3d^{10} 4p^5$

7.17 a. Ne b. Ti c. S

7.18 a. H, He, Li, Be b. Ti, V c. He, Li, Be, B, C, N, O, F, Ne

7.19 a. 1.00 b. 0.400 c. 8/30 = 0.267

7.20 a. excited b. ground c. excited d. impossible
e. excited f. excited

7.21

	1s	2s	2p	3s	3p	4s
a.	(↑↓)	(↑↓)	(↑↓)(↑↓)(↑↓)	(↑↓)	(↑)(↑)( )	
b.	(↑↓)	(↑↓)	(↑↓)(↑↓)(↑↓)	(↑↓)	(↑↓)(↑↓)(↑↓)	(↑↓)

3d
(↑)(↑)(↑)(↑)(↑)

	1s	2s	2p	3s	3p
c.	(↑↓)	(↑↓)	(↑↓)(↑↓)(↑↓)	(↑↓)	(↑)( )( )
d.	(↑↓)	(↑↓)	(↑↓)(↑↓)(↑↓)		

7.22 a. Ni b. Fe c. As

7.23 a. 3 b. 0 c. 2

7.24 a. 2, 8 b. 1, 3, 7 c. 4, 6 d. 5

7.25

n	1	1	2	2	2	2
ℓ	0	0	0	0	1	1
m_ℓ	0	0	0	0	1	0
m_s	$+\frac{1}{2}$	$-\frac{1}{2}$	$+\frac{1}{2}$	$-\frac{1}{2}$	$+\frac{1}{2}$	$+\frac{1}{2}$

7.26 a. $n = 4$, $\ell = 0$, $m_\ell = 0$, $m_s = +\frac{1}{2}$

b.

n	3	3	3	3	3	3	3
ℓ	2	2	2	2	2	2	2
m_ℓ	2	2	1	1	0	-1	-2
m_s	$+\frac{1}{2}$	$-\frac{1}{2}$	$+\frac{1}{2}$	$-\frac{1}{2}$	$+\frac{1}{2}$	$+\frac{1}{2}$	$+\frac{1}{2}$

c.

n	2	2	2	2	2	2	3	3	3	3
ℓ	1	1	1	1	1	1	1	1	1	1
m_ℓ	1	1	0	0	-1	-1	1	1	0	-1
m_s	$+\frac{1}{2}$	$-\frac{1}{2}$	$+\frac{1}{2}$	$-\frac{1}{2}$	$+\frac{1}{2}$	$-\frac{1}{2}$	$+\frac{1}{2}$	$-\frac{1}{2}$	$+\frac{1}{2}$	$+\frac{1}{2}$

7.27 a. 1 b. 6 c. 0

7.28 a. no 2d electrons c. no 3f electrons d. m_s cannot be 1
e. m_ℓ cannot be 2 when $\ell = 0$

7.29 a. no two electrons can have the same four quantum numbers
b. when orbitals of equal energy are available, electrons tend to enter singly with parallel spins
c. light at a discrete wavelength formed as result of electron transition
d. number (1, 2, 3, - -) designating the principal energy level

7.30 a. directly proportional to ν
b. $-E$ is inversely proportional to n^2
c. 4s fills before 3d

CHAPTER 8

BASIC SKILLS

1. Having located a maingroup element in the Periodic Table, state its outer electron configuration.
2. Predict trends in the Periodic Table with respect to atomic radius, ionization energy, and metallic character.
3. Describe, with the aid of balanced equations, the reactions of the metals in Groups 1 and 2 with nonmetals and with water.
4. Compare and contrast the properties of transition metals to those of the metals in Groups 1 and 2.

LECTURE NOTES

This chapter divides naturally into two parts, each of which can be covered in one lecture. The first portion of the chapter (Sections 8.1-8.4) deals with the Periodic Table and the general properties of metals. The second portion (Sections 8.5 and 8.6) is descriptive in nature and deals with the chemistry of the Group 1 and Group 2 metals and some of the general properties of the transition metals. Included here are several examples reviewing material from Chapters 5-7; many of the end-of-chapter problems are also of a review nature.

Much of the material in this chapter is included for interest and general information. It need not be covered in lecture unless you wish to do so. Topics in this category include:
- the historical development of the Periodic Table (Section 8.1)
- the physical properties of metals and alloys (Section 8.4)
- the properties and uses of some of the compounds of sodium and calcium (Section 8.5).

LECTURE 1

I The Periodic Table

A. Correlation with Electronic Structure

1. All the elements in a given group have the same outer electron configuration.

Group	1	2	3	4	5	6	7	8
	ns^1	ns^2	ns^2np^1	ns^2np^2	ns^2np^3	ns^2np^4	ns^2np^5	ns^2np^6

where n = quantum no. of outermost level
= period number of element

Outer electron configuration of iodine? 5th period, Group 7
hence, $5s^2 5p^5$

2. Elements in Groups 1 and 2 fill s sublevels
(e.g., Na and Mg fill 3s sublevel)
Elements in Groups 3-8 fill p sublevels
(e.g., Al-Ar fill 3p sublevel)
Transition metals (10 in each series) fill d sublevels
Lanthanides (14) fill 4f; actinides (14) fill 5f sublevel

B. Atomic Radius

1. In general, atomic radius decreases going across a period from left to right, increases going down group

Na 0.186 nm	Mg 0.160 nm	S 0.104 nm	Cl 0.099 nm
K 0.231 nm			Br 0.114 nm

2. Trends can be expressed in terms of effective nuclear charge experienced by outermost electron(s).

Eff. nuclear charge = Z - S, where Z is nuclear charge, S is number of electrons in inner, complete levels. Electrons in outer level do not shield one another effectively.

Na 11 - 10 = +1   Mg 12 - 10 = +2   Al 13 - 10 = +3
K 19 - 18 = +1

C. Ionization Energy

1. First ionization energy = energy that must be absorbed to convert an atom to a +1 ion:

$Na(g) \longrightarrow Na^+(g) + e^-$

Increases → in Periodic Table (atoms get smaller, making it more difficult to remove electron)
Decreases ↓ in Periodic Table (atoms get larger)

2. Successive ionization energies increase steadily; more difficult to remove electrons as + charge increases. Large jump in ionization energy occurs when electron is removed from inner level:

$Mg(1s^2 2s^2 2p^6 3s^2) \longrightarrow Mg^+ (1s^2 2s^2 2p^6 3s^1)$ $\Delta E_1 = +738$ kJ
$Mg^+(1s^2 2s^2 2p^6 3s^1) \longrightarrow Mg^{2+}(1s^2 2s^2 2p^6)$ $\Delta E_2 = +1450$ kJ
$Mg^{2+}(1s^2 2s^2 2p^6) \longrightarrow Mg^{3+}(1s^2 2s^2 2p^5)$ $\Delta E_3 = +7731$ kJ

D. Metallic Character

Decreases → in Periodic Table, increases ↓ . Metals are found below and to the left of the diagonal stairway, nonmetals above and to the right. Elements along the stairway (B, Si, Ge, - -) are referred to as metalloids and are often semiconductors.

LECTURE 2

I Chemistry of the Alkali and Alkaline Earth Metals

A. In general, these metals, except for Be, exist as cations in their compounds (+1 for Group 1, +2 for Group 2).

Group 1 halides: MX — Group 2 halides: MX_2

Group 1 sulfides: M_2S — Group 2 sulfides: MS

Group 1 nitrides: M_3N — Group 2 nitrides: M_3N_2

B. Reaction with Hydrogen

$2\ Na(s) + H_2(g) \longrightarrow 2\ NaH(s)$

$Ca(s) + H_2(g) \longrightarrow CaH_2(s)$

What volume of H_2 at 1.00 atm, 200°C is required to react with 1.00 g Na?

$$\text{no. moles } H_2 = 1.00\text{ g Na} \times \frac{1\text{ mol Na}}{22.99\text{ g Na}} \times \frac{1\text{ mol } H_2}{2\text{ mol Na}} = 0.0217\text{ mol } H_2$$

$$\text{volume} = \frac{(0.0217)(0.0821)(473)}{1.00}\text{ L} = 0.843\text{ L}$$

C. Reaction with Water

$2\ Na(s) + 2\ H_2O(l) \longrightarrow H_2(g) + 2\ Na^+(aq) + 2\ OH^-(aq)$

$Ca(s) + 2\ H_2O(l) \longrightarrow H_2(g) + Ca^{2+}(aq) + 2\ OH^-(aq)$

Calculate ΔH for the reaction of sodium with water, using heats of formation.

$$\Delta H = 2\,\Delta H_f\ Na^+(aq) + 2\,\Delta H_f\ OH^-(aq) - 2\,\Delta H_f\ H_2O(l)$$
$$= 2(-239.7\text{ kJ}) + 2(-229.9\text{ kJ}) - 2(-285.8\text{ kJ}) = -367.6\text{ kJ}$$

(enough heat to raise t of a kilogram of water to the boiling point)

D. Reaction with Oxygen

1. Li and the Group 2 metals $\longrightarrow$ normal oxides (O^{2-} ion)
2. Na (Ba) $\longrightarrow$ peroxides (O_2^{2-} ion)
3. K, Rb, Cs $\longrightarrow$ superoxides (O_2^- ion)

Write balanced equations for reactions of O_2 with Mg, Na, K

$2\ Mg(s) + O_2(g) \longrightarrow 2\ MgO(s)$

$2\ Na(s) + O_2(g) \longrightarrow Na_2O_2(s)$

$K(s) + O_2(g) \longrightarrow KO_2(s)$

II Transition Metals

A. Generally less reactive toward nonmetals (H_2, N_2, O_2) or water than are the Group 1 and Group 2 metals. Explained, at least in part, by higher ionization energies.

B. Often form more than one cation (e.g., Fe^{2+}, Fe^{3+}). This is explained by fact that transition metal contain several electrons of roughly the same energy

	E_1	E_2	E_3
Ca	590	1145	4912
Fe	759	1561	2957

$Ca(g) \longrightarrow Ca^{2+}(g) + 2\,e^-$; $\Delta E = 1735$ kJ

$Ca(g) \longrightarrow Ca^{3+}(g) + 3\,e^-$; $\Delta E = 6647$ kJ

$Fe(g) \longrightarrow Fe^{2+}(g) + 2\,e^-$; $\Delta E = 2320$ kJ

$Fe(g) \longrightarrow Fe^{3+}(g) + 3\,e^-$; $\Delta E = 5277$ kJ

easier to form Ca^{2+} than Fe^{2+}; more difficult to form Ca^{3+} than Fe^{3+}

C. Form many colored compounds. Energy required to promote electron is about equivalent to that of visible light.

DEMONSTRATIONS

1. Reaction of Na with water: Test. Dem. 5; J. Chem. Educ. 58 506 (1981), 61 635 (1984)
2. Reaction of Mg with steam: Test. Dem. 9, 58, 127, 142
3. Flame tests: Test. Dem. 22, 91
4. Slaking of lime: Test. Dem. 29
5. Low melting alloys (Wood's metal) Test. Dem. 40, 93
6. Reaction of Mg with CO_2: Test. Dem. 29; J. Chem. Educ. 55 450 (1978)

QUIZZES

Quiz 1

1. Give the outer electron configuration of
   a. selenium b. bismuth c. xenon
2. Write balanced equations for the reaction of lithium with
   a. nitrogen b. oxygen c. chlorine d. water

Quiz 2

1. Consider the four elements oxygen, fluorine, sulfur, and chlorine.
   a. Which one has the largest atomic radius?
   b. Which one has the largest first ionization energy?
   c. Which one(s) react with Group 1 metals to form ionic compounds?
   d. Which one(s) have the outer electron configuration ns^2np^4?
2. The red color observed when lithium salts are heated in a flame is due to a spectral line at 670.8 nm. Calculate the energy difference and the frequency corresponding to this line.
   ($E_{hi} - E_{lo} = \frac{1.196 \times 10^5}{\lambda} \frac{kJ.nm}{mol}$; $\nu = \frac{2.998 \times 10^8}{\lambda}$ m/s

Quiz 3

1. Give the outer electron configuration of
   a. strontium b. antimony c. polonium
2. Write balanced equations for the reaction of calcium with
   a. oxygen b. iodine c. nitrogen d. water

Quiz 4

1. Give the atomic number of the element that just fills the
   a. 6s sublevel b. 4d sublevel c. 5p sublevel

2. Explain, in your own words, why atomic radius decreases going across a period.

3. Give the formulas of the ions present in
   a. KO_2 b. Mg_3N_2 c. BaO_2

Quiz 5

1. Where would you expect to find the largest increase in successive ionization energies for Al? Explain your reasoning.

2. Write a balanced equation for the reaction of lithium with water. What volume of hydrogen gas (718 mm Hg, 22°C) is formed when 1.16 g of lithium reacts?

PROBLEMS 1-30

8.1 a. vertical family of elements b. Group 7 element
c. Group 1 metal d. elements filling 5f sublevel
e. Group 8 element

8.2 118

8.3 a. 38 b. 87 c. 51 d. 57

8.4 a. $6s^2$ b. $5s^2 5p^5$ c. $6s^2 6p^2$ d. $6s^2 6p^3$

8.5 a. Br < Se < K b. K < Se < Br c. K > Se > Br

8.6 a. O b. Na c. Na

8.7 a. metal b. metalloid c. nonmetal d. metal e. metal

8.8 a. Cs, Fr b. Ne, Ar, Kr, Xe, Rn c. Tm, Yb
d. Cs, Ba, Tl, Pb, Bi

8.9 5.00 g/cm^3

8.10 a. Li_3N b. Li_2S c. LiH d. LiOH e. LiI

8.11 a. $K_2O_2(s) + 2\ H_2O \longrightarrow H_2O_2(aq) + 2\ K^+(aq) + 2\ OH^-(aq)$
b. $2\ Sr(s) + O_2(g) \longrightarrow 2\ SrO(s)$
c. $Cs(s) + O_2(g) \longrightarrow CsO_2(s)$
d. $SrH_2(s) + 2\ H_2O \longrightarrow 2\ H_2(g) + Sr^{2+}(aq) + 2\ OH^-(aq)$

8.12 $2\ Ca(s) + O_2(g) \longrightarrow 2\ CaO(s)$
$CaO(s) + H_2O(l) \longrightarrow Ca(OH)_2(s)$
$Ca(s) + 2\ H_2O(l) \longrightarrow H_2(g) + Ca(OH)_2(s)$

8.13 a. Na b. Cs c. Na d. K

8.14 a. $CaCO_3(s) \longrightarrow CaO(s) + CO_2(g)$
b. $CaSO_4 \cdot 2H_2O(s) \longrightarrow CaSO_4 \cdot \frac{1}{2}H_2O(s) + 3/2\ H_2O(g)$

8.14 c. $4\ KO_2(s) + 2\ H_2O(g) \longrightarrow 3\ O_2(g) + 4\ KOH(s)$

d. $Mg(s) + H_2O(g) \longrightarrow MgO(s) + H_2(g)$

8.15 a. $Li(s) + H_2O(l) \longrightarrow Li^+(aq) + OH^-(aq) + \frac{1}{2}\ H_2(g)$

b. for one mole of Li:

$$\Delta H = \Delta H_f\ Li^+(aq) + \Delta H_f\ OH^-(aq) - \Delta H_f\ H_2O(l)$$
$$= -278.5\ kJ - 229.9\ kJ + 285.8\ kJ = -222.6\ kJ$$

$$\Delta H = 2.50\ g\ Li \times \frac{1\ mol\ Li}{6.941\ g\ Li} \times \frac{-222.6\ kJ}{1\ mol\ Li} = -80.2\ kJ$$

8.16 $BaO_2(s) \longrightarrow BaO(s) + \frac{1}{2}\ O_2(g)$

$$\Delta H \text{ per mole} = 1\ mol\ BaO_2 \times \frac{169.33\ g}{1\ mol\ BaO_2} \times \frac{0.503\ kJ}{1.18\ g\ BaO_2} = 72.2\ \frac{kJ}{mol}$$

$$72.2\ kJ = -558\ kJ - \Delta H_f\ BaO_2(s)$$

$$\Delta H_f\ BaO_2(s) = -558\ kJ - 72\ kJ = -630\ kJ$$

8.17 for equation as written:

$$\Delta H = \Delta H_f\ CaSO_4 . 2H_2O(s) - 3/2\ \Delta H_f\ H_2O(l) - \Delta H_f CaSO_4 \cdot \tfrac{1}{2}H_2O(s)$$
$$= -2021\ kJ - 1.5(-286\ kJ) + 1575\ kJ = -17\ kJ$$

$$\Delta H = 2.50 \times 10^3\ g\ CaSO_4 \cdot 2H_2O \times \frac{1\ mol\ CaSO_4 \cdot 2H_2O}{172.16\ g\ CaSO_4 . 2H_2O} \times \frac{-17\ kJ}{1\ mol}$$
$$= -2.5 \times 10^2\ kJ$$

8.18 a. $$1.000\ g\ MgO \times \frac{1\ mol\ MgO}{40.30\ g\ MgO} \times \frac{-601.8\ kJ}{1\ mol\ MgO} = -14.93\ kJ$$

b. $$14{,}930\ J = 4.18\ \frac{J}{g \cdot {}^oC} \times 452\ g \times \Delta t$$

$$\Delta t = 7.90^oC$$

8.19 $$n\ O_2 = 12.0\ g\ BaO_2 \times \frac{1\ mol\ BaO_2}{169.33\ g\ BaO_2} \times \frac{1\ mol\ O_2}{2\ mol\ BaO_2} = 0.0354\ mol\ O_2$$

$$P = \frac{(0.0354\ mol)(0.0821\ L \cdot atm/mol \cdot K)(293\ K)}{0.500\ L} = 1.70\ atm$$

8.20 a. $Na(s) + H_2O(l) \longrightarrow Na^+(aq) + OH^-(aq) + \frac{1}{2}\ H_2(g)$

b. $$n\ H_2 = 3.05\ g\ Na \times \frac{1\ mol\ Na}{22.99\ g\ Na} \times \frac{1\ mol\ H_2}{2\ mol\ Na} = 0.0663\ mol\ H_2$$

$$V = \frac{(0.0663\ mol)(0.0821\ L \cdot atm/mol.K)(298\ K)}{742/760\ atm} = 1.66\ L$$

8.21 a. $$n\ H_2O = \frac{(116 \times 0.062\ L)(752/760\ atm)}{(0.0821\ L \cdot atm/mol \cdot K)(310\ K)} = 0.280\ mol\ H_2O$$

b. $$0.280\ mol\ H_2O \times \frac{4\ mol\ KO_2}{2\ mol\ H_2O} \times \frac{71.10\ g\ KO_2}{1\ mol\ KO_2} = 39.8\ g\ KO_2$$

mass left = 216 g - 40 g = 176 g

8.22 $1s^2 2s^2 2p^6 3s^2 3p^6 4s^2 3d^{10} 4p^6 5s^2$

1s (↑↓) 2s (↑↓) 2p (↑↓)(↑↓)(↑↓) 3s (↑↓) 3p (↑↓)(↑↓)(↑↓) 4s (↑↓)

3d (↑↓)(↑↓)(↑↓)(↑↓)(↑↓) 4p (↑↓)(↑↓)(↑↓) 5s (↑↓)

8.23 a. $n = 5, \ell = 0, m_\ell = 0, m_s = +\frac{1}{2}$

b. $n = 6, \ell = 0, m_\ell = 0, m_s = +\frac{1}{2}, -\frac{1}{2}$

c. $n = 7, \ell = 0, m_\ell = 0, m_s = +\frac{1}{2}, -\frac{1}{2}$

8.24 $E_{hi} - E_{lo} = \frac{1.196 \times 10^5}{\lambda \text{ (in nm)}} = \frac{1.196 \times 10^5}{780.0} = 153.3$ kJ/mol

$= \frac{1.196 \times 10^5}{420.2} = 284.6$ kJ/mol

8.25 $\lambda = \frac{1.986 \times 10^{-25}}{3.368 \times 10^{-19}}$ m $= 5.897 \times 10^{-7}$ m $= 589.7$ nm

8.26 See discussion, Section 8.1

8.27 a. outer s electrons are being removed
b. inner (3p) electron is being removed with Ca

8.28 high electrical conductivity, high thermal conductivity, ductility, malleability, luster, formation of positive ions

8.29 a. have closely spaced energy levels, partially occupied by electrons
b. principal energy level, in which outer electron is located, increases
c. have two outer electrons, which are readily removed

8.30 a. increases going across period
b. ns^2np^5
c. at least one metal (Fe + C)

CHAPTER 9

BASIC SKILLS

1. State the electron configurations of monatomic ions, including those derived from transition metals.
2. Compare the radii of monatomic ions to those of the corresponding atoms.
3. Write Lewis structures for molecules and polyatomic ions.
4. Predict whether a species is likely to show resonance, given or having derived its Lewis structure.
5. Using only the Periodic Table, describe electronegativity trends and compare the relative polarity of different bonds.
6. Compare bond distances and bond energies for polar vs nonpolar bonds; multiple vs single bonds.
7. Use a table of bond energies to calculate ΔH for gas phase reactions.

LECTURE NOTES

This is the first of two chapters dealing with chemical bonding and molecular structure. After a brief review of ionic bonding, the chapter focuses on Lewis structures. Resonance structures, non-octet structures, and bond polarity are among the other topics covered. The chapter ends with a discussion of bond energy and its use in calculating ΔH. Some points to keep in mind:

1. The most important skill to be obtained from this chapter is writing Lewis structures. Students find this relatively easy to do, provided they get sufficient practice. Note that the ability to write Lewis structures is essential to an understanding of many of the topics in Chapter 10 (molecular geometry, polarity, hybridization).

2. Non-octet structures are simply mentioned here; they will be considered in greater detail in later chapters.

3. The calculation of ΔH from bond energies, in previous editions, appeared in the thermochemistry chapter. It would seem to fit more logically here, when students are better able to identify the bonds in molecules.

This material is readily covered in two lectures. It helps

to spread Lewis structures over both lectures, assuming students review their notes from the previous lecture before attending the next one.

LECTURE 1

I <u>Ionic Bonding</u> Electrostatic attraction between ions of opposite charge.

A. Formulas of Ionic Compounds (recall Chapter 3)

	X^-	X^{2-}	X^{3-}
M^+	MX	M_2X	M_3X
M^{2+}	MX_2	MX	M_3X_2
M^{3+}	MX_3	M_2X_3	MX

B. Electron Configurations, Monatomic Ions

1. Noble gas structures: Group 1 and 2 metals, Group 6 and 7 nonmetals. Ions with argon configuration:

$$S^{2-},\ Cl^-,\ K^+,\ Ca^{2+},\ Sc^{3+}$$

2. Transition metal cations Outer s electrons lost when cation is formed.

$$Fe\ [{}_{18}Ar]\ 4s^2 3d^6 \qquad Fe^{2+}\ [{}_{18}Ar]\ 3d^6 \qquad Fe^{3+}\ [{}_{18}Ar]\ 3d^5$$

C. Sizes of Ions

Cations are smaller than corresponding atoms; anions are larger:

Na 0.186 nm   Cl 0.099 nm

Na^+ 0.095 nm Cl^- 0.181 nm

II Covalent Bonding; Lewis Structures

A covalent bond consists of an electron pair shared between two nonmetal atoms

$$H\cdot + H\cdot \longrightarrow H-H$$

A Lewis structure shows the distribution of outer (valence) electrons in an atom, molecule, or polyatomic ion. Unshared electrons are shown as dots, bonds as straight lines.

$$H\cdot + \underset{\cdot\cdot}{\overset{\cdot\cdot}{F}}\!: \longrightarrow H-\underset{\cdot\cdot}{\overset{\cdot\cdot}{F}}\!:$$

$$2H\cdot + \cdot\underset{\cdot\cdot}{\overset{\cdot\cdot}{O}}\cdot \longrightarrow H-\underset{\cdot\cdot}{\overset{\cdot\cdot}{O}}-H$$

In HF and H_2O, as in most molecules and polyatomic ions, nonmetal atoms (except H) are surrounded by 8 electrons, an octet. In this sense, each atom has a noble gas structure.

Lewis structures are written following a stepwise procedure. To illustrate that procedure, consider the species OCl^-, CH_3OH, and SO_3^{2-}.

1. Count valence electrons available. Number of valence electrons contributed by nonmetal atom is equal to its group number in the Periodic Table (1 for H). Add electrons to account for negative charge.

OCl^- ion: $6 + 7 + 1 = 14$ valence e^-

CH_3OH molecule: $4 + 4(1) + 6 = 14$ valence e^-

SO_3^{2-} ion: $6 + 3(6) + 2 = 26$ valence e^-

2. Draw skeleton structure, using single bonds

$$O - Cl \qquad H - \overset{\displaystyle H}{\underset{\displaystyle H}{C}} - O - H \qquad O - \overset{}{\underset{\displaystyle O}{S}} - O$$

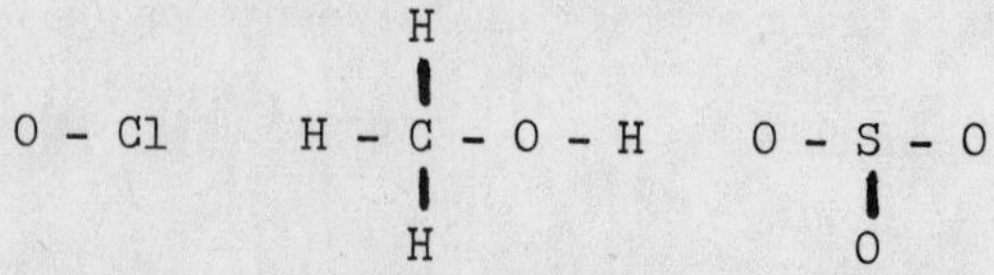

Note that carbon virtually always forms four bonds. In molecules and polyatomic ions, oxygen atoms are ordinarily bonded to a central nonmetal atom rather than to each other.

3. Deduct two electrons for each single bond in the skeleton.

OCl^- ion: $14 - 2 = 12$ valence e^- left

CH_3OH molecule: $14 - 10 = 4$ valence electrons left

SO_3^{2-} ion: $26 - 6 = 20$ valence e^- left

4. Distribute these electrons to give each atom a noble gas structure, if possible.

$$(:\!\ddot{O} - \ddot{C}l\!:)^- \qquad H - \overset{\displaystyle H}{\underset{\displaystyle H}{C}} - \ddot{O} - H \qquad (:\!\ddot{O} - \underset{\displaystyle :\!\ddot{O}\!:}{S} - \ddot{O}\!:)^{2-}$$

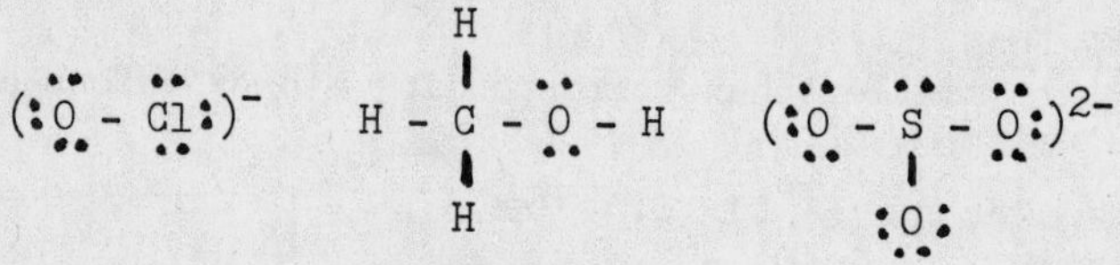

LECTURE 2

I <u>Lewis Structures</u> (cont.)

A. Structure of NO_3^- ion?

no. valence $e^- = 5 + 18 + 1 = 24$

skeleton: $\quad O - \overset{}{\underset{\displaystyle O}{N}} - O$

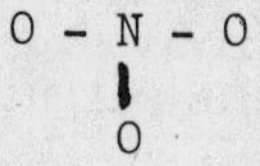

valence e^- left: $24 - 6 = 18$

If all electrons distributed as unshared pairs: $:\!\ddot{O} - N - \ddot{O}\!:$ with $:\!\ddot{O}\!:$ above

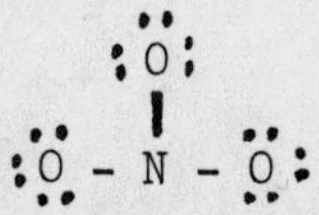

N surrounded by only 6 valence electrons. To remedy a deficiency of two electrons, form a double bond:

$$:\!\ddot{O} - N - \ddot{O}\!: \quad \underset{\displaystyle :\!\ddot{O}\!:}{\overset{\displaystyle \|}{}}$$

N_2 structure? 10 valence e^- $\quad :\!N \equiv N\!:$

B. <u>Resonance Forms</u>

In the NO_3^- ion, all three bonds are the same length. To explain this, invoke the concept of resonance.

$$:\ddot{O} - N - \ddot{O}: \longleftrightarrow :\ddot{O} - N = \ddot{O} \longleftrightarrow \ddot{O} = N - \ddot{O}:$$
$$\overset{\displaystyle \|}{\underset{:\ddot{O}:}{}} \qquad\qquad \overset{\displaystyle |}{\underset{:\ddot{O}:}{}} \qquad\qquad \overset{\displaystyle |}{\underset{:\ddot{O}:}{}}$$

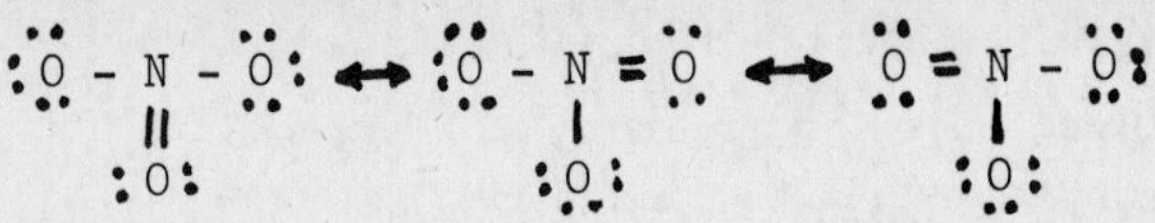

True structure of NO_3^- ion is a "hybrid" of these three forms. Note that:
1. Resonance forms obtained by moving electrons, not atoms.
2. Resonance can be expected when it is possible to draw more than one structure that follows the octet rule.

C. Nonoctet Structures
1. NO (11 valence e^-), NO_2 (17 valence e^-)

$$\cdot \ddot{N} = \ddot{O}: \quad , \quad \cdot N \overset{\displaystyle \nearrow^O}{\searrow_{\ddot{O}:}}$$

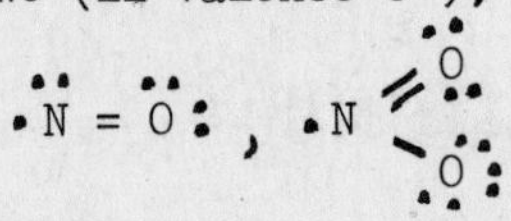

2. Central atom surrounded by less than four pairs of electrons

BeF_2 $\quad :\ddot{F} - Be - \ddot{F}:$ $\qquad BF_3$ $\quad :\ddot{F} - B - \ddot{F}:$
$$\overset{\displaystyle |}{\underset{:\ddot{F}:}{}}$$

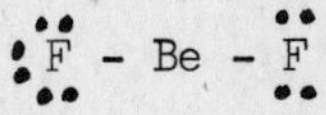

3. Central atom surrounded by more than four electron pairs:
PF_5, SF_6

II Bond Properties
A. Polarity
Bond between two identical atoms, as in H - H, Cl - Cl, is nonpolar; electrons equally shared. If atoms differ, as in H- F, the bond is polar; bonding electrons shifted toward more electronegative atom.

$$\begin{array}{c} H \\ 2.1 \end{array}$$

Li	Be	B	C	N	O	F
1.0	1.5	2.0	2.5	3.0	3.5	4.0

Extent of polarity depends upon difference in electronegativity

C-H bond only very slightly polar (Δ EN = 0.4)
H-F bond strongly polar (Δ EN = 1.9)
Li-F bond ionic (Δ EN = 3.0)

B. Bond Energy = ΔH when one mole of bonds is broken in the gas state.

$Cl_2(g) \longrightarrow 2\ Cl(g)$; ΔH = B.E. Cl - Cl = 243 kJ
$N_2(g) \longrightarrow 2\ N(g)$; ΔH = B.E. N $\equiv$ N = 941 kJ

In general ,multiple bonds are stronger than single bonds:

C - C 347 kJ $\qquad$ C = C 612 kJ $\qquad$ C $\equiv$ C 820 kJ

Estimation of ΔH from bond energies. Proceed in two steps:
1. Break up reactant molecules into atoms. Since bonds are broken:

$\Delta H_1 = \Sigma$ bond energies reactants

2. Combine atoms to form new bonds in products. Since bonds

are formed:

$\Delta H_2 = -\Sigma$ bond energies products

Example: $N_2(g) + 3\ H_2(g) \longrightarrow 2\ NH_3(g)$

ΔH_1 = B.E. N≡N + 3 B.E. H-H = 941 kJ + 3(436 kJ) = 2249 kJ

ΔH_2 = -6 B.E. N-H = -6(389 kJ) = -2334 kJ

$\Delta H = \Delta H_1 + \Delta H_2 = -85$ kJ

(actual ΔH, calculated from heats of formation, is -92 kJ)

DEMONSTRATIONS

1. Paramagnetism of liquid O_2: J. Chem. Educ. 57 373 (1980)

QUIZZES

Quiz 1

1. Give the abbreviated electron configuration of

   a. Sc^{3+} b. Mg^{2+} c. Co^{2+} d. Co^{3+}

2. Draw Lewis structures for

   a. PCl_3 b. O_3 c. ClO_4^-

Quiz 2

1. Draw Lewis structures for:

   a. BrCl b. C_2H_5OH

2. Using the table of bond energies in your text, calculate ΔH for:

   $C_2H_5OH(g) + HCl(g) \longrightarrow C_2H_5Cl(g) + H_2O(g)$

Quiz 3

1. Give the electron configuration of each monatomic ion and the Lewis structure of each polyatomic ion in:

   a. $(NH_4)_2S$ b. $Fe_2(SO_4)_3$ c. $Ni(NO_3)_2$

Quiz 4

1. Arrange the following bonds in order of increasing polarity:

   Si-Cl, O-Cl, Cl-Cl, Na-Cl, F-Cl

2. Draw Lewis structures for:

   a. CN^- b. PO_4^{3-} c. CH_2O

Quiz 5

1. Choose the smaller member of each pair:

   a. Tl^+ ot Tl^{3+} b. Cd or Cd^{2+} c. I or I^- d. F^- or Cl^-

2. Draw Lewis structures for:

   a. SO_4^{2-} b. N_2F_2 c. CO

PROBLEMS 1-30

9.1 a. K^+, Ca^{2+}, Sc^{3+} b. S^{2-}, Cl^- c. K_2S, KCl, CaS, $CaCl_2$, Sc_2S_3, $ScCl_3$

9.2 a. Li_3N b. $Ba(NO_3)_2$ c. $FeSO_4$ d. Cr_2O_3

9.3 a. tellurium b. hydrogen c. zinc d. nitrogen

9.4 NaF, KCl, RbBr, CsI

9.5 a. $1s^22s^22p^63s^2$; $1s^22s^22p^6$

b. $1s^22s^22p^63s^23p^4$; $1s^22s^22p^63s^23p^6$

c. $1s^22s^22p^63s^23p^64s^23d^2$; $1s^22s^22p^63s^23p^63d^2$

d. $1s^22s^22p^63s^23p^63d^4$; $1s^22s^22p^63s^23p^63d^3$

9.6 a. $3d^6$ (4 unp. e^-) b. $3d^5$ (5 unp. e^-) c. $3d^8$ (2 unp. e^-) d. 0

9.7 a. Mg atom b. S^{2-} ion c. Ti atom d. Cr^{2+} ion

9.8 K > Na > Mg > Mg^{2+}

9.9

$$Cs(s) + Br(g) \longrightarrow Cs(s) + \tfrac{1}{2}Br_2(l) \quad ; \Delta H = -191 \text{ kJ}$$
$$Cs^+(g) + Br^-(g) \longrightarrow Cs(g) + Br(g) \quad ; \Delta H = -50 \text{ kJ}$$
$$Cs(s) + \tfrac{1}{2}Br_2(l) \longrightarrow CsBr(s) \quad ; \Delta H = -395 \text{ kJ}$$
$$Cs^+(g) + Br^-(g) \longrightarrow CsBr(s) \quad ; \Delta H = -636 \text{ kJ}$$

9.10 a. 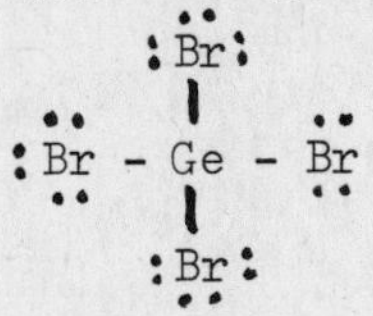 b. 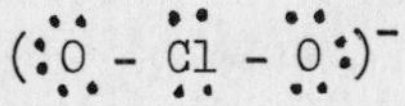 c. 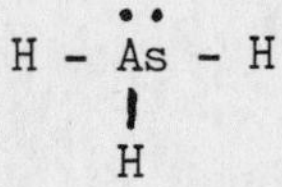

d. 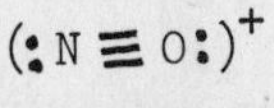

9.11 a. 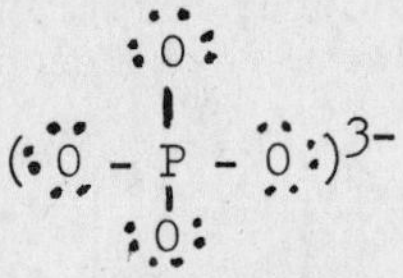 b. 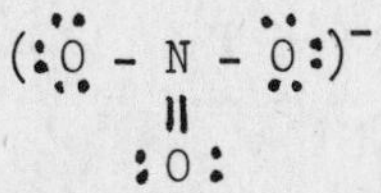 c. 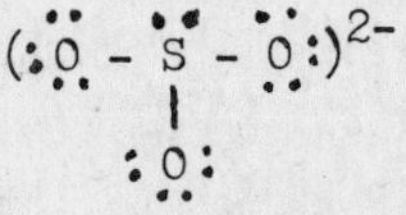

d. 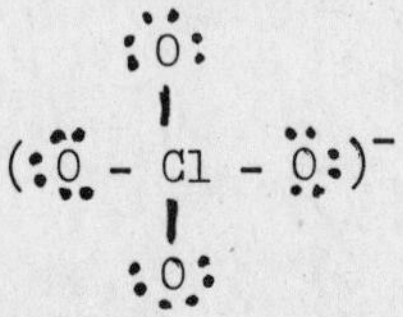

9.12 a. 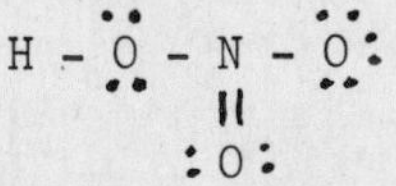 b. 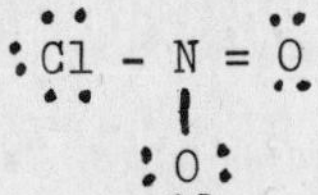 c 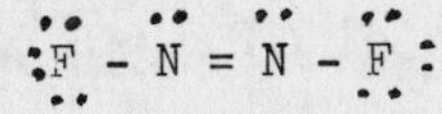

9.13 (Lewis structures: $H_2C = \ddot{O} - \ddot{O}:$ or cyclic structure with H_2C, O, O)

9.14 $H - \underset{\underset{H}{|}}{\overset{\overset{H}{|}}{C}} - \ddot{O} - \underset{\underset{H}{|}}{\overset{\overset{H}{|}}{C}} - H$, $H - \underset{\underset{H}{|}}{\overset{\overset{H}{|}}{C}} - \underset{\underset{H}{|}}{\overset{\overset{H}{|}}{C}} - \ddot{O} - H$

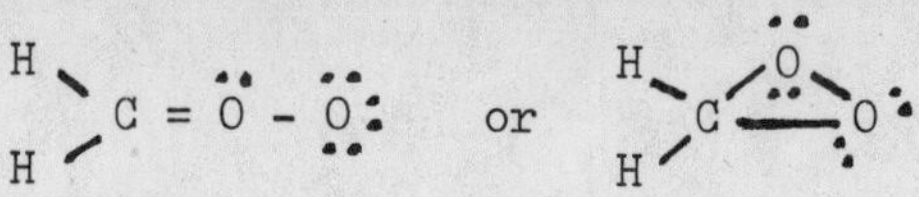

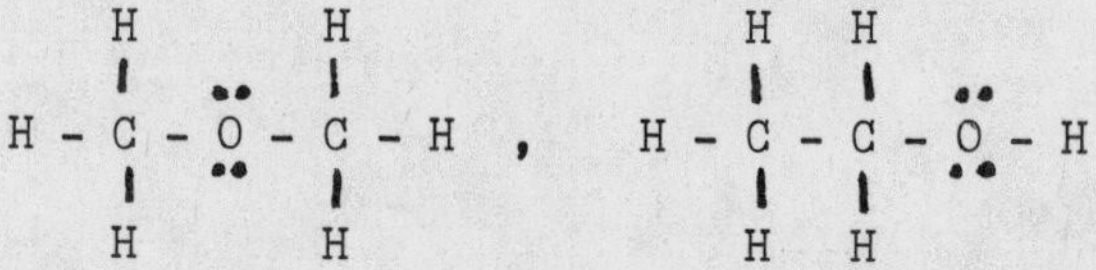

9.15 a. Cl_2 b. H_2SO_4 c. CH_4 d. CCl_4

9.16 a. $(:\ddot{S} = C = \ddot{N}:)^-$ b. $(:\ddot{S} - \overset{\overset{:\ddot{O}:}{|}}{\underset{\underset{:\ddot{O}:}{|}}{S}} - \ddot{O}:)^{2-}$ c. $\underset{H}{\overset{H}{>}} N - \ddot{O} - H$

d. $(:\ddot{O} - \overset{\overset{:\ddot{O}:}{|}}{\underset{\underset{:\ddot{O}:}{|}}{P}} - \ddot{O} - \overset{\overset{:\ddot{O}:}{|}}{\underset{\underset{:\ddot{O}:}{|}}{P}} - \ddot{O}:)^{4-}$

9.17 a. $\cdot N = \ddot{O}:$ b. $(:\ddot{O} - \dot{S} - \ddot{O}:)^-$ c. $:\ddot{Cl} - \overset{\overset{:\ddot{Cl}:}{|}}{B} - \ddot{Cl}:$

d. $(\cdot C \equiv O:)^+$

9.18 a. (resonance structures of SO_2/sulfur with oxygens) $\longleftrightarrow \longleftrightarrow$

b. $(:\ddot{O} - \ddot{N} = \ddot{O})^- \longleftrightarrow (\ddot{O} = \ddot{N} - \ddot{O}:)^-$

c. $(:\ddot{S} = C = \ddot{N}:)^- \longleftrightarrow (:\ddot{S} \equiv C - \ddot{N}:)^-$ $(:\ddot{S} - C \equiv N:)^-$

9.19 a. $\underset{H}{\overset{}{\diagdown}} N \equiv N - \ddot{N}:$, $\underset{H}{\overset{}{\diagdown}} \ddot{N} - N \equiv N:$ b. no; different skeleton

9.20 P-P $<$ P-C $<$ P-N $<$ P-O

9.21 I-Cl

9.22 a. Cl b. O c. F d. Cl

9.23 longest in N_2H_4, shortest in N_2

9.24 N-N bond in N_2H_4 is weakest; N$\equiv$N bond in N_2 is strongest

9.25 a. $\Delta H = 2$ B.E. O-H = 928 kJ
     b. $\Delta H = 3$ B.E. N-H = 1167 kJ
     c. $\Delta H = 4$ B.E. C-H = 1656 kJ

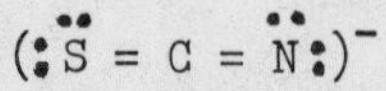

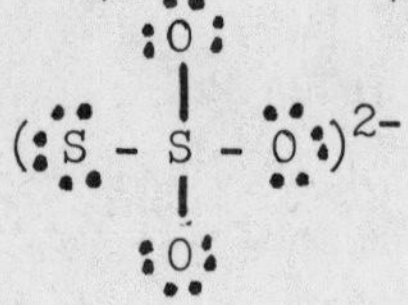

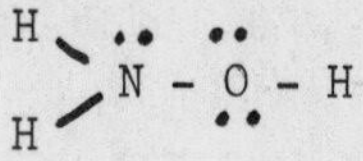

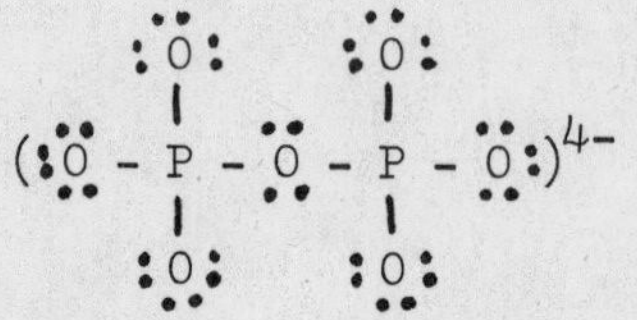

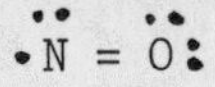

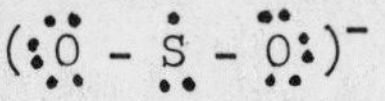

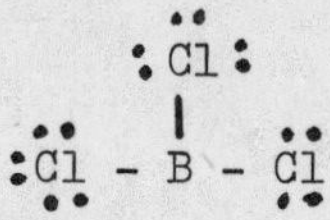

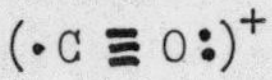

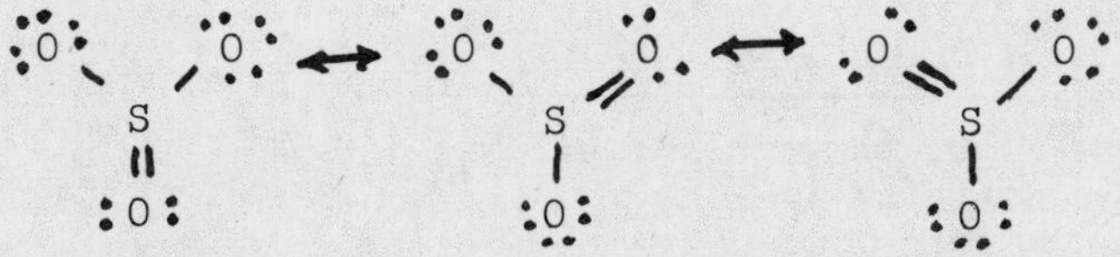

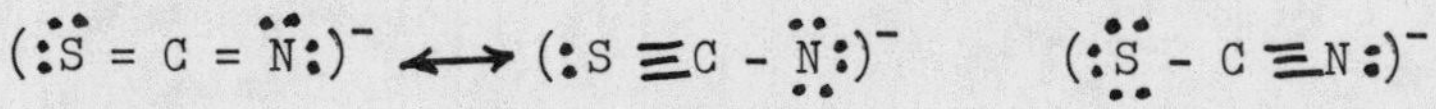

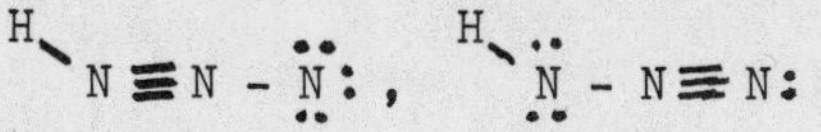

9.26 $Cl-Cl + 3\,F-F \longrightarrow 2\,F-\underset{\displaystyle F}{\underset{|}{Cl}}-F$

Break 1 Cl-Cl bond, 3 F-F bonds; form 6 Cl-F bonds

ΔH = B.E. Cl-Cl + 3 B.E. F-F - 6 B.E. Cl-F

= 243 kJ + 3(153 kJ) - 6(255 kJ) = -828 kJ

9.27 $H-C\equiv C-H + 2\,Br-Br \longrightarrow H-\underset{\displaystyle Br}{\overset{\displaystyle Br}{C}}-\underset{\displaystyle Br}{\overset{\displaystyle Br}{C}}-H$

Break 1 C≡C, 2 Br-Br; form 1 C-C, 4 C-Br

ΔH = B.E. C≡C + 2 B.E. Br-Br - B.E. C-C - 4 B.E. C-Br

= 820 kJ + 386 kJ - 347 kJ - 4(276 kJ) = -245 kJ

9.28 a. Fe^{2+}: $1s^2 2s^2 2p^6 3s^2 3p^6 3d^6$ Cl^-: $1s^2 2s^2 2p^6 3s^2 3p^6$

b. Fe^{3+}: $1s^2 2s^2 2p^6 3s^2 3p^6 3d^5$ 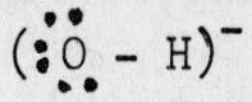

c. Ca^{2+}: $1s^2 2s^2 2p^6 3s^2 3p^6$ PO_4^{3-} 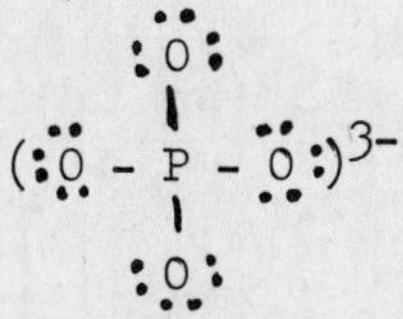

9.29 a. increases b. closer c. electron pairs
d. decreases

9.30 a. extra electron increases electron repulsion
b. energy evolved when oppositely charged ions come together
c. covalent bond forms
d. H-F bond stronger than average of H-H and F-F bonds

CHAPTER 10

BASIC SKILLS

1. Given or having derived the Lewis structure of a molecule or polyatomic ion, apply VSEPR principles to deduce its geometry.
2. Given or having derived the geometry of a molecule or polyatomic ion, predict whether it will be a dipole.
3. Starting with the Lewis structure of a molecule or polyatomic ion, give the hybridization of each atom.
4. Predict the number of sigma and pi bonds in a species, knowing its Lewis structure.
5. Predict the hybridization and geometry of a molecule in which a central atom is surrounded by 5 or 6 electron pairs.

LECTURE NOTES

This is the second of two chapters emphasizing molecular structure. It starts with the discussion of two basic properties of molecules: geometry and polarity. Valence bond theory is then applied to simple molecules, introducing the concepts of hybridization, sigma and pi bonds. The last two sections of the chapter deal with expanded octets and molecular orbital theory. Many instructors choose to omit these sections; if you do this, you will need only about $1\frac{1}{2}$ lectures for this chapter. If you cover the entire chapter, it will run about $2\frac{1}{2}$ lectures.

Several observations concerning the coverage of material:

1. We choose to deemphasize the effect of lone pairs on molecular geometry (i.e., the decrease in bond angle). In our experience, students, in their first exposure to molecular geometry, have considerable trouble with the basic principles.

2. Table 10.1 is a useful way of summarizing all that is said about molecular geometry in Section 10.1. The point of view is slightly different; instead of talking about type of bonds, we consider the number of atoms surrounding the central atom.

3. Students often get the mistaken idea that promotion of electrons is necessary for hybridization to occur. They also have trouble distinguishing between sigma and pi bonds.

4. If you intend to cover Section 10.4 on expanded octets, you may wish to integrate it into earlier sections, covering

geometry in Section 10.1 and hybridization in Section 10.3.

LECTURE 1

I Molecular Geometry

A. Will restrict discussion to simple molecules in which a central atom is surrounded by 2, 3 or 4 electron pairs, which may or may not be shared. The general principle is that these electron pairs tend to be oriented so as to be as far apart as possible.

B. No Unshared Pairs

1. XY_2: linear, 180° bond angle

   :F̤̈ - Be - F̤̈:   Ö = C = Ö

   Note that, as far as geometry is concerned, multiple bond behaves like single bond.

2. XY_3: equilateral triangle, 120° bond angles

   Again, double bond behaves like single bond

3. XY_4: tetrahedral, 109.5° bond angle CH_4

C. Unshared Pairs

1. Atoms occupy roughly the same positions as before

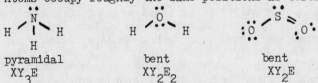

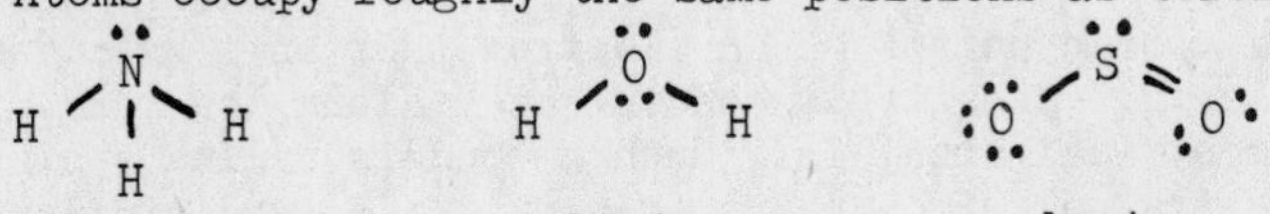

   pyramidal XY_3E — bent XY_2E_2 — bent XY_2E

   Note that in describing geometry, refer only to positions of atoms.

2. Bond angles ordinarily slightly smaller; lone pair spreads out over more space. Bond angle in H_2O is 105° rather than 109°.

II Molecular Polarity

A. Diatomic molecules: polar if atoms differ

H - Cl	Cl - Cl
polar	nonpolar

HCl molecules line up in electric field, Cl_2 molecules don't

B. Polyatomic Molecules Even though bonds are polar, molecule may be nonpolar if it is symmetrical.

F ← Be → F	H → O ← H	CH_4	CH_3Cl
nonpolar	polar	nonpolar	polar

III Atomic Orbitals; Hybridization

In molecules, the orbitals occupied by electron pairs are seldom "pure" s or p orbitals. Instead, they are "hybrid" orbitals, formed by combining s and p orbitals.

A. s orbital + p orbital → two "sp" hybrid orbitals

	2s	2p		
Be in BeF_2	(↑↓)	(↑↓)		

s orbital + two p orbitals → three "sp^2" hybrid orbitals

	2s	2p	2p	
B in BF_3	(↑↓)	(↑↓)	(↑↓)	

s orbital + three p orbitals → four "sp^3" hybrid orbitals

	2s	2p	2p	2p
C in CH_4	(↑↓)	(↑↓)	(↑↓)	(↑↓)

LECTURE 2

I Hybridization (cont.)

A. sp hybridization in BeF_2, sp^2 in BF_3, sp^3 in CH_4

B. Unshared pairs can be hybridized. In both NH_3 and H_2O, hybridization is sp^3.

C. Only one of the electron pairs in a multiple bond can be hybridized.

O = S - O sp^2 hybridization for sulfur

O = C = O   sp hybridization for carbon

D. When a bond consists of an electron pair in a hybrid orbital, the electron density is concentrated along the bond axis and is symmetrical about it. Such a bond is called a sigma bond. The "extra" electron pairs in a multiple bond are located in unhybridized orbitals which are not concentrated along the bond axis. Such bonds are called pi bonds.

CO_2	two sigma bonds, two pi bonds
SO_2	two sigma bonds, one pi bond
CH_4	four sigma bonds
N_2	one sigma bond, two pi bonds

II Expanded Octets

A. In certain molecules, a central atom is surrounded by 5 or 6 electron pairs rather than 4. Most of these consist of a central nonmetal atom (3rd, 4th or 5th period) surrounded by halogen atoms. For such a molecule, the number of electrons around the central atom is found by applying the rule:

no. of valence e^- = x + y

where x = group no. central atom, y = no. halogen atoms

SF_6 : 6 + 6 = 12; 6 electron pairs

XeF_4 : 8 + 4 = 12; 6 electron pairs

PF_5 : 5 + 5 = 10; 5 electron pairs

SF_4: 6 + 4 = 10; 5 electron pairs

B. Hybridization

5 electron pairs: sp^3d

6 electron pairs: sp^3d^2

Note that expanded octets do not occur with atoms in the 2nd period (e.g., N, O, F), since there are no 2d orbitals available for hybridization.

C. Geometry

XY_5 trigonal bypyramid (two triangular pyramids fused through the base)

XY_6 octahedron (two square pyramids fused through the base)

Other possible geometries; see Figure 10.10

LECTURE $2\frac{1}{2}$

I Molecular Orbitals

A. Formed by combining atomic orbitals

two atomic orbitals → high energy, antibonding MO
two atomic orbitals → low energy, bonding MO

Each MO can hold two electrons; MO's fill in order of increasing energy.

two 1s orbitals	σ^b_{1s}	$+\sigma^*_{1s}$	
H_2	(↑↓)	( )	1 bond
He_2	(↑↓)	(↑↓)	0 bonds

B. Molecular orbitals formed by 2nd period elements

two 2s orbitals $\sigma^b_{2s} + \sigma^*_{2s}$

two p_x orbitals $\sigma^b_{2p} + \sigma^*_{2p}$

two p_y orbitals $\pi^b_{2p} + \pi^*_{2p}$

two p_z orbitals $\pi^b_{2p} + \pi^*_{2p}$

Arrangement: $\sigma^b_{2s}, \sigma^*_{2s}, \pi^b_{2p}, \pi^b_{2p}, \sigma^b_{2p}, \pi^*_{2p}, \pi^*_{2p}, \sigma^*_{2p}$

Hund's rule is obeyed:

	σ^b_{2s}	σ^*_{2s}	π^b_{2p}	π^b_{2p}	σ^b_{2p}	π^*_{2p}	π^*_{2p}	σ^*_{2p}
N_2 (10 valence e^-)	(↑↓)	(↑↓)	(↑↓)	(↑↓)	(↑↓)			
O_2 (12 valence e^-)	(↑↓)	(↑↓)	(↑↓)	(↑↓)	(↑↓)	(↑)	(↑)	
F_2 (14 valence e^-)	(↑↓)	(↑↓)	(↑↓)	(↑↓)	(↑↓)	(↑↓)	(↑↓)	
Ne_2 (16 valence e^-)	(↑↓)	(↑↓)	(↑↓)	(↑↓)	(↑↓)	(↑↓)	(↑↓)	(↑↓)

	number of bonds	number unpaired electrons
N_2	3	0
O_2	2	2

F_2	1	0
Ne_2	0	0

Explains how O_2 can have double bond, two unpaired electrons

QUIZZES

Quiz 1

1. For each of the following species, draw the Lewis structure, describe the geometry, and state whether the species is a dipole.

   a. NO_3^- b. SCN^- c. O_3

2. State the hybridization shown by carbon in

   a. CH_4 b. CO_2 c. CH_2O

Quiz 2

1. State the number of electron pairs surrounding the central atom in:

   a. BCl_3 b. CCl_4 c. PCl_3 d. PCl_5 e. SCl_4

2. Draw the Lewis structure of the C_3H_6 molecule and state all the bond angles.

Quiz 3

1. Draw the Lewis structure and give the hybridization of each atom in

   a. CO   b. $COCl_2$ c. N_2O

2. Explain how, using MO theory, there can be a double bond and two unpaired electrons in the O_2 molecule.

Quiz 4

1. Give the hybridization of sulfur in

   a. H_2S b. SO_2 c. SO_4^{2-}

2. State the bond angles in each species in (1) and state whether it is a dipole.

Quiz 5

1. State the number of sigma and pi bonds in

   a. CO_3^{2-} b. BF_3 c. CN^- d. NO_2^-

2. Describe the geometry of each species in (1).

PROBLEMS 1-30

10.1 a. tetrahedral   b. linear   c. bent   d. pyramidal

10.2 a. triangular   b. bent   c. pyramidal   d. linear

10.3 a. 109° around C at left; 180° around triple bonded carbon
b. 120° around C atom; 109° around O atom
c. 120°

10.4 a. $:\ddot{F} - \ddot{N} - \ddot{F}:$ b. $H - \overset{\displaystyle H}{\underset{\displaystyle :\ddot{Cl}:}{C}} - H$ c. $(:\ddot{N} = N = \ddot{N}:)^{-}$

$\qquad\quad :\ddot{F}:$ $\qquad\qquad\qquad\qquad\qquad\qquad$ linear

$\qquad$ pyramidal $\qquad\qquad$ tetrahedral

$\qquad\qquad\quad :\ddot{Cl}:$

$\qquad (:\ddot{Cl} - \overset{\displaystyle |}{P} - \ddot{Cl}:)^{+}$

$\qquad\qquad\quad :\ddot{Cl}:$

$\qquad$ tetrahedral

10.5 a. $H - \ddot{O} - \ddot{Cl} - \ddot{O}:$ b. $(:\ddot{O} - \overset{\displaystyle |}{P} - \ddot{O}:)^{3-}$ c. $:\ddot{Cl} - \overset{\displaystyle |}{As} - \ddot{Cl}:$

$\quad$ bent $(109^{\circ}$ angles) $\qquad\qquad\qquad :\ddot{O}: \qquad\qquad\qquad\qquad :\ddot{Cl}:$

$\qquad\qquad\qquad\qquad\qquad\qquad$ pyramidal $\qquad\qquad$ pyramidal

$\qquad :\ddot{O} \diagup \overset{\displaystyle \ddot{O}}{\diagdown} \ddot{O}:$

$\qquad$ bent

10.6 a. CO_3^{2-}: 0 unshared pairs; NO_2^{-}: 1 unshared pair
$\qquad ClO_3^{-}$: 1 unshared pair ; SCN^{-}: 0 unshared pairs

$\quad$ b. CO_3^{2-}: 3 NO_2^{-}: 2 ClO_3^{-}: 3 SCN^{-}: 2

$\quad$ c. CO_3^{2-}: 120° NO_2^{-}: 120° ClO_3^{-}: 109° SCN^{-}: 180°

10.7 a. $H - \overset{\displaystyle H}{\underset{\displaystyle H}{C}} - \overset{\displaystyle :O:}{C} - \ddot{O} - \ddot{O} - \overset{\displaystyle :O:}{C} - \overset{\displaystyle H}{\underset{\displaystyle H}{C}} - H$

$\quad$ b. All bond angles 109° except around double bonded C atoms, where bond angles are 120°.

10.8 a. 109° b. 120° around double bonded C atoms; 109° in CH_3 groups

$\quad$ c. 180°

10.9 effect of unshared pairs makes bond angle less than 109.5° in PH_3, H_2S.

10.10 all

10.11 b, c, d

10.12 1st molecule is polar (unsymmetrical)

10.13 a. sp^3 b. sp c. sp^2 d. sp^3

10.14 a. sp^2 for carbon; sp^3 for single bonded O; sp^2 for double bonded O

$\quad$ b. sp^2 for nitrogen and O atoms at right; sp^3 for O at left

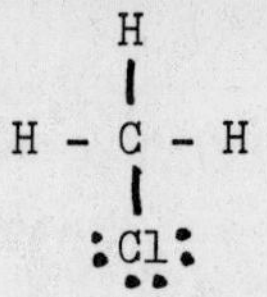

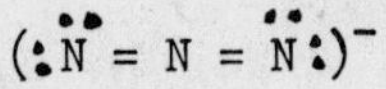

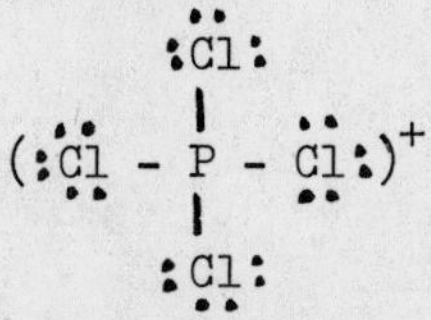

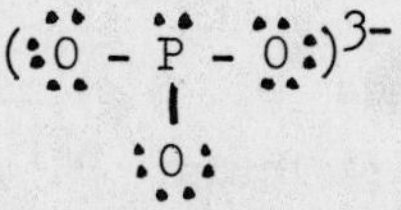

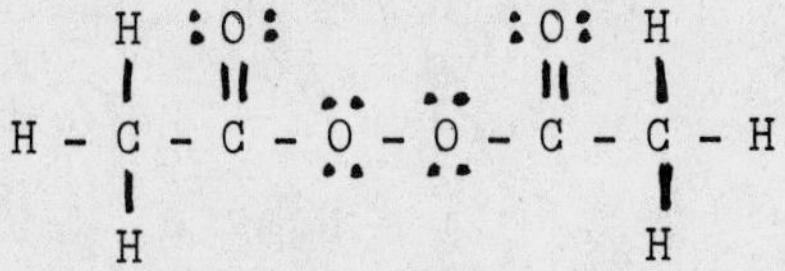

10.14 c. sp^3

d. sp for C; sp^2 for S, N

10.15 sp^2 around all C atoms and N atom; sp^3 around single bonded O; sp^2 around double bonded O

10.16 a. sp^3 b. sp^2 c. sp d. sp^2

10.17 a. 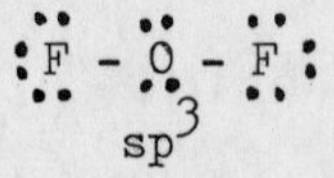 b. 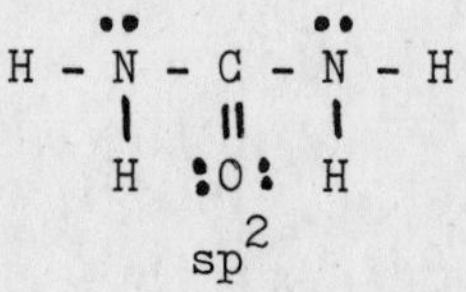 c. 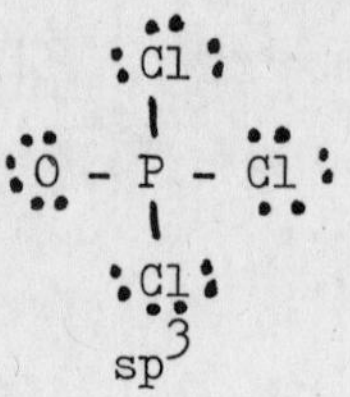

10.18 a. 4 sigma bonds b. 3 sigma bonds, 1 pi bond

c. 2 sigma, 2 pi bonds d. 4 sigma bonds, 1 pi bond

10.19 a. NH_4^+ b. NO_2^- c. C_2H_4, CO_3^{2-}, - -

10.20 a. 6, sp^3d^2 b. 5, sp^3d c. 5, sp^3d d. 5, sp^3d

10.21 a. octahedral b. linear c. distorted tetrahedron
d. T-shaped

10.22 a. octahedral b. square pyramid c. square, planar

10.23 a. 6, sp^3d^2 b. 6, sp^3d^2 c. 5, sp^3d

10.24

	σ^b_{2s}	σ^*_{2s}	π^b_{2p}	π^b_{2p}	σ^b_{2p}	π^*_{2p}	π^*_{2p}	σ^*_{2p}	Bonds	Unpaired e^-
a. CO	2	2	2	2	2				3	0
b. C_2^-	2	2	2	2	1				$2\frac{1}{2}$	1
c. F_2^-	2	2	2	2	2	2	2	1	$\frac{1}{2}$	1

10.25

+1 ion	Li_2^+	Be_2^+	B_2^+	C_2^+	N_2^+	O_2^+	F_2^+
bonds	$\frac{1}{2}$	$\frac{1}{2}$	$\frac{1}{2}$	3/2	5/2	5/2	3/2
unp. e^-	1	1	1	1	1	1	1

10.26 a. 10 valence e^-; 3 bonds
b. 8 valence e^-; 2 bonds
c. 14 valence e^-; 1 bond

10.27 a. consider bond polarity and molecular symmetry
b. consider bonding and unshared pairs around atom
c. add group no. of central atom to number of halogens bonded to it.

10.28 a. 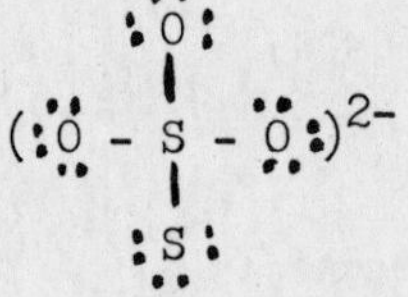 b. tetrahedral, 109° c. yes d. sp^3

10.29 a. 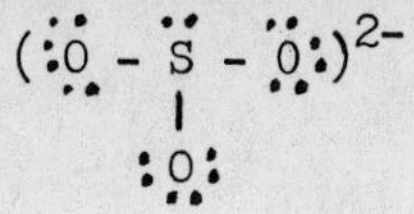 b. pyramidal c. yes d. sp^3

10.30

XY_2E_2	2	2	bent
XY_3	3	0	triangular
XY_4E_2	4	2	square planar
XY_5	5	0	trigonal bypyramid

CHAPTER 11

BASIC SKILLS

1. Use the Clausius-Clapeyron equation (Equation 11.2) to calculate one of the three quantities P_2, T_2, or H_{vap}, given the values of the other two and P_1 at T_1.
2. Draw a phase diagram for a pure substance, given appropriate data, and state what phases are present at any given point on the diagram.
3. Distinguish among ionic, molecular, network covalent, and metallic solids with regards to both particle structure and physical properties (melting point, conductivity, water solubility).
4. Compare different molecular substances with respect to physical properties (melting point, boiling point) and types of intermolecular forces (dipole forces, hydrogen bonds, dispersion forces).
5. Given the type of unit cell for a substance (simple cubic, face-centered cubic, or body-centered cubic), relate the cell dimensions to atomic radius.

LECTURE NOTES

This chapter requires more lecture time than most. Many of the concepts are relatively abstract; most are likely to be new to the student. We suggest you allow $2\frac{1}{2}$, possibly 3, lectures for this chapter. Among the more difficult topics are the following:

- the idea of vapor pressure. Perhaps surprisingly, students have less trouble with problems involving the Clausius-Clapeyron equation (e.g., Problems 11.4 and 11.5) than they do with problems such as 11.1, 11.2 and 11.7, which test their understanding of what vapor pressure really means.

- trends in melting and boiling points. Students have a great deal of difficulty making comparisons between structurally different types of substances (e.g., NaCl vs Cl_2, CO_2 vs SiO_2). At this point, they should be able to classify a substance as ionic or molecular, given its formula, but many of them still haven't learned to do this. Students should be expected to recognize some common network covalent substances such as C, Si, SiC, and SiO_2.

- the geometry of different types of unit cells. Here models using styrofoam balls are a great help.

<u>LECTURE 1</u>

I Liquid-Vapor Equilibria

A. Vapor Pressure

When a sample of liquid is placed in a closed container, it reaches equilibrium with its vapor:

$$\text{liquid} \rightleftharpoons \text{vapor}$$

The pressure exerted by the vapor at equilibrium is referred to as the vapor pressure of the liquid. Vapor pressure depends upon the identity of the liquid and the temperature:

vp water at $25^{\circ}C$ = 24 mm Hg; vp water at $100^{\circ}C$ = 760 mm Hg
vp benzene $25^{\circ}C$ = 92 mm Hg; vp benzene $80^{\circ}C$ = 760 mm Hg

Vapor pressure does not depend on volume. Suppose one has a sample of $H_2O(l)$ in equilibrium with $H_2O(g)$ at $25^{\circ}C$, 24 mm Hg. If volume of container is increased, more liquid vaporizes, maintaining a constant pressure. If volume is decreased, some vapor condenses.

Dependence of vapor pressure on temperature:

$$\log_{10} \frac{P_2}{P_1} = \frac{\Delta H_{vap}(T_2 - T_1)}{2.30\ RT_2T_1}$$

where P_2 = vapor pressure at temperature T_2 (Kelvin)
P_1 = vapor pressure at temperature T_1 (Kelvin)
ΔH_{vap} = heat of vaporization in joules per mole
$R = 8.31$ J/K

ΔH_{vap} of benzene is 30.8 kJ/mol; vp = 92 mm Hg at $25^{\circ}C$. Calculate vp at $50^{\circ}C$.

$$\log_{10} \frac{P_2}{P_1} = \frac{30{,}800(323 - 298)}{(2.30)(8.31)(323)(298)} = 0.419$$

$$P_2/P_1 = 2.62;\ P_2 = 2.62(92\ \text{mm Hg}) = 240\ \text{mm Hg}$$

Note that pressure more than doubles when T rises from 25 to $50^{\circ}C$, reflecting the fact that more benzene vaporizes (pressure of ideal gas would increase by less than 10%).

B. Boiling Point - temperature at which vapor bubbles form in liquid.

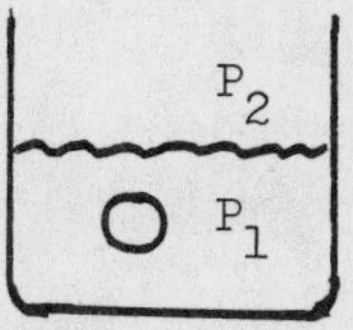

$$P_1 \geq P_2$$

Hence, boiling point varies with applied pressure, P_2. When P_2 = 760 mm Hg, bp H_2O = $100^{\circ}C$. If P_2 = 1075 mm Hg, bp H_2O = $110^{\circ}C$ (pressure cooker). If P_2 = 5 mm Hg, bp H_2O = $0^{\circ}C$.

C. Critical Temperature - temperature above which liquid cannot exist.

Critical pressure = vp liquid at critical temperature. O_2: crit. T = $-119^{\circ}C$. Liquid O_2 cannot exist at room T, regardless of pressure. C_3H_8: crit. T = $97^{\circ}C$. Propane stored as liquid under pressure at room T.

II Phase Diagrams

Graph showing temperatures and pressures at which liquid, solid, and vapor phases of a substance can exist.

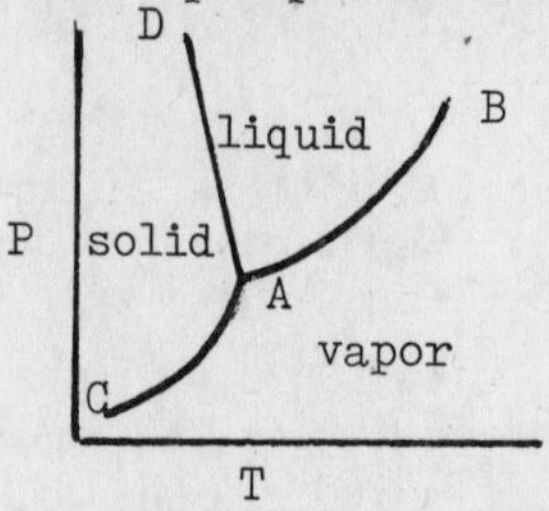

AB = vapor pressure curve of liquid
AC = vapor pressure curve of solid
AD = melting (or freezing) point curve
A = triple point; all three phases in equilibrium

Note that:

- solid sublimes (passes directly to vapor) below triple point (0°C, 5 mm Hg for water; 115°C, 90 mm Hg for I_2).

- if line AD inclines toward P axis, melting point decreases as P increases. This behavior is observed for water, where the liquid is the more dense phase. More often, the solid is the more dense, AD tilts away from the P axis, and the melting point increases with pressure.

LECTURE 2

I Types of Solids

A. Ionic ($NaCl$, KNO_3, etc.)

M^+ X^- M^+
X^- M^+ X^-
M^+ X^- M^+

High melting (strong attractive forces between + and - ions). Nonconducting as solids, but conduct when melted. Often soluble in polar solvents such as water.

B. Molecular (Cl_2, CH_4, etc.)

MX MX MX
MX MX MX
MX MX MX

Low melting (weak attractive forces between molecules). Nonconducting (no charged particles). Seldom water-soluble.

C. Network covalent (C, SiC, SiO_2, etc.)

```
M - X - M
|   |   |
X - M - X
|   |   |
M - X - M
```

High melting (covalent bonds must be broken). Nonconducting. Insoluble in water or other common solvents.

D. Metallic

M^+ e^- M^+
e^- M^+ e^-
M^+ e^- M^+

Conduct electricity and heat through movement of electrons. Cover wide range of melting points. Insoluble in water.

II Properties of Molecular Substances

A. Trends in melting point, boiling point

1. Usually increase with molar mass

$F_2 < Cl_2 < Br_2 < I_2$; $CH_4 < C_2H_6 < C_3H_8 < C_4H_{10}$

2. Slightly higher for polar molecules than for nonpolar molecules of comparable molar mass.

bp $ICl = 97^\circ C$, $Br_2 = 59^\circ C$

3. Unusually high for molecules where hydrogen is bonded to a small, strongly electronegative atom (N, O, F)

	H_2O	H_2S	H_2Se	H_2Te
bp	$100^\circ C$	$-61^\circ C$	$-42^\circ C$	$-2^\circ C$

B. Types of intermolecular forces

1. Dipole forces = electrical attractive forces between + end of polar molecule and - end of adjacent molecule.

dipole force

2. Hydrogen bond - unusually strong dipole force. H atom is very small and differs greatly in electronegativity from F, O, or N. Note that water has many unusual properties in addition to high boiling point. Open structure of ice, a result of H bonding, accounts for its low density.

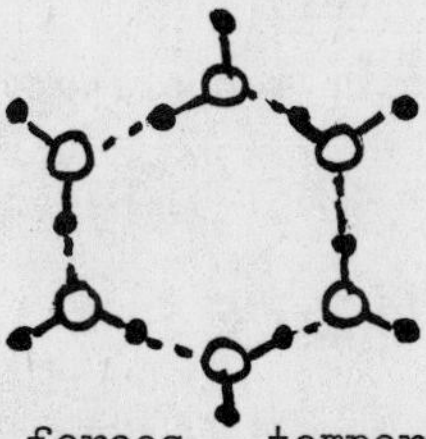

3. Dispersion forces - temporary dipoles formed in nonpolar molecules. Strength depends upon how readily electrons are "polarized". Increase with molecular size or molar mass.

LECTURE $2\frac{1}{2}$

I Unit Cells in Solids

Unit cell = simplest unit which, repeated over and over again in three dimensions, generates the solid lattice.

A. Simple cubic Unit cell consists of eight atoms at the corners of a cube.

$$2r = s$$

where r = atomic radius, s = side of cube

B. Face-centered cubic Atoms at corners of cube and in center of each face. Atoms at corners of cube do not touch each other; instead, atoms touch along face diagonal.

$$4r = s\sqrt{2}$$

Typical of many metals and LiCl (Cl^- ions form face-centered cubic lattice with Li^+ ions fitting between them).

C. Body-centered cubic Atoms at corners of cube and one atom in center of cube. Contact is along a body diagonal, which has a length of $s\sqrt{3}$.

$$4r = s\sqrt{3}$$

Found in many metals. In CsCl, Cs^+ ion is at center of cube touching 8 Cl^- ions at corners.

Suppose a certain metal crystallizes in a BCC pattern with s = 0.400 nm. Value of r?

$$r = \frac{0.400 \times 1.73}{4} = 0.173 \text{ nm}$$

DEMONSTRATIONS

1. Phase Changes: Test. Dem. 64
2. Vapor pressure of liquids: Test. Dem. 65; J. Chem. Educ. 58 725 (1981)
3. Critical temperature: Test. Dem. 66; J. Chem. Educ. 56 614 (1979)
4. Sublimation of solids: Test. Dem. 71
5. Dependence of vapor pressure on temperature: J. Chem. Educ. 56 474 (1979); 57 667 (1980)

QUIZZES

Quiz 1

1. The vapor pressure of a certain liquid doubles when the temperature rises from 12 to 25°C. Calculate its heat of vaporization.
2. State which member of the following pairs will have the higher boiling point and explain your reasoning.

   a. NaCl, I_2 b. SiO_2, CO_2 c. Br_2, I_2

Quiz 2

1. A sample of 225 mL of hydrogen is collected over water at 20°C and a total pressure of 752 mm Hg. The vapor pressure of water at 20°C is 17.5 mm Hg.
   a. What is the mole fraction of water in the wet gas?
   b. What is the mass of water in the wet gas?
2. Give two examples of:
   a. macromolecular solids
   b. molecules showing hydrogen bonding
   c. molecules showing no type of intermolecular force except dispersion

Quiz 3

1. A certain metal crystallizes with a body-centered cubic unit cell 0.318 nm on an edge. What is the atomic radius of the metal? Explain your reasoning.

2. A certain solid is low melting and almost completely insoluble in water. What structural type of solid might it be (ionic, molecular, network covalent, or metallic)? What further experiments would you carry out to determine the structural type?

Quiz 4

1. A certain substance has a triple point at 52°C, 12 mm Hg, and a critical point at 218°C, 18.0 atm. The liquid is less dense than the solid. Draw a phase diagram for the substance and state what phase is present in each area of the diagram.

2. Which is stronger:
   a. hydrogen bond or covalent bond?
   b. dispersion force in CH_4 or in C_2H_6?
   c. ionic bond or dispersion force?
   d. dipole force or polar bond?

Quiz 5

1. What is the normal boiling point (1.00 atm pressure) of a liquid which has a vapor pressure of 142 mm Hg at 28°C and a heat of vaporization of 56.2 kJ/mol?

2. State the strongest type of force present in a crystal of

   a. $CaCl_2$ b. I_2 c. SiC d. Fe

PROBLEMS 1-30

11.1 a. $390 \text{ mm Hg} \times \frac{363 \text{ K}}{373 \text{ K}} = 380 \text{ mm Hg}$

$390 \text{ mm Hg} \times \frac{353 \text{ K}}{373 \text{ K}} = 369 \text{ mm Hg}$

b. 380 mm Hg < 526 mm Hg; no condensation at 90°C
369 mm Hg > 355 mm Hg; condensation at 80°C

c. 380 mm Hg at 90°C; 355 mm Hg at 80°C

11.2 a. $n\ H_2O = \frac{(18/760 \text{ atm})(3.5 \times 10^4 \text{ L})}{(0.0821 \text{ L·atm/mol·K})(293 \text{ K})} = 34.5 \text{ mol}$

$\text{mass } H_2O = 34.5 \text{ mol} \times \frac{18.02 \text{ g}}{1 \text{ mol}} = 6.2 \times 10^2 \text{ g}$

b. 800 g: P = 18 mm Hg

400 g: $P = \frac{(400/18.02 \text{ mol})(0.0821 \text{ L·atm/mol·K})(293 \text{ K})}{3.5 \times 10^4 \text{ L}}$

$= 0.0153 \text{ atm} = 12 \text{ mm Hg}$

11.3 a. 105°C b. 93°C c. 66°C

11.4 a. $\log_{10} \frac{184}{113} = \frac{\Delta H_{vap}(293\ K - 282\ K)}{(2.30)(8.31)(293\ K)(282\ K)} = 0.212$

$\Delta H_{vap} = 3.0 \times 10^4\ J$

b. $\log_{10} \frac{P_2}{184} = \frac{3.0 \times 10^4(323 - 293)}{(2.30)(8.31)(323)(293)} = 0.498$

$P_2/184\ mm\ Hg = 3.14;\ P_2 = 5.8 \times 10^2\ mm\ Hg$

11.5 a. $\log_{10} \frac{P_2}{17.3} = \frac{59400(613 - 473)}{(2.30)(8.31)(613)(473)} = 1.50$

$P_2/17.3\ mm\ Hg = 32;\ P_2 = 5.5 \times 10^2\ mm\ Hg$

b. $\log_{10} \frac{760}{17.3} = \frac{59400(T_2 - 473)}{(2.30)(8.31)(473)T_2}$

$6.57\frac{(T_2 - 473)}{T_2} = 1.64;\ \ 4.93\ T_2 = 6.57(473)$

$T_2 = 630\ K = 357^\circ C$

11.6

$\log_{10}$ vp	0.778	1.064	1.328	1.572
1/T	0.003534	0.003413	0.003300	0.003195

slope $\approx -2.34 \times 10^3$

$\Delta H_{vap} = (2.30)(8.31)(2.34 \times 10^3) = 4.5 \times 10^4\ J$

11.7 a. false; it cannot contain $SO_2(l)$
b. true
c. true (greater than $32^\circ C$)

11.8 a. vapor b. vapor, liquid c. liquid

11.9 a. vapor b. vapor c. vapor

11.10 a, b

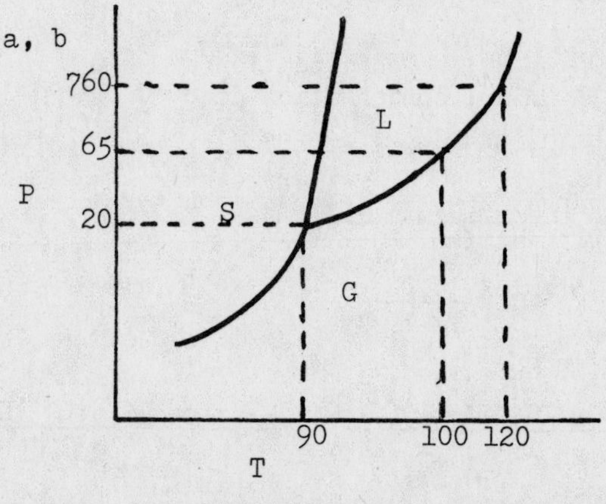

c. move from liquid through liquid-gas mixture to gas

11.11 a. ionic b. molecular c. metallic (Na, K, - -)

11.12 a. metallic, network covalent, most molecular substances
b. ionic, network covalent
c. metallic

11.13 a. CO_2 (dry ice); very high MM hydrocarbons; $C_{12}H_{22}O_{11}$
b. Na_2CO_3
c. C, SiC
d. steel

11.14 a. molecular b. network covalent c. molecular d. ionic
e. metallic

11.15 a. ions, electrons b. ions c. molecules d. atoms

11.16 $I_2 > Br_2 > Cl_2 > F_2$

11.17 a. dispersion b. dispersion, dipole c. dispersion, dipole
d. dispersion, dipole e. dispersion

11.18 b, d

11.19 a. dispersion forces stronger in C_2H_6
b. hydrogen bonds in C_2H_5OH
c. NaF is ionic
d. stronger dispersion forces in O_2
e. dipole forces in CO

11.20 d. $N_2O_4(g) \longrightarrow 2\ NO_2(g)$

11.21 a. F_2; weaker dispersion forces
b. ClO_2; molecular
c. C_3H_8; molecular
d. HCl; weaker dispersion forces

11.22 a. H bonds b. dispersion forces c. ionic bonds
d. dispersion forces

11.23 $4r = s\sqrt{3}$; $s = \dfrac{4r}{\sqrt{3}} = \dfrac{4(0.217\ \text{nm})}{1.732} = 0.501\ \text{nm}$

11.24 $4r = s\sqrt{2}$; $r = \dfrac{s\sqrt{2}}{4} = \dfrac{(0.390\ \text{nm})(1.414)}{4} = 0.138\ \text{nm}$

11.25 a. $0.513\ \text{nm} - 2(0.181\ \text{nm}) = 0.151\ \text{nm}$
b. $2r\ Na^+ = 0.190$ nm; no
$2r\ K^+ = 0.266$ nm; no

11.26 $2r$ cation + $2r$ anion = $s\sqrt{3}$

$$s = \frac{2(r\ \text{cation} + r\ \text{anion})}{\sqrt{3}} = 1.155(r\ \text{cation} + r\ \text{anion})$$

11.27 a. $200.59\ \text{g Hg} \times \dfrac{1\ \text{cm}^3}{13.6\ \text{g Hg}} = 14.7\ \text{cm}^3$

b. $V = \dfrac{(1.00\ \text{mol})(0.0821\ \text{L·atm/mol·K})(293\ \text{K})}{(0.0012/760\ \text{atm})} = 1.5 \times 10^7\ \text{L}$

c. $V = \dfrac{4}{3}\pi(1.55 \times 10^{-8}\ \text{cm})^3 \times 6.022 \times 10^{23} = 9.39\ \text{cm}^3$

d. liquid: $100 \times 9.39/14.7 = 63.9$

gas: $100 \times 9.39/1.5 \times 10^{10} = 6.3 \times 10^{-8}$

11.28 a. true

b. false; above triple point, solid will melt

c. false; NaF melts higher because it is ionic

d. false; $r = \frac{s\sqrt{2}}{4}$ for FCC, $\frac{s\sqrt{3}}{4}$ for BCC

11.29 a. intermolecular forces = forces between adjacent molecules (dispersion, dipole, H bonds); intramolecular forces = forces within a molecule (covalent bonds)

b. molecular solid contains small, discrete units (molecules) while network covalent solid does not.

c. hydrogen bond is a very strong dipole force between H atom of one molecule and F, O or N atom of another molecule.

d. ionic solid consists of oppositely charged ions; molecular solid contains molecules

11.30 a. food is frozen, evacuated to remove water by sublimation

b. vapor pressure lower at low T

c. lower mass of air at higher altitude

d. sublimation

CHAPTER 12

BASIC SKILLS

1. Write balanced equations to represent the solution process for an electrolyte or nonelectrolyte.
2. Given the amounts (grams or moles) of the components of a solution, calculate mole fraction or molality.
3. Describe how to prepare a solution to a desired molarity, starting either with a pure solute or a concentrated solution.
4. Relate the molarity of an electrolyte solution to the molarities of its ions.
5. Given appropriate data, convert between molarity, molality, mole fraction, and mass percent of solute.
6. Predict the effect of changes in temperature or pressure upon solubilities of solids and gases.
7. Use Equation 12.10 to relate vapor pressure lowering to solute mole fraction.
8. Use Equation 12.11 to relate osmotic pressure to molarity.
9. Use Equations 12.14 and 12.15 to obtain
   a. the boiling point or freezing point of a nonelectrolyte solution, knowing or having calculated the molality.
   b. the molar mass of a nonelectrolyte.
10. Compare the colligative properties of electrolytes to those of nonelectrolytes.

LECTURE NOTES

This chapter can be covered in two lectures. The first deals largely with concentration units, the second with colligative properties. Note that molarity, which is covered here in considerable detail, was introduced in Chapter 2. Discussed here for the first time are the calculations involved in preparing a solution of known molarity from a more concentrated solution. The important point to get across here is that the number of moles of solute stays the same.

LECTURE 1

I Types of Solutes

A. Nonelectrolytes; dissolve as molecules

$$CH_3OH(l) \longrightarrow CH_3OH(aq)$$

B. Electrolytes: dissolve as ions

$NaCl(s) \longrightarrow Na^{+}(aq) + Cl^{-}(aq)$ 2 mol ions per mole solute

$Ca(NO_3)_2(s) \longrightarrow Ca^{2+}(aq) + 2\ NO_3^{-}(aq)$ 3 mol ions per mole solute

II Concentrations of Solutes

A. Mole Fraction: $X_a = \dfrac{n_a}{n_a + n_b}$

Dissolve 12.0 g of CH_3OH in 100.0 g of water. Mole fraction of CH_3OH?

$n\ CH_3OH = 12.0\ g \times \dfrac{1\ mol}{32.0\ g} = 0.375$; $n\ H_2O = 100.0\ g \times \dfrac{1\ mol}{18.0\ g} = 5.56\ mol$

$X\ CH_3OH = \dfrac{0.375}{0.375 + 5.56} = 0.0631$; $X\ H_2O = 1 - 0.0631 = 0.9369$

B. Molality $m = \dfrac{\text{no. moles solute}}{\text{no. kg solvent}}$

Molality of solution of CH_3OH referred to above?

$m = 0.375/0.1000 = 3.75$

C. Molarity $M = \dfrac{\text{no. moles solute}}{\text{no. liters solution}}$

1. How prepare 35.0 mL of 0.200 M $Al(NO_3)_3$ from solid?

   molar mass = (27.0 + 42.0 + 144.0)g/mol = 213.0 g/mol

   no. moles required = 0.0350 L x 0.200 mol/L = 7.00×10^{-3} mol

   mass required = $7.00 \times 10^{-3}\ mol \times \dfrac{213.0\ g}{1\ mol} = 1.49\ g$

   Dissolve 1.49 g $Al(NO_3)_3$ in enough water to form 35.0 mL of solution.

2. How prepare 35.0 mL of 0.200 M $Al(NO_3)_3$ from 0.500 M solution? Note that number of moles remains constant upon dilution.

   $0.0350\ L \times 0.200\ \dfrac{mol}{L} = 0.500\ \dfrac{mol}{L} \times V$

   V = 0.0140 L = 14.0 mL; dilute to 35.0 mL with water

3. Concentration of Al^{3+}, NO_3^{-} in 0.200 M $Al(NO_3)_3$?

   $Al(NO_3)_3(s) \longrightarrow Al^{3+}(aq) + 3\ NO_3^{-}(aq)$

   conc. Al^{3+} = 0.200 M, conc. NO_3^{-} = 0.600 M

4. Molality of 0.200 M $Al(NO_3)_3$ solution (d = 1.012 g/mL)?

   Work with one liter of solution.

   Mass = $1000\ mL \times \dfrac{1.012\ g}{1\ mL} = 1012\ g$

$$\text{Mass } Al(NO_3)_3 = 0.200 \text{ mol} \times \frac{213.0 \text{ g}}{1 \text{ mol}} = 42.6 \text{ g}$$

$$\text{Mass water} = 1012 \text{ g} - 43 \text{ g} = 969 \text{ g}$$

$$\text{Molality} = \frac{0.200 \text{ mol}}{0.969 \text{ kg}} = 0.206$$

LECTURE 2

I Principles of Solubility

A. Nature of solute and solvent. Most nonelectrolytes that are appreciably soluble in water are hydrogen bonded (CH_3OH, H_2O_2, sugars). Other types of nonelectrolytes are generally more soluble in nonpolar or slightly polar solvents (C_6H_6, CCl_4, $CHCl_3$).

B. Effect of temperature: increase in T favors endothermic process:

solid + water $\longrightarrow$ solution; $\Delta H > 0$ (usually); increase in T increases solubility

gas + water $\longrightarrow$ solution; $\Delta H < 0$ (usually); solubility decreases as T increases

C. Effect of pressure: negligible, except for gases, where solubility is directly proportional to the partial pressure of gas. Carbonated beverages.

II Colligative properties of nonelectrolytes

Depend primarily upon concentration of solute particles rather than type.

A. Vapor pressure lowering

$VPL = X_2P_1^{\circ}$ where X_2 = mole fraction solute, P_1° = vapor pressure of pure solvent

B. Osmosis, osmotic pressure

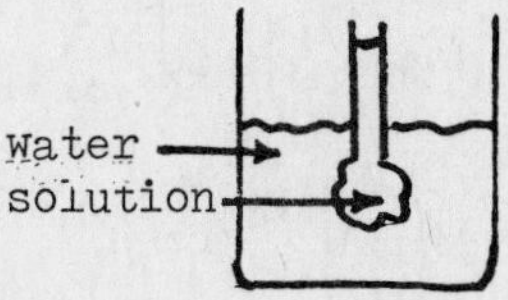

Water passes through semipermeable membrane to region where its vapor presure is lower. Osmotic pressure = pressure which must be applied to solution to prevent osmosis.

$\pi = MRT$; 1 M solution at 25°C has osmotic pressure of 24.5 atm

C. Boiling point elevation, freezing point lowering

1. Results from vapor pressure lowering

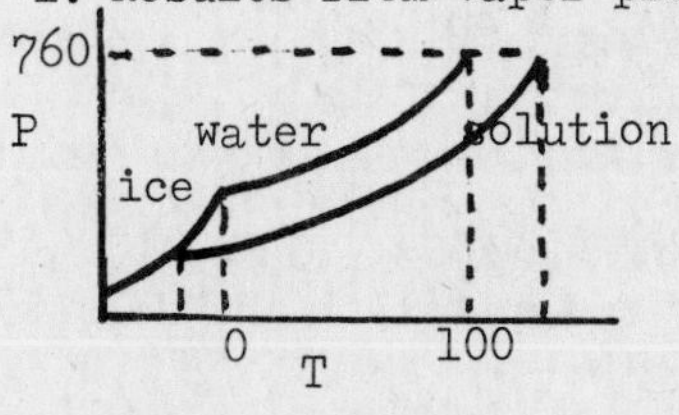

2. For water solutions:

$$\Delta T_f = 1.86^\circ C \times m \; ; \; \Delta T_b = 0.52^\circ C \times m$$

Freezing point and boiling point of solution containing 20.0 g of ethylene glycol (molar mass = 62.0 g/mol) in 50.0 g water?

$$m = \frac{20.0/62.0}{0.050} = 6.45$$

$$\Delta T_f = 6.45 \times 1.86^\circ C = 12.0^\circ C; \quad T_f = 0 - 12.0^\circ C = -12.0^\circ C$$

$$\Delta T_b = 6.45 \times 0.52^\circ C = 3.4^\circ C; \quad T_b = 103.4^\circ C$$

3. Use in determining molar mass. Suppose solution is prepared by dissolving 0.100 g of nonelectrolyte in 1.00 g of water and freezing point is found to be -1.00°C. Molar mass?

$$\text{molality} = \frac{1.00}{1.86} = \frac{0.100/MM}{0.00100}$$

$$MM = \frac{1.86 \times 10^{-1}}{1.00 \times 10^{-3}} = 186 \text{ g/mol}$$

III <u>Electrolytes</u>

Colligative effects are greater because of increased number of particles.

$$\Delta T_f = 1.86^\circ C \times m \times i$$

where i is approximately equal to the number of moles of ions per mole of solute (2 for NaCl, 3 for $CaCl_2$, etc.)

DEMONSTRATIONS

1. Supersaturation: J. Chem. Educ. <u>57</u> 152 (1980)
2. Conductivity of water solutions: Test. Dem. 15, 74, 129
3. Osmosis: Test. Dem. 16, 66
4. Raoult's Law: Test. Dem. 128, 145, 195; J. Chem. Educ. <u>53</u> 303 (1976)
5. Conductivity HCl in water, benzene: J. Chem. Educ. <u>43</u> A539 (1966)

QUIZZES

<u>Quiz 1</u>

1. A solution is prepared by dissolving 5.82 g of urea, $CO(NH_2)_2$, in 24.0 g of water. Calculate:
   a. the molality of urea
   b. the mole fraction of urea
   c. the freezing point of the solution (K_f = 1.86°C)
   d. the molarity of urea, taking the density of the solution to be 1.010 g/cm^3.

Quiz 2

1. You are given 225 mL of 0.200 M $Cr_2(SO_4)_3$ solution.
   a. What is the molarity of Cr^{3+}? of SO_4^{2-}?
   b. What volume of 0.150 M $Cr_2(SO_4)_3$ solution can be prepared from this solution?

2. A solution containing 1.80 g of a nonelectrolyte in 10.0 g of water freezes at -3.45°C. What is the molar mass of the nonelectrolyte? (K_f = 1.86°C).

Quiz 3

1. In dilute nitric acid, the concentration of HNO_3 is 6.00 M and the density is 1.19 g/cm³. What is:
   a. the mass percent of HNO_3?
   b. the mole fraction of HNO_3?

2. What is the freezing point of a solution containing 12.0 g of naphthalene, $C_{10}H_8$, in 100.0 g of benzene? (K_f = 5.10°C; fp of pure benzene = 5.50°C).

Quiz 4

1. How would you prepare 218 mL of 0.169 M K_2CrO_4 starting with
   a. pure K_2CrO_4?
   b. a 0.414 M solution of K_2CrO_4?

2. A solution containing 0.125 g of a nonelectrolyte in 10.0 mL of solution has an osmotic pressure of 3.12 atm at 20°C. What is the molar mass of the nonelectrolyte?

Quiz 5

1. A solution is prepared by dissolving 16.0 g of ethylene glycol, $C_2H_6O_2$, in 50.0 g of water. Calculate:
   a. the molality of ethylene glycol
   b. the mole fraction of ethylene glycol
   c. the freezing point of the solution (K_f = 1.86°C)
   d. the molarity of ethylene glycol, taking the density of the solution to be 0.965 g/cm³.

PROBLEMS 1-30

12.1 Add a crystal of KNO_3. If it dissolves, the solution was unsaturated. If excess solute crystallizes, it was supersaturated. If nothing happens, the solution was saturated.

12.2 a. $MgI_2(s) \longrightarrow Mg^{2+}(aq) + 2\ I^-(aq)$

b. $KClO_4(s) \longrightarrow K^+(aq) + ClO_4^-(aq)$

c. $RbHCO_3(s) \longrightarrow Rb^+(aq) + HCO_3^-(aq)$

d. $Sc_2(SO_4)_3(s) \longrightarrow 2\ Sc^{3+}(aq) + 3\ SO_4^{2-}(aq)$

12.3 a. 3 x 0.10 = 0.30 mol   b. 4 x 0.10 mol = 0.40 mol

c. 5 x 0.10 mol = 0.50 mol   d. 2 x 0.10 mol = 0.20 mol

12.4 a. $\frac{1.25}{1.25 + 11.6} \times 100 = 9.73$ b. $\frac{11.6}{1.25 + 11.6} \times 100 = 90.3$

12.4 c. MM K_2CrO_4 = (78.20 + 52.00 + 64.00)g/mol = 194.20 g/mol

$$m = \frac{1.25/194.20}{0.0116} = 0.555\ M$$

12.5 $$\text{moles } CH_2O = 40\ cm^3 \times \frac{0.82\ g}{1\ cm^3} \times \frac{1\ mol}{30.03\ g} = 1.1\ mol$$

$$\text{mass water} = 100\ cm^3 \times \frac{1.00\ g}{1\ cm^3} = 100\ g$$

$$m\ CH_2O = \frac{1.1\ mol}{0.100\ kg\ water} = 11$$

12.6 MM CS_2 = 76.13 g/mol; MM $CHCl_3$ = 119.38 g/mol

$$X \text{ of } CS_2 = \frac{50.0/76.13}{50.0/76.13 + 50.0/119.38} = \frac{0.657}{0.657 + 0.419} = 0.611$$

X of $CHCl_3$ = 1.000 - 0.611 = 0.389

12.7 MM $C_6H_{12}O_6$ = 180.16 g/mol

$$12.5\ g \times \frac{1\ mol}{180.16\ g} = 0.0694\ mol;\quad M = \frac{0.0694\ mol}{0.219\ L} = 0.317\ mol/L$$

$$1.08\ mol \times \frac{180.16\ g}{1\ mol} = 195\ g;\quad V = 1.08\ mol \times \frac{1\ L}{0.519\ mol} = 2.08\ L$$

$$\frac{1.08\ mol}{1\ L} \times 1.62\ L = 1.75\ mol;\quad 1.75\ mol \times \frac{180.16\ g}{1\ mol} = 315\ g$$

12.8 a. $$\text{mass KOH} = 0.220\ L \times \frac{0.500\ mol}{1\ L} \times \frac{56.11\ g}{1\ mol} = 6.17\ g$$

dissolve 6.17 g of KOH in enough water to form 220 mL of solution

b. $$\text{moles KOH} = 0.220\ L \times \frac{0.500\ mol}{1\ L} = 0.110\ mol$$

$$V \text{ of } 1.25\ M\ KOH = 0.110\ mol \times \frac{1\ L}{1.25\ mol} = 0.0880\ L$$

Measure out 88.0 mL of 1.25 M KOH, dilute with water to 220 mL

12.9 a. $$\text{moles } Ca(NO_3)_2 = 0.150\ L \times \frac{0.210\ mol}{1\ L} = 0.0315\ mol$$

$$M\ Ca(NO_3)_2 = \frac{0.0315\ mol}{0.450\ L} = 0.0700\ mol/L$$

b. 0.0700 M in Ca^{2+}; 0.140 M in NO_3^-

c. orig. soln.: $$0.100\ L \times \frac{0.420\ mol\ NO_3^-}{1\ L} = 0.0420\ mol\ NO_3^-$$

dil. soln.: $$0.100\ L \times \frac{0.140\ mol\ NO_3^-}{1\ L} = 0.0140\ mol\ NO_3^-$$

12.10 a. $$1.50\ L \times \frac{0.250\ mol}{1\ L} = 0.375\ mol$$

$$V = 0.375\ mol \times \frac{1\ L}{16.0\ mol} = 0.0234\ L$$

12.10 b. 1.50 L - 0.0234 L = 1.48 L

12.11 1000 g water, 1.62 x 342.30 g = 555 g sugar

a. $X \text{ sugar} = \dfrac{1.62 \text{ mol}}{1.62 \text{ mol} + \dfrac{1000}{18.02} \text{ mol}} = 1.62/57.11 = 0.0284$

$X \text{ water} = 1.0000 - 0.0284 = 0.9716$

b. mass % sugar = 100 x 555/1555 = 35.7; mass % water = 64.3

12.12 1st solution: one liter weighs 1050 g, contains 1.32 x 40.00 g = 52.8 g NaOH and 1050 g - 53 g = 997 g water

molality = 1.32/0.997 = 1.32
mass % = 100 x 52.8/1050 = 5.03

2nd solution: 100 g solution contains 20.0 g NaOH and 80.0 g water, has a volume of $100 \text{ g} \times \dfrac{1 \text{ mL}}{1.22 \text{ g}} = 82.0 \text{ mL}$

$\text{molarity} = \dfrac{20.0/40.00}{0.0820} = 6.10 \text{ M}; \ m = \dfrac{20.0/40.00}{0.0800} = 6.25$

3rd solution: contains 1000 g water and 11.8 x 40.00 g = 472 g NaOH. It has a volume of $1472 \text{ g} \times \dfrac{1 \text{ mL}}{1.35 \text{ g}}$

= 1090 mL

$\text{molarity} = \dfrac{11.8 \text{ mol}}{1.09 \text{ L}} = 10.8 \text{ M}; \ \text{mass \%} = 100 \times \dfrac{472}{1472} = 32.1$

12.13 a. NH_3 (hydrogen bonding)

b. ethylene glycol (hydrogen bonding)

c. NaOH (ionic vs molecular)

12.14 a. greater (solubility increases as T drops and as P rises)
b. less (solubility decreases as T rises and as P drops)
c. greater
d. less

12.15 a. 0.0014 mol/L b. 0.0013 mol/L c. 0.013 mol/L

12.16 a. vpl = 0.0100(23.76 mm Hg) = 0.238 mm Hg
vp water = 23.76 mm Hg - 0.24 mm Hg = 23.52 mm Hg

b. vpl = 0.100(23.76 mm Hg) = 2.38 mm Hg
vp water = 23.76 mm Hg - 2.38 mm Hg = 21.38 mm Hg

c. vpl = 0.200(23.76 mm Hg) = 4.75 mm Hg
vp water = 23.76 mm Hg - 4.75 mm Hg = 19.01 mm Hg

12.17 mole fraction I_2 = 1.00/114 = 0.00877

$\text{moles } CCl_4 = 1.00 \times 10^3 \text{ mL} \times \dfrac{1.60 \text{ g}}{1 \text{ mL}} \times \dfrac{1 \text{ mol}}{153.81 \text{ g}} = 10.4 \text{ mol}$

$0.00877 = n\ I_2/(n\ I_2 + 10.4)$; solving, $n\ I_2 = 0.0920$

$\text{mass } I_2 = 0.0920 \text{ mol} \times \dfrac{253.8 \text{ g}}{1 \text{ mol}} = 23.3 \text{ g}$

12.18 MM $C_{12}H_{22}O_{11}$ = 342.30 g/mol

a. $\pi = \frac{10.0}{342.30} \frac{mol}{L} \times 0.0821 \frac{L \cdot atm}{mol \cdot K} \times 298\ K = 0.715\ atm$

b. $\pi = \frac{25.0}{342.30} \frac{mol}{L} \times 0.0821 \frac{L \cdot atm}{mol \cdot K} \times 298\ K = 1.79\ atm$

c. $\pi = \frac{155}{342.30} \frac{mol}{L} \times 0.0821 \frac{L \cdot atm}{mol \cdot K} \times 298\ K = 11.1\ atm$

12.19 a. $m = \frac{14.9/180.16}{0.100} = 0.827$

$\Delta T_b = 0.827 \times 0.52^\circ C = 0.43^\circ C$; $T_b = 100.43^\circ C$

$\Delta T_f = 0.827 \times 1.86^\circ C = 1.54^\circ C$; $T_f = -1.54^\circ C$

b. $m = \frac{6.00/60.05}{0.0800} = 1.25$

$\Delta T_b = 1.25 \times 0.52^\circ C = 0.65^\circ C$; $T_b = 100.65^\circ C$

$\Delta T_f = 1.25 \times 1.86^\circ C = 2.32^\circ C$; $T_f = -2.32^\circ C$

12.20 a. mass ethylene glycol = $4.00 \times 10^3\ mL \times \frac{1.12\ g}{1\ mL} = 4480\ g$

mass water = $6.00 \times 10^3\ mL \times \frac{1.00\ g}{1\ mL} = 6.00 \times 10^3\ g$

$m = \frac{4480/62.07}{6.00} = 12.0$

b. $\Delta T_f = 12.0(1.86^\circ C) = 22.3^\circ C$; $T_f = -22.3^\circ C$

12.21 a. $\Delta T_f = 0.20(5.10^\circ C) = 1.0^\circ C$; $T_f = 5.50^\circ C - 1.0^\circ C = 4.5^\circ C$

b. $\Delta T_f = 0.20(20.2^\circ C) = 4.0^\circ C$; $T_f = 6.5^\circ C - 4.0^\circ C = 2.5^\circ C$

c. $\Delta T_f = 0.20(40.0^\circ C) = 8.0^\circ C$; $T_f = 178^\circ C - 8.0^\circ C = 170^\circ C$

12.22 $m = \frac{1.42}{1.86} = \frac{6.70/MM}{0.0500}$; solving, MM = 176 g/mol

In one mole of vitamin C, there are:

$(176 \times 0.409)\ g\ C \times \frac{1\ mol\ C}{12.01\ g\ C} = 6.00\ mol\ C$

$(176 \times 0.0458)\ g\ H \times \frac{1\ mol\ H}{1.008\ g\ H} = 8.00\ mol\ H$ $\quad C_6H_8O_6$

$(176 \times 0.545)\ g\ O \times \frac{1\ mol\ O}{16.00\ g\ O} = 6.00\ mol\ O$

12.23 $\Delta T_f = 5.50^\circ C - 5.15^\circ C = 0.35^\circ C$

$m = \frac{0.35}{5.10} = \frac{0.180/MM}{0.0500}$; solving, MM = 52 g/mol

12.24 $\frac{4.60}{760} = \frac{3.27/MM}{0.200} \times 0.0821 \times 293$; solving, $MM = 6.50 \times 10^4\ \frac{g}{mol}$

12.25 a. $\Delta T_f = (0.10)(3)(1.86^\circ C) = 0.56^\circ C$; $T_f = -0.56^\circ C$

b. $\Delta T_f = (0.10)(2)(1.86^\circ C) = 0.37^\circ C$; $T_f = -0.37^\circ C$

12.25 c. $\Delta T_f = (0.10)(4)(1.86^\circ C) = 0.74^\circ C$; $T_f = -0.74^\circ C$

12.26 $\Delta T_f = 1.86^\circ C \times m \times i$

$i = 0.38/(1.86 \times 0.20) = 1.0$; HF molecules

12.27 MM = 342.30 g/mol

a. $M = \frac{338/342.30 \text{ mol}}{1.00 \text{ L}} = 0.987 \text{ mol/L}$

b. One liter of solution has a mass of 1127 g, contains 1127 g - 338 g = 789 g water

$m = \frac{338/342.30}{0.789} = 1.25$

c. $\pi = 0.987 \frac{\text{mol}}{\text{L}} \times 0.0821 \frac{\text{L} \cdot \text{atm}}{\text{mol} \cdot \text{K}} \times 293 \text{ K} = 23.7 \text{ atm}$

d. $\Delta T_f = 1.25(1.86^\circ C) = 2.32^\circ C$; $T_f = -2.32^\circ C$

12.28 a. H^+, Cl^- ions in water; HCl molecules in liquid

b. Na^+ and Cl^- ions free to move in solution

c. In very dilute solution, there is about 1000 g of water in one liter solution.

d. Must exceed osmotic pressure to make reverse osmosis occur

12.29 a. may contain very little solute if solubility is small
b. may increase if ΔH of solution is negative
c. not true, especially in concentrated solution
d. 3/2 as great
e. 1/2 as great for sucrose

12.30 a. solution formed has freezing point below $0^\circ C$
b. contains dissolved solute
c. infinitely soluble in water because of hydrogen bonding
d. depends on concentration of solute, not type

CHAPTER 13

BASIC SKILLS

1. Describe the structure (molecular or network covalent) of the various allotropic forms of oxygen, sulfur, phosphorus, and carbon.
2. Describe different methods of preparation of the hydrogen halides.
3. Give the molecular formulas and Lewis structures of the several oxides of carbon, nitrogen, phosphorus, and sulfur.
4. Identify nonmetal halides in which the central atom has an expanded octet.
5. Distinguish between alkanes, alkenes, alkynes, and aromatic hydrocarbons.
6. Identify the several structural isomers of an alkane or related organic compound.

LECTURE NOTES

This is a descriptive chapter; the only new terms introduced are allotropy and isomerism. The lecture time devoted to this chapter can be adjusted to your schedule. Two lectures seems about right, but, with selective omissions, this could easily be reduced to 1½ lectures or even to a single lecture.

<u>LECTURE 1</u>

I <u>Structures of the Nonmetals</u>

A. <u>Oxygen</u> O_2 and O_3 are allotropic forms of the element (different properties, same physical state. Structure of O_3:

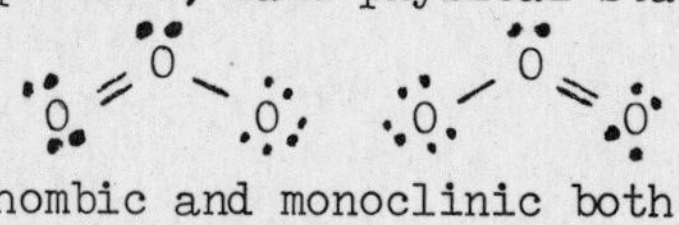

B. <u>Sulfur</u> Rhombic and monoclinic both contain S_8 molecules; different types of packing give different crystal structures.

rhombic $\rightleftharpoons$ monoclinic ; triple point at 96°C, 0.0043 mm Hg

Increase in T favors monoclinic (higher enthalpy); increase in P favors rhombic (higher density). Plastic sulfur contains long chain molecules randomly oriented.

C. <u>Phosphorus</u> White allotrope, P_4, is toxic, reactive

$P_4(s) \longrightarrow P_4O_6(s) \longrightarrow P_4O_{10}(s)$; phosphorescence

Red allotrope is much less reactive; has a network structure; used in safety matches, with $KClO_3$.

C. Carbon Both graphite and diamond are network covalent and hence very high melting. Diamond has a 3-d, tetrahedral structure, accounting for its hardness. In graphite, C atoms are bonded within layers; weak dispersion forces between layers. Since diamond has a higher density than graphite (3.51 vs 2.26 g/cm^3), it is formed at high pressures.

II Nonmetal Hydrides

A.

Group 5	Group 6	Group 7
XH_3	H_2X	HX
and: N_2H_4, P_2H_4, HN_3	H_2O_2	

Lewis structure of N_2H_4:

```
    ..  ..
H - N - N - H
    |   |
    H   H
```

HN_3: $H-\ddot{N}=N=\ddot{N}:$ and resonance forms

B. Preparation of hydrogen halides

1. $X_2(g) + H_2(g) \longrightarrow 2\,HX(g)$ Suitable for HCl, HBr. Not used for HF (too violent) or HI (too slow)

2. $NaX(s) + H_2SO_4(l) \longrightarrow HX(g) + NaHSO_4(s)$

   Suitable for HF, HCl. For HBr, HI, use H_3PO_4 to avoid oxidation of Br^-, I^- ions to Br_2, I_2.

III Nonmetal Oxides

A. Sulfur: SO_2 and SO_3. Sulfur dioxide formed when sulfur or metal sulfides are burned in air:

$S(s) + O_2(g) \longrightarrow SO_2(g)$

$2\,ZnS(s) + 3\,O_2(g) \longrightarrow 2\,ZnO(s) + 2\,SO_2(g)$

Sulfur trioxide formed by slow reaction of SO_2 with O_2.

Lewis structures, resonance forms of SO_2, SO_3 (Ch. 9)

B. Nitrogen (Figure 13.9, p. 402)

Molecular formulas: N_2O_5, N_2O_4, N_2O_3, N_2O_2, NO_2, NO, N_2O

Consider N_2O_4

Lewis structure?

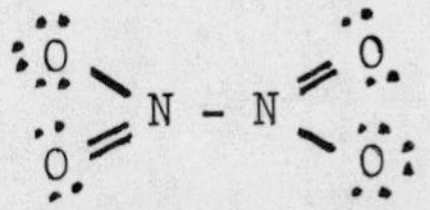

Bond angle? 120° Polarity? nonpolar

Hybridization of N? sp^2 Types of bonds? 5 sigma, 2 pi

I <u>Nonmetal Oxides</u> (cont.)

  A. Carbon CO and CO_2; Lewis structures, resonance forms CO_2.

    CO formed by burning carbon or hydrocarbon in limited amount of air:

$$C(s) + \tfrac{1}{2} O_2(g) \longrightarrow CO(g)$$

$$CH_4(g) + 3/2\ O_2(g) \longrightarrow CO(g) + 2\ H_2O(l)$$

    CO_2 formed in excess air or in pure O_2:

$$C(s) + O_2(g) \longrightarrow CO_2(g)$$

$$CH_4(g) + 2\ O_2(g) \longrightarrow CO_2(g) + 2\ H_2O(l)$$

II <u>Nonmetal Halides</u>

Halogen atoms (F, Cl, Br or I) bonded to a central nonmetal atom (Group 4, 5, 6, 7 or 8). Examples: CCl_4, PF_3, SF_4, IF_5, XeF_4. No. of electrons around central atom = Group no. central atom + no. of halogen atoms bonded to central atom

CCl_4: 4 + 4 = 8; octet structure

PF_3 : 5 + 3 = 8; octet structure

SF_4 : 6 + 4 = 10; expanded octet, sp^3d hybridization

IF_5 : 7 + 5 = 12; expanded octet, sp^3d^2 hybridization

XeF_4: 8 + 4 = 12; expanded octet, sp^3d^2 hybridization

III Hydrocarbons

  A. Saturated (alkanes) All single bonds. General formula = C_nH_{2n+2}: CH_4, C_2H_6, C_3H_8, C_4H_{10}, - -

    Alkanes containing four or more carbon atoms per molecule show isomerism; two or more different compounds have the same molecular formula

```
   H   H   H   H                 H   H   H
   |   |   |   |                 |   |   |
H- C - C - C - C -H          H - C - C - C - H
   |   |   |   |                 |   |   |
   H   H   H   H                 H   C   H
                                    /|\
                                   H H H
```

    butane (bp = -1°C)   2-methylpropane (bp = -10°C)

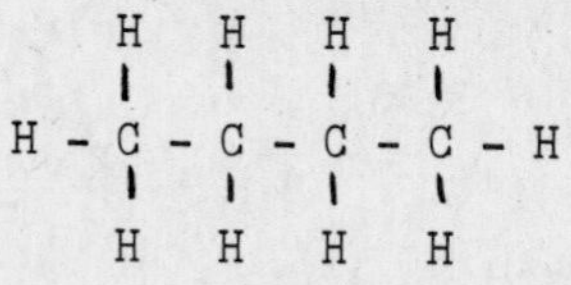

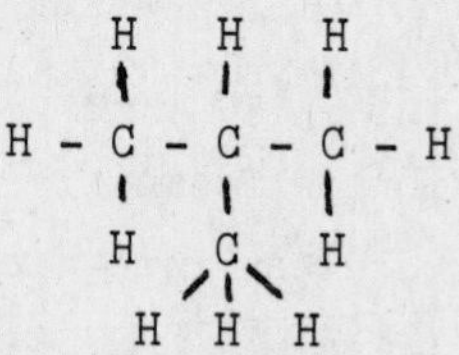

Isomers of C_6H_{14}?

6 carbon chain: C - C - C - C - C - C

```
5 carbon chains:  C - C - C - C - C       C - C - C - C - C
                          |                       |
                          C                       C
```

```
                      C
                      |
4 carbon chains:  C - C - C - C       C - C - C - C
                      |                   |   |
                      C                   C   C
```

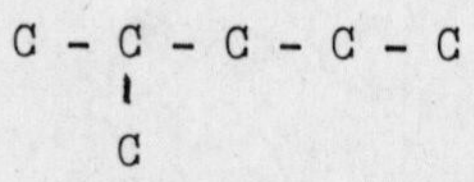

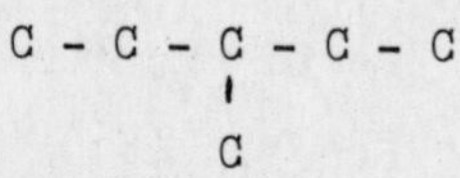

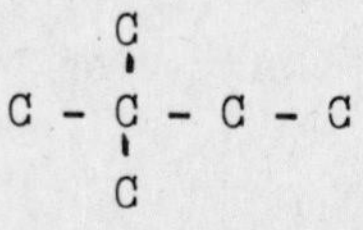

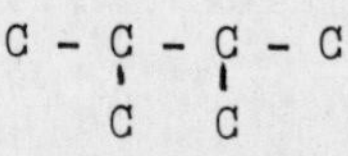

B. Unsaturated

1. Alkenes: 1 double bond 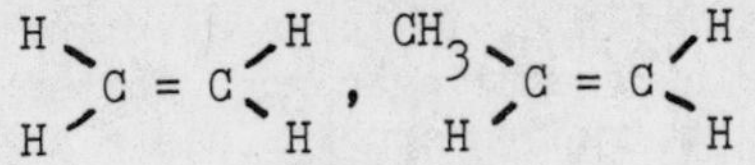

   120° bond angle

2. Alkynes: 1 triple bond $H - C \equiv C - H$, $CH_3 - C \equiv C - H$

   180° bond angle

C. Aromatic - contain benzene ring. Benzene = C_6H_6

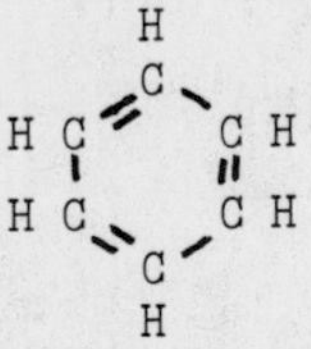 or  extra three pairs of electrons spread over entire molecule

Isomers of $C_6H_4Cl_2$? 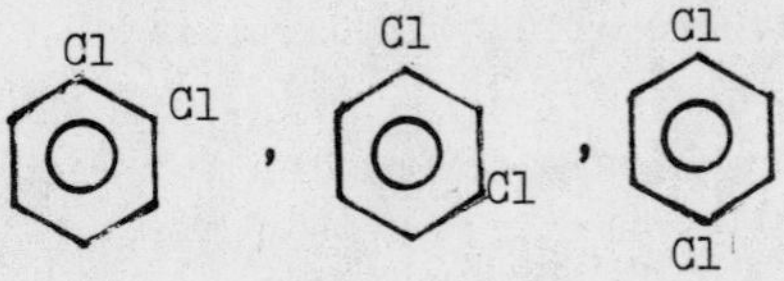

DEMONSTRATIONS

1. Etching glass with HF: Test. Dem. 34
2. White and red phosphorus: Test. Dem. 39, 177
3. PH_3; smoke rings: Test. Dem. 39, 100
4. Monoclinic and plastic sulfur: Test. Dem. 41, 101
5. Reaction of acid with halide ions: Test. Dem. 44
6. Properties of H_2O_2: Test. Dem. 62
7. Preparation of oxides of nitrogen: Test. Dem. 99
8. Phosphorus in matches: Test. Dem. 100
9. Reaction of xenon with fluorine: J. Chem. Educ. <u>43</u> 202 (1966)

QUIZZES

<u>Quiz 1</u>

1. Draw Lewis structures for:

   a. O_3 b. N_2O c. H_2O_2 d. C_2H_4

2. Explain, in a few well-chosen words, what is meant by each of the following terms:

   a. allotropy b. isomerism c. expanded octet d. alkene

Quiz 2

1. Explain, in terms of structure, why
   a. graphite is much softer than diamond
   b. red phosphorus has a higher melting point than white phosphorus
   c. there are three different compounds with the molecular formula C_5H_{12}
2. Consider the N_2O_2 molecule
   a. Draw its Lewis structure
   b. Are there resonance forms of N_2O_2? Explain.
   c. Could there be isomers of N_2O_2? Explain.

Quiz 3

1. Draw all the isomers of $C_4H_8Cl_2$.
2. Explain what is meant by

   a. allotropy b. resonance c. network covalent structure
   d. aromatic hydrocarbon

Quiz 4

1. Draw Lewis structures for each of the following oxides of nitrogen:

   a. NO b. NO_2 c. N_2O_3
2. Choose one of the molecules in (1) and:
   a. draw at least one resonance form
   b. state the bond angles in the molecule
   c. indicate whether the molecule is polar or nonpolar
   d. give the hybridization of the nitrogen atom(s)

Quiz 5

1. Draw the structural formulas of all the isomers of $C_6H_3Cl_2Br$, which is derived from benzene by substituting halogen atoms for hydrogen.
2. Write chemical equations for:
   a. the formation of a gas with a choking odor when sulfur burns in air.
   b. the reaction that occurs when C_3H_8 burns in a limited amount of air.

PROBLEMS 1-30

13.1 a. O_2 vs O_3
b. crystal structure
c. arrangement of atoms
d. molecular vs network covalent

13.2 a. $2\ H_2O_2(l) \longrightarrow 2\ H_2O(l) + O_2(g)$
b. $N_2H_4(l) + 2\ H_2O_2(l) \longrightarrow N_2(g) + 4\ H_2O(g)$
c. $CaF_2(s) + H_2SO_4(l) \longrightarrow 2\ HF(g) + CaSO_4(s)$
d. $NaI(s) + H_3PO_4(l) \longrightarrow HI(g) + NaH_2PO_4(s)$

13.3 a. react O_2 with NO at room temperature
b. heat NH_4NO_3
c. burn sulfur in air
d. burn phosphorus in limited supply of O_2

13.4 a. XeF_2, XeF_4, XeF_6 b. SF_2, SF_4, SF_6
c. CCl_4, C_2Cl_4, C_2Cl_6 d. ClF, BrF

13.5 a. S b. O c. Br

13.6 a. true b. false; paramagnetic c. false; graphite to diamond d. false; add As instead of Ge e. false; different boiling points

13.7 a. allotropy refers to two different forms of an element; isomerism refers to two different compounds with the same molecular formula

b. there is one double bond in an alkene molecule, one triple bond in an alkyne

c. graphite has a layer structure; in diamond, the C atoms are in a tetrahedral pattern

d. butane has a straight chain; 2-methylpropane has a branched chain

13.8 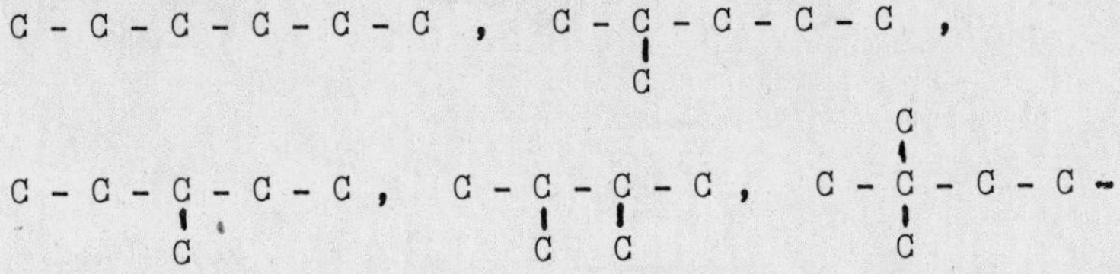

13.9 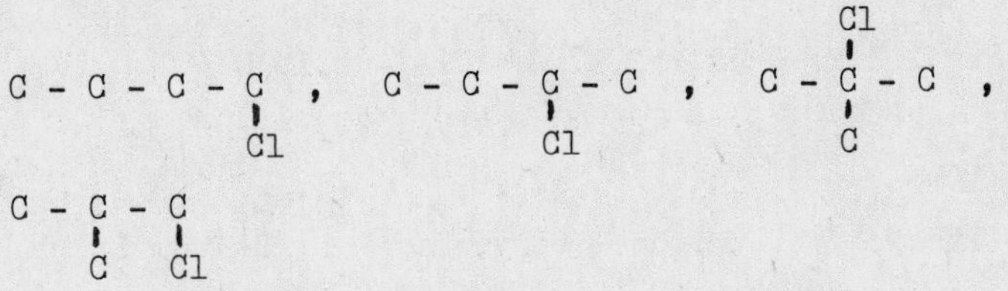

13.10

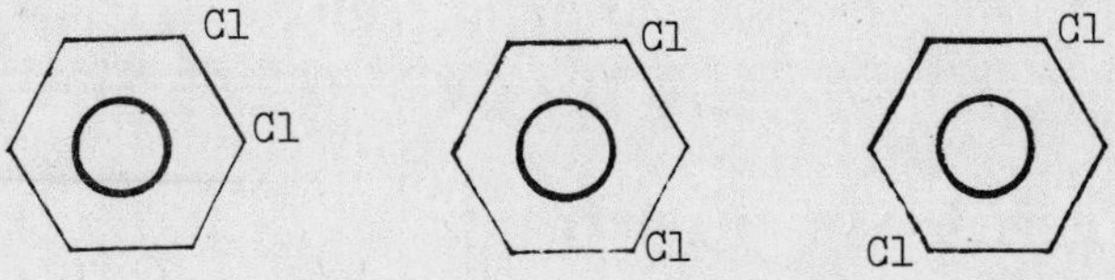

13.11 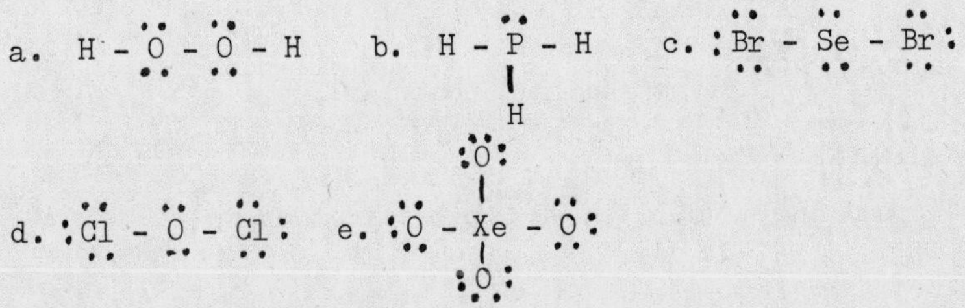

13.12 a. $:\!\ddot{O} = C = \ddot{O}\!:$ b. $:\!\ddot{O} - \ddot{S} = \ddot{O}\!:$ c. $:\!N \equiv N - \ddot{O}\!:$

d. $:\!\ddot{O}\!:\!\!\diagdown\!\! N - N\!\!\diagup\!\!:\!\ddot{O}\!:$, $:\!\ddot{O}\!:\!\!\diagup\!\!\qquad\diagdown\!\!:\!\ddot{O}\!:$

13.13 a. $:\!\ddot{Cl} - Xe - \ddot{Cl}\!:$ b. $:\!\ddot{F}\!:$ ... $P - \ddot{F}\!:$ c. $:\!\ddot{F} - \ddot{Cl} - \ddot{F}\!:$

d. $:\!\ddot{F}\!:$ ... $I - \ddot{F}\!:$

13.14 a. $:\!\ddot{O} = C = \ddot{O}\!: \longleftrightarrow :\!O \equiv C - \ddot{O}\!: \longleftrightarrow :\!\ddot{O} - C \equiv O\!:$

b. $:\!\ddot{O} - \ddot{S} = \ddot{O}\!: \longleftrightarrow :\!\ddot{O} = \ddot{S} - \ddot{O}\!:$

c. $:\!N \equiv N - \ddot{O}\!: \longleftrightarrow :\!\ddot{N} = N = \ddot{O}\!: \longleftrightarrow :\!\ddot{N} - N \equiv O\!:$

d. $\overset{O}{\underset{O}{\diagdown}}N - N\overset{O}{\underset{O}{\diagup}} \longleftrightarrow \overset{O}{\underset{O}{\diagup}}N - N\overset{O}{\underset{O}{\diagdown}} \longleftrightarrow \overset{O}{\underset{O}{\diagdown}}N - N\overset{O}{\underset{O}{\diagdown}} \longleftrightarrow \overset{O}{\underset{O}{\diagup}}N - N\overset{O}{\underset{O}{\diagup}}$

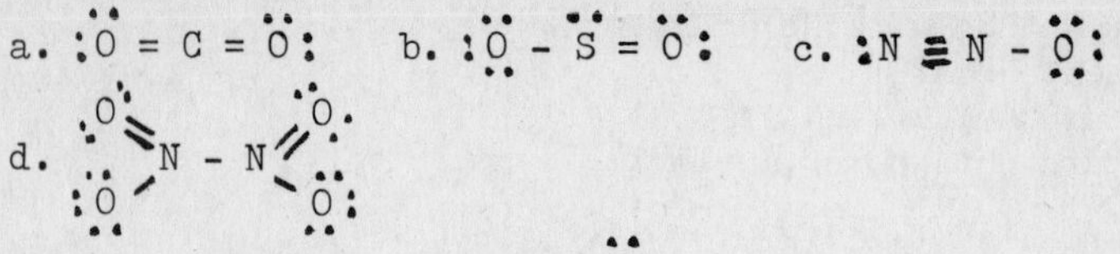

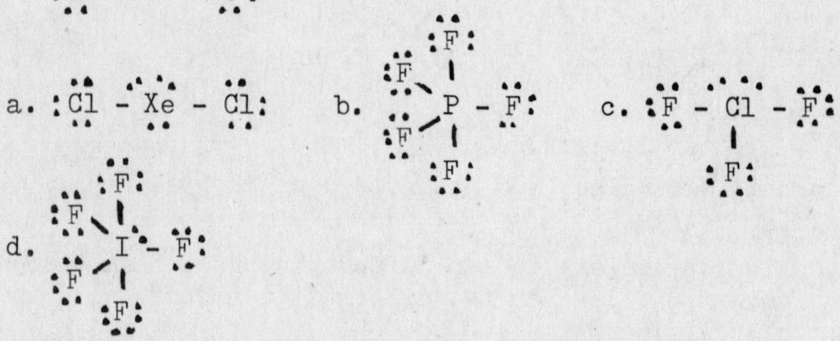

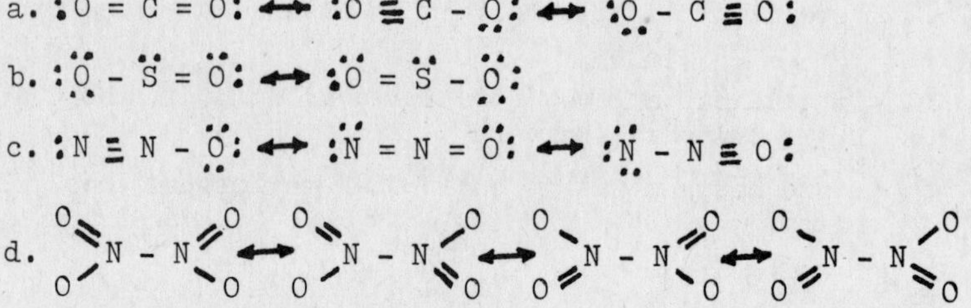

13.15 $Xe(g) + 2 F_2(g) \longrightarrow XeF_4(g)$.

$-215 \text{ kJ} = -4 \text{ B.E. Xe-F} + 2 \text{ B.E. F} - F$

$\text{B.E. Xe-F} = \dfrac{215 \text{ kJ} + 2(153 \text{ kJ})}{4} = +130 \text{ kJ/mol}$

13.16 $\Delta H = 2 \text{ B.E. C} \equiv O + \text{B.E. } O = O - 4 \text{ B.E. C} = O$

$= 2(1075 \text{ kJ}) + 498 \text{ kJ} - 4(715 \text{ kJ}) = -212 \text{ kJ}$

13.17 a. 109° b. 109° c. 109° d. 109° e. 109°

13.18 a. linear   b. trigonal bipyramid   c. T-shaped   d. square pyramid

13.19 a. polar   b. polar   c. polar   d. polar   e. nonpolar

13.20 a. sp^3d b. sp^3d c. sp^3d d. sp^3d^2

13.21 all are sp^2, except for N_2O, where the hybridization is sp

13.22 120°; polar

13.23 $:\!\ddot{F} - \ddot{S} - \ddot{F}\!:$; bent; polar; sp^3

$:\!\ddot{F}\!:$ ... S ... $:\!\ddot{F}\!:$; octahedral; nonpolar; sp^3d^2

13.24 bond angle would be 90°, hybridization sp^2

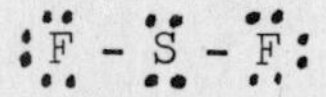

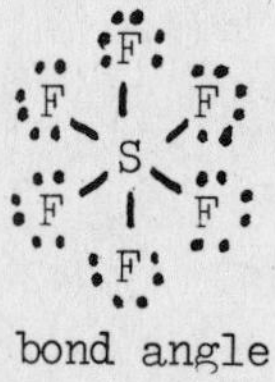

13.25 a. increase
b. heat between 96 and 119°C
c. monoclinic, liquid, vapor; vapor; monoclinic

13.26 $\log_{10} \frac{0.133}{0.0254} = \frac{\Delta H_{subl}(313 - 293)}{(2.30)(8.31)(313)(293)} = 0.719$

$\Delta H_{subl} = 6.30 \times 10^4$ J/mol

13.27 a. dispersion force b. H bond c. covalent bond
d. dispersion force

13.28 SF_6 is a compact molecule; dispersion forces weaker

13.29 consider 100 g of solution:

$m = \frac{10.0/34.02}{0.0900} = 3.27$

Volume of solution = $100 \text{ g} \times \frac{1 \text{ cm}^3}{1.035 \text{ g}} = 96.6 \text{ cm}^3$

$M = \frac{10.0/34.02}{0.0966} = 3.04$

13.30 a. no. moles HBr = $0.350 \text{ L} \times \frac{1.00 \text{ mol}}{1 \text{ L}} = 0.350 \text{ mol}$

Volume = $0.350 \text{ mol} \times \frac{1 \text{ L}}{2.50 \text{ mol}} = 0.140 \text{ L}$

b. $0.225 \text{ L} \times \frac{2.50 \text{ mol HBr}}{1 \text{ L}} \times \frac{2 \text{ mol ions}}{1 \text{ mol HBr}} = 1.12 \text{ mol}$

CHAPTER 14

BASIC SKILLS

1. Use Tables 14.1-14.4 to calculate ΔH and ΔS^o for a reaction.
2. Use the Gibbs-Helmholtz equation (Equation 14.13) to calculate the free energy change for a reaction.
3. Given or having calculated ΔH and ΔS^o, determine the temperature at which a reaction is at equilibrium at 1 atm.
4. Describe how the signs of ΔH, ΔS^o and ΔG^o relate to the spontaneity of a reaction.
5. Use the Second Law equation, Equation 14.16, to determine the maximum efficiency of a heat engine.

LECTURE NOTES

This chapter takes an empirical approach to the thermodynamic functions H, S and G. Emphasis is placed on the effect of ΔH, ΔS^o and ΔG^o on reaction spontaneity. We do not attempt to "derive" the Gibbs-Helmholtz equation; neither do we discuss how molar entropies are determined or how ΔS and ΔG vary with pressure or concentration. These aspects of thermodynamics can safely be deferred to a later course. It is important that students, in their first exposure, appreciate the power of thermodynamics, even at the expense of its rigor or elegance.

Perhaps surprisingly, students find this material quite straightforward. It is readily covered in two lectures, $1\frac{1}{2}$ if you omit Section 14.4 on the Second Law. Note, in Section 14.3, that ΔG^o is calculated from ΔH and ΔS^o, rather than from free energies of formation. A table of ΔG_f^o values is available in Appendix 1.

<u>LECTURE 1</u>

I <u>Enthalpy Changes</u> (review Ch. 5)

A. $\Delta H = H_{products} - H_{reactants}$ = heat flow in reaction at constant T, P

$\Delta H = \Sigma \Delta H_f$ products $- \Sigma \Delta H_f$ reactants

$Fe_2O_3(s) + 3\ H_2(g) \longrightarrow 2\ Fe(s) + 3\ H_2O(g)$

$\Delta H = 3\ \Delta H_f\ H_2O(g) - \Delta H_f\ Fe_2O_3(s) = +96.8$ kJ

B. If ΔH is negative, reaction is usually spontaneous at 25°C, 1 atm. However, certain endothermic phase changes are spontaneous:

$$H_2O(s) \longrightarrow H_2O(l); \Delta H = +6.0 \text{ kJ}; \text{ spontaneous above } 0^\circ C$$

Also, endothermic reactions often become spontaneous at high T. This is true of the reaction of iron(III) oxide with hydrogen, for which $\Delta H = +96.8$ kJ

II <u>Entropy Changes</u>

A. $\Delta S = S_{products} - S_{reactants}$; measure of change in order or randomness

solid ⟶ liquid; ΔS positive
liquid ⟶ gas ; ΔS positive

ΔS is usually positive for a reaction in which the number of moles of gas increases.

$$2\ SO_3(g) \longrightarrow 2\ SO_2(g) + O_2(g) \ ; \ \Delta n_g = +1; \Delta S \text{ positive}$$

$$N_2(g) + 3\ H_2(g) \longrightarrow 2\ NH_3(g); \ \Delta n_g = -2; \ \Delta S \text{ negative}$$

B. Calculation of ΔS° (ΔS at 1 atm, 1 M conc.)

$$\Delta S^\circ = \sum S^\circ \text{ products } - \sum S^\circ \text{ reactants}$$

$$Fe_2O_3(s) + 3\ H_2(g) \longrightarrow 2\ Fe(s) + 3\ H_2O(g)$$

$$\Delta S^\circ = 2S^\circ Fe(s) + 3S^\circ H_2O(g) - S^\circ Fe_2O_3(s) - 3S^\circ H_2(g)$$
$$= 2(27.2 \text{ J/K}) + 3(188.7 \text{ J/K}) - 90.0 \text{ J/K} - 3(130.6 \text{ J/K})$$
$$= 138.7 \text{ J/K}$$

Note that S° is a positive quantity for all substances, including elements.

C. Reactions for which ΔS° is positive tend to be spontaneous, at least at high temperatures.

$$H_2O(s) \longrightarrow H_2O(l) \ ; \ \Delta S^\circ +$$
$$H_2O(l) \longrightarrow H_2O(g) \ ; \ \Delta S^\circ +$$
$$Fe_2O_3(s) + 3\ H_2(g) \longrightarrow 2\ Fe(s) + 3\ H_2O(g); \Delta S^\circ +$$

Note that all these reactions are endothermic (ΔH +). They become spontaneous at high temperatures.

III <u>Free Energy Changes</u>

A. $\Delta G^\circ = \Delta H - T\Delta S^\circ$

Note that ΔG, like ΔS, is dependent on P. Unlike ΔH and ΔS°, ΔG° is strongly temperature dependent becuse of T in equation.

If ΔG° is -, reaction spontaneous at standard conditions
If ΔG° is +, reaction nonspontaneous at standard conditions
If $\Delta G^\circ = 0$, reaction at equilibrium at standard conditions

LECTURE 2

I Gibbs-Helmholtz Equation

$\Delta G^{\circ} = \Delta H - T\Delta S^{\circ}$

ΔH +, ΔS° -; ΔG° + at all T; nonspontaneous at 1 atm

ΔH -, ΔS° +; ΔG° - at all T; spontaneous at 1 atm

ΔH +, ΔS° +; ΔG° + at low T, becomes negative at high T

ΔH -, ΔS° -; ΔG° - at low T, becomes positive at high T

II Calculation of ΔG°

A. $Fe_2O_3(s) + 3\ H_2(g) \longrightarrow 2\ Fe(s) + 3\ H_2O(g)$

$\Delta H = +96.8$ kJ; $\Delta S^{\circ} = +138.7$ J/K = +0.1387 kJ/K

ΔG° (in kJ) = +96.8 - 0.1387 T

at 25°C: $\Delta G^{\circ} = +96.8 - 0.1387(298) = +55.5$ kJ
nonspontaneous at 25°C, 1 atm

at 500°C: $\Delta G^{\circ} = +96.8 - 0.1387\ (773) = -10.4$ kJ
spontaneous at 500°C, 1 atm

B. At what temperature does the reduction of Fe_2O_3 become spontaneous at 1 atm?

$\Delta G^{\circ} = 0$; $\Delta H = T\Delta S^{\circ}$

$T = \Delta H/\Delta S^{\circ} = 96.8/0.1387 = 698$ K (425°C)

Consider the vaporization of water:

$H_2O(l) \longrightarrow H_2O(g)$; $\Delta H = +44.0$ kJ, $\Delta S^{\circ} = +118.8$ J/K

T = 44.0/0.1188 = 370 K (bp water at 1 atm)

III Second Law

Heat cannot be completely converted to work in a constant temperature, cyclical process. Consequences:

1. Maximum efficiency engine:

$$\frac{w}{q} = \frac{T_2 - T_1}{T_2}$$

where w = work done in cycle
q = heat absorbed at high T

For steam engine operating between 100°C, 25°C

$$\text{Efficiency} = \frac{373 - 298}{373} = 0.20$$

2. Entropy function $\Delta S_{total} = \Delta S_{system} + \Delta S_{surroundings}$

$\Delta S_{total} > 0$ for a spontaneous process

3. Free Energy function

ΔG system < 0 for spontaneous process

ΔG system $= -w_{max}$ (i.e., maximum work obtained from process)

A spontaneous process is one from which it is possible to obtain useful work.

DEMONSTRATIONS

1. Spontaneous endothermic reactions: J. Chem. Educ. 46 A55 (1969); 51 A178 (1974)

QUIZZES

(Tables 14.1 - 14.4 should be available)

Quiz 1

1. Consider the reaction:

   $SnO_2(s) + 2\ H_2(g) \longrightarrow Sn(s) + 2\ H_2O(g)$

   calculate:

   a. ΔH b. ΔS° c. ΔG° at 25°C d. ΔG° at 200°C

2. For which of the following processes would you expect ΔS to be positive?

   a. osmosis b. $NO(g) + \frac{1}{2}\ O_2(g) \longrightarrow NO_2(g)$

   c. compressing a gas

Quiz 2

1. For the reaction: $N_2(g) + 3\ H_2(g) \longrightarrow 2\ NH_3(g)$, calculate:

   a. ΔH b. ΔS° c. ΔG° at 200°C

   d. the temperature at which $\Delta G^\circ = 0$

2. Of the three functions ΔH, ΔS, and ΔG, which one:

   a. varies least with pressure?
   b. varies most with temperature?
   c. is always negative for a spontaneous reaction?

Quiz 3

1. Consider the reaction:

   $CaSO_4(s) \longrightarrow CaO(s) + SO_2(g) + \frac{1}{2}\ O_2(g)$

   At what temperature does this reaction become spontaneous?

2. Which of the following are spontaneous at 25°C, 1 atm?

   a. dissolving HCl in water to form a 1 M solution.
   b. rusting of iron
   c. decomposition of water to the elements

Quiz 4

1. Consider the reaction between Fe_3O_4 and hydrogen gas to form iron and water vapor. Write a balanced equation for the reaction and calculate ΔG° at 300°C.

2. What are the signs of ΔH and ΔS for a reaction that:

   a. is spontaneous at all temperatures?
   b. becomes spontaneous as temperature increases?

Quiz 5

1. The heat of vaporization of benzene is 30.8 kJ/mol. The standard entropy of liquid benzene is 181.9 J/K; that of the vapor is 269.2 J/K, Calculate:

   a. ΔG^o_{vap} benzene at 0°C b. ΔG^o_{vap} benzene at 100°C

   c. the normal boiling point of benzene

2. Predict the sign of ΔH, ΔS^o, and ΔG^o at 25°C for the reaction:

   $2\ Fe(s) + 3/2\ O_2(g) \longrightarrow Fe_2O_3(s)$

PROBLEMS 1-30

14.1 a. $\Delta H = 2\Delta H_f\ H_2O(g) + 2\Delta H_f\ SO_2(g) - 2\Delta H_f\ H_2S(g)$
$= 2(-241.8\ kJ) + 2(-296.1\ kJ) - 2(-20.1\ kJ) = -1035.6\ kJ$

b. $\Delta H = 4\Delta H_f\ NO(g) + 6\Delta H_f\ H_2O(l) - 4\Delta H_f\ NH_3(g)$
$= 4(+90.4\ kJ) + 6(-285.8\ kJ) - 4(-46.2\ kJ) = -1168.4\ kJ$

c. $\Delta H = 2\Delta H_f\ NH_3(g) + 2\Delta H_f\ H_2O(l) - 2\Delta H_f\ NH_4^+(aq)$
$- 2\Delta H_f\ OH^-(aq) = 2(-46.2\ kJ) + 2(-285.8\ kJ)$
$-2(-132.8\ kJ) - 2(-229.9\ kJ) = +61.4\ kJ$

14.2 a. $\Delta H = \Delta H_f\ CaO(s) + \Delta H_f\ H_2O(g) - \Delta H_f\ Ca(OH)_2(s)$
$= -635.5\ kJ - 241.8\ kJ + 986.6\ kJ = +109.3\ kJ$

b. $\Delta H = \Delta H_f\ NO_2(g) + \Delta H_f\ H_2O(l) + \Delta H_f\ Ag^+(aq) - \Delta H_f\ NO_3^-(aq)$
$= +33.9\ kJ - 285.8\ kJ + 105.9\ kJ + 206.6\ kJ = +60.6\ kJ$

c. $\Delta H = \Delta H_f\ NO(g) + 2\Delta H_f\ H_2O(l) + 3\Delta H_f\ Ag^+(aq) - \Delta H_f\ NO_3^-(aq)$
$= +90.4\ kJ + 2(-285.8\ kJ) + 3(105.9\ kJ) + 206.6\ kJ$
$= +43.1\ kJ$

14.3 a. $NH_4Cl(s) \longrightarrow NH_4^+(aq) + Cl^-(aq)$
$\Delta H = \Delta H_f\ NH_4^+(aq) + \Delta H_f\ Cl^-(aq) - \Delta H_f\ NH_4Cl(s)$
$= -132.8\ kJ - 167.4\ kJ + 315.4\ kJ = +15.2\ kJ$

b. $Ag^+(aq) + Cl^-(aq) \longrightarrow AgCl(s)$
$\Delta H = \Delta H_f\ AgCl(s) - \Delta H_f\ Ag^+(aq) - \Delta H_f\ Cl^-(aq)$
$= -127.0\ kJ - 105.9\ kJ + 167.4\ kJ = -65.5\ kJ$

c. $C_2H_6(g) + 7/2\ O_2(g) \longrightarrow 2\ CO_2(g) + 3\ H_2O(l)$
$\Delta H = 2\Delta H_f\ CO_2(g) + 3\Delta H_f\ H_2O(l) - \Delta H_f\ C_2H_6(g)$
$= 2(-393.5\ kJ) + 3(-285.8\ kJ) + 84.7\ kJ = -1559.7\ kJ$

14.4 a. $75.0\ g\ Mg_3N_2 \times \frac{1\ mol\ Mg_3N_2}{100.92\ g\ Mg_3N_2} \times \frac{-691\ kJ}{1\ mol\ Mg_3N_2} = -514\ kJ$

b. $-691\ kJ = 3(-924.7\ kJ) + 2(-46.2\ kJ) - 6(-285.8\ kJ)$
$-\Delta H_f\ Mg_3N_2$

Solving, $\Delta H_f\ Mg_3N_2(s) = -461\ kJ$

14.5 a. + b. - c. + d. +

14.6 a. + b. - c. -

14.7 a. $\Delta S^\circ = 4\ S^\circ\ PCl_3(g) - S^\circ\ P_4(s) - 6\ S^\circ\ Cl_2(g)$
$= 4(311.7\ J/K) - 177.4\ J/K - 6(222.9\ J/K) = -268.0\ J/K$

b. $\Delta S^\circ = 2\ S^\circ\ Fe(s) + 3\ S^\circ\ H_2O(l) - 3\ S^\circ\ H_2(g) - S^\circ\ Fe_2O_3(s)$
$= 2(27.2\ J/K) + 3(69.9\ J/K) - 3(130.6\ J/K) - 90.0\ J/K$
$= -217.7\ J/K$

c. $\Delta S^\circ = 3\ S^\circ\ Mn(s) + 2\ S^\circ\ Al_2O_3(s) - 4\ S^\circ\ Al(s) - 3\ S^\circ\ MnO_2(s)$
$= 3(31.8\ J/K) + 2(51.0\ J/K) - 4(28.3\ J/K) - 3(53.1\ J/K)$
$= -75.1\ J/K$

14.8 a. $\Delta S^\circ = S^\circ\ Zn^{2+}(aq) + S^\circ\ H_2(g) - S^\circ\ Zn(s)$
$= -106.5\ J/K + 130.6\ J/K - 41.6\ J/K = -17.5\ J/K$

b. $\Delta S^\circ = S^\circ\ H_2O(l) - S^\circ\ OH^-(aq)$
$= 69.9\ J/K + 10.5\ J/K = +80.4\ J/K$

c. $\Delta S^\circ = S^\circ\ NH_4^+(aq) + S^\circ\ OH^-(aq) - S^\circ\ NH_3(g) - S^\circ\ H_2O(l)$
$= +112.8\ J/K - 10.5\ J/K - 192.5\ J/K - 69.9\ J/K$
$= -160.1\ J/K$

14.9 a. $\Delta S^\circ = 2\ S^\circ\ H_2O(g) + 2\ S^\circ\ SO_2(g) - 2\ S^\circ\ H_2S(g) - 3\ S^\circ\ O_2(g)$
$= 2(188.7\ J/K) + 2(248.5\ J/K) - 2(205.6\ J/K)$
$- 3(205.0\ J/K) = -151.8\ J/K$

b. $\Delta S^\circ = 4\ S^\circ\ NO(g) + 6\ S^\circ\ H_2O(l) - 5\ S^\circ\ O_2(g) - 4\ S^\circ\ NH_3(g)$
$= 4(210.6\ J/K) + 6(69.9\ J/K) - 5(205.0\ J/K)$
$- 4(192.5\ J/K) = -533.2\ J/K$

c. $\Delta S^\circ = 2\ S^\circ\ H_2O(l) + 2\ S^\circ\ NH_3(g) - 2\ S^\circ\ OH^-(aq)$
$- 2\ S^\circ\ NH_4^+(aq) = 2(69.9\ J/K) + 2(192.5\ J/K)$
$- 2(-10.5\ J/K) - 2(112.8\ J/K) = +320.2\ J/K$

14.10 a. $\Delta S^\circ = S^\circ\ NH_4^+(aq) + S^\circ\ Cl^-(aq) - S^\circ\ NH_4Cl(s)$
$= 112.8\ J/K + 55.1\ J/K - 94.6\ J/K = +73.3\ J/K$

b. $\Delta S^\circ = S^\circ\ AgCl(s) - S^\circ\ Ag^+(aq) - S^\circ\ Cl^-(aq)$
$= 96.1\ J/K - 73.9\ J/K - 55.1\ J/K = -32.9\ J/K$

c. $\Delta S^\circ = 2\ S^\circ\ CO_2(g) + 3\ S^\circ\ H_2O(l) - 7/2\ S^\circ\ O_2(g)$
$- S^\circ\ C_2H_6(g) = 2(213.6\ J/K) + 3(69.9\ J/K)$
$- 7/2(205.0\ J/K) - 229.5\ J/K = -310.1\ J/K$

14.11 a. $\Delta G^\circ = -109.0\ kJ - 298(0.0278)kJ = -117.3\ kJ$

b. $\Delta G^\circ = +842\ kJ + 298(0.1162)kJ = +877\ kJ$

c. $\Delta G^\circ = +82.9\ kJ - 298(0.115)kJ = +48.6\ kJ$

14.12 a. $\Delta G^\circ = -109.0\ kJ - 773(0.0278)kJ = -130.5\ kJ$

b. $\Delta G^\circ = +842\ kJ + 773(0.1162)kJ = +932\ kJ$

14.12 c. ΔG° = +82.9 kJ - 773(0.115)kJ = -6.0 kJ

14.13 a. ΔG° = -1035.6 kJ + 298(0.1518)kJ = -990.4 kJ

b. ΔG° = -1168.4 kJ + 298(0.5332)kJ = -1009.5 kJ

c. ΔG° = +61.4 kJ - 298(0.3202)kJ = -34.0 kJ

14.14 a. ΔG° = +15.2 kJ b. ΔG° = -65.5 kJ c. ΔG° = -1559.7 kJ

14.15 a. $Na(s) + \frac{1}{2} Cl_2(g) \longrightarrow NaCl(s)$

$\Delta H = \Delta H_f$ NaCl(s) = -411.0 kJ

$\Delta S^\circ = S^\circ$ NaCl(s) - S° Na(s) - $\frac{1}{2}$ S° $Cl_2(g)$

= 72.4 J/K - 51.0 J/K - 111.4 J/K = -90.0 J/K

ΔG° = -411.0 kJ + 298(0.0900)kJ = -384.2 kJ

b. $Cu(s) + S(s) + 2\ O_2(g) \longrightarrow CuSO_4(s)$

$\Delta H = \Delta H_f$ $CuSO_4(s)$ = -769.9 kJ

$\Delta S^\circ = S^\circ$ $CuSO_4(s)$ - S° Cu(s) - S° S(s) - 2 S° $O_2(g)$

= (113.4 - 33.3 -31.9 - 410.0)J/K = -361.8 J/K

ΔG° = -769.9 kJ + 298(0.3618)kJ = -662.1 kJ

c. $2\ Fe(s) + \frac{3}{2}\ O_2(g) \longrightarrow Fe_2O_3(s)$

$\Delta H = \Delta H_f$ $Fe_2O_3(s)$ = -822.2 kJ

$\Delta S^\circ = S^\circ$ $Fe_2O_3(s)$ - 2 S° Fe(s) - 3/2 S° $O_2(g)$

= (90.0 - 54.4 - 307.5)J/K = -271.9 J/K

ΔG° = -822.2 kJ + 298(+0.2719)kJ = -741.2 kJ

14.16 a. $\Delta S^\circ = \dfrac{\Delta H - \Delta G^\circ}{T} = \dfrac{-126.0\ \text{kJ} + 173.8\ \text{kJ}}{298\ \text{K}}$ = +0.160 kJ/K

Would expect ΔS° to be positive, since gas is formed

b. ΔS° = 4 S° NaOH(s) + S° $O_2(g)$ - 2 S° $H_2O(l)$- 2 S° $Na_2O_2(s)$

160 J/K = 209.2 J/K + 205.0 J/K - 139.8 J/K - 2$S^\circ$$Na_2O_2(s)$

solving, S° $Na_2O_2(s)$ = 57 J/K

c. $\Delta H = 4\ \Delta H_f$ NaOH(s) - 2 ΔH_f $Na_2O_2(s)$ - 2 ΔH_f $H_2O(l)$

-126.0 kJ = -1706.8 kJ + 571.6 kJ - 2 ΔH_f $Na_2O_2(s)$

solving, ΔH_f $Na_2O_2(s)$ = -504.6 kJ/mol

14.17 a. ΔG° = -55.6 kJ - 298(0.1331)kJ = -95.3 kJ

b. $\Delta H = \Delta H_f$ $H_2O(l)$ + 2 ΔH_f $Cl^-(aq)$ + ΔH_f $Cu_2O(s)$

- 2 ΔH_f $OH^-(aq)$ - 2 ΔH_f CuCl(s)

-55.6 kJ = -285.8 kJ - 334.8 kJ - 166.7 kJ + 459.8 kJ

- 2 ΔH_f CuCl(s)

solving, ΔH_f CuCl(s) = -136.0 kJ

14.17 c. $\Delta S^o = S^o\ H_2O(l) + 2\ S^o\ Cl^-(aq) + S^o\ Cu_2O(s) - 2\ S^o\ OH^-(aq)$
$- 2\ S^o\ CuCl(s)$

$+133.1\ J/K = 69.9\ J/K + 110.2\ J/K + 100.8\ J/K + 21.0\ J/K$
$- 2\ S^o\ CuCl(s)$

solving, $S^o\ CuCl(s) + +84.4\ J/K$

14.18 a. ΔG^o is + at low T; becomes - at high T
reaction is nonspontaneous at low T, becomes spontaneous as T is raised

b. ΔG^o is - at all T; reaction spontaneous at all T

c. ΔG^o is - at low T, becomes + at high T
reaction is spontaneous at low T, becomes nonspontaneous as T is raised

14.19 a. 839.4/0.203 = 4130 K; in practice, ΔG^o is + at all reasonable temperatures

b. At no temperature is ΔG^o zero

c. 1897.8/0.318 = 5970 K; in practice, ΔG^o is - at all reasonable temperatures

14.20 a. $\Delta H = 2\ \Delta H_f\ NO_2(g) - 2\ \Delta H_f\ NO(g)$
$= 2(+33.9\ kJ) - 2(90.4\ kJ) = -113.0\ kJ$

b. $\Delta S^o = 2\ S^o\ NO_2(g) - 2\ S^o\ NO(g) - S^o\ O_2(g)$
$= (481.0 - 421.2 - 205.0) J/K = -145.2\ J/K$

c. $T = \frac{113.0}{0.1452}\ K = 778\ K$

14.21 a. $\Delta H = -2\ \Delta H_f\ Ag_2O(s) = +61.2\ kJ$

$\Delta S^o = 4\ S^o\ Ag(s) + S^o\ O_2(g) - 2\ S^o\ Ag_2O(s)$
$= +170.8\ J/K + 205.0\ J/K - 243.4\ J/K = +132.4\ J/K$

$\Delta G^o = +61.2 - 0.1324\ T$

T	100 K	200 K	300 K	400 K	500 K
ΔG^o	+48.0 kJ	+34.7 kJ	+21.5 kJ	+8.2 kJ	-5.0 kJ

b. $T = \Delta H/\Delta S^o = 61.2/0.1324 = 462\ K$

14.22 a. zero

b. $\Delta S^o = \frac{\Delta H}{T} = \frac{31.3\ kJ}{334\ K} = 0.0937\ kJ/K = 93.7\ J/K$

$93.7\ J/K = S^o\ CHCl_3(g) - 202.9\ J/K$

$S^o\ CHCl_3(g) = 296.6\ J/K$

14.23 red P $\longrightarrow$ white P ; $\Delta H = +17.6\ kJ$; $\Delta S^o = 18.29\ J/K$

T = 17.6/0.01829 = 962 K

14.24 $\frac{T_2 - 308}{T_2} = 0.18$; $T_2 = \frac{308\ K}{0.82} = 380\ K \approx 100^oC$

14.25 a. $\Delta H = 2\,\Delta H_f\ H_2O(g) = -483.6$ kJ

$\Delta S^\circ = 2\ S^\circ\ H_2O(g) - 2\ S^\circ\ H_2(g) - S^\circ\ O_2(g)$

$= 377.4$ J/K $- 261.2$ J/K $- 205.0$ J/K $= -88.8$ J/K

$\Delta G^\circ = -483.6$ kJ $+ 298(0.0888)$kJ $= -457.1$ kJ

can obtain 457.1 kJ of work

b. $\Delta G^\circ = -483.6$ kJ $+ 1273(0.0888)$kJ $= -370.6$ kJ

can obtain 370.6 kJ of work

14.26 a. spontaneous b. nonspontaneous c. nonspontaneous d. spontaneous

14.27 only (a)

14.28 a. reaction is not spontaneous at standard concentrations
b. not generally true; usually true at room T
c. becomes spontaneous at high T
d. 500 kJ is maximum amount of useful work at 1 atm

14.29 a. ΔS° may be negative
b. solid more ordered
c. $\Delta G^\circ = \Delta H - T\,\Delta S^\circ$

14.30 a. 0 b. 0

CHAPTER 15

BASIC SKILLS

1. Given the balanced equation for a reaction involving gases, write the corresponding expression for K_c.
2. For a given equation, calculate K_c, knowing:
   a. the equilibrium concentrations of all species, or
   b. the original concentrations of all species and the equilibrium concentration of one species.
3. Given the value of K_c, predict:
   a. the direction in which a chemical system will move to reach equilibrium.
   b. the equilibrium concentration of one species, given those of all other species.
   c. the equilibrium concentrations of all species, given their original concentrations.
4. Using Le Chatelier's Principle, predict the effect of a change in the number of moles, volume, or temperature upon the position of an equilibrium.
5. Given one of the two quantities, K_c or K_p, for a system involving gases, calculate the other quantity.
6. Relate the standard free energy change for a reaction, ΔG°, to the equilibrium constant.

LECTURE NOTES

In many ways, this is the most difficult chapter covered to this point. Many students find equilibrium calculations difficult to follow, let alone carry out on their own. This is particularly true for the most important calculation: determination of equilibrium from original concentrations (skill 3c above). It helps to go through a couple of calculations of this sort slowly and carefully, emphasizing the chemical reasoning rather than the algebra.

This chapter will require $2\frac{1}{2}$ - 3 lectures, depending upon how much time you wish to devote to K_p. In the outline below, we assume that K_c will be emphasized with K_p referred to only in passing.

<u>LECTURE 1</u>

I <u>The Equilibrium Constant</u>, Kc

A. $2\ HI(g) \rightleftharpoons H_2(g) + I_2(g)$

To study this equilibrium at 520°C, put pure HI or a mixture of H_2 and I_2 in a sealed container at this temperature. Take samples over a period of time until concentrations become constant. At that point, system is at equilibrium.

	Orig. Conc.			Final Conc.		
	HI	H_2	I_2	HI	H_2	I_2
Expt. 1	0.200	0.000	0.000	0.160	0.020	0.020
Expt. 2	0.000	0.100	0.100	0.160	0.020	0.020
Expt. 3	0.100	0.100	0.100	0.240	0.030	0.030

Note that:

a) Δconc. HI = $-2\,\Delta$conc. H_2 = $-2\,\Delta$conc. I_2 (required by coefficients of equation)

b) $\frac{[H_2] \times [I_2]}{[HI]^2} = 0.016$ in all three experiments

In general, at any temperature: $\frac{[H_2] \times [I_2]}{[HI]^2}$ = constant = K_c

K_c depends only on T; independent of original concentrations, volume, or pressure.

II General Expression for K_c

A. Only gases involved

$N_2O_4(g) \rightleftharpoons 2\ NO_2(g)$; $K_c = \frac{[NO_2]^2}{[N_2O_4]}$

$N_2(g) + 3\ H\ (g) \rightleftharpoons 2\ NH_3(g)$; $K_c = \frac{[NH_3]^2}{[N_2] \times [H_2]^3}$

$aA(g) + bB(g) \rightleftharpoons cC(g) + dD(g)$; $K_c = \frac{[C]^c \times [D]^d}{[A]^a \times [B]^b}$

Note that products (right side of equation) appear in numerator, reactants in denominator.

B. Solids or liquids as well as gases

$CaCO_3(s) \rightleftharpoons CaO(s) + CO_2(g)$; $K_c = [CO_2]$

Terms for solids or liquids do not appear. Adding or removing a solid or liquid does not affect position of equilibrium. For example, conc. CO_2 is the same regardless of how much $CaCO_3$ or CaO is present.

III Calculation of K_c

A. $N_2O_4(g) \rightleftharpoons 2\ NO_2(g)$

At 100°C, 0.240 mol NO_2 and 0.080 mol N_2O_4 are present at equilibrium in a 2.00 L container. K_c?

$$K_c = \frac{(0.240/2.00)^2}{(0.080/2.00)} = 0.36$$

B. At a different T, start with 1.00 mol N_2O_4 in a 4.00 L container; end with 0.400 mol N_2O_4 at equilibrium. K_c?

	orig.	change	equil.
conc. N_2O_4	0.250 M -	0.150 M =	0.100 M
conc. NO_2	0.000 M +	0.300 M =	0.300 M

$$K_c = \frac{(0.100)^2}{0.300} = 0.033$$

LECTURE 2

I Applications of K_c

In general, if K_c is large, equilibrium lies far to right (products favored). If K_c is very small, equilibrium favors reactants.

A. Determination of direction of reaction

Compare Q, original concentration quotient, to K_c, the concentration quotient at equilibrium. If $Q < K_c$, system moves to right (products formed). If $Q > K_c$, system moves to left (reactants formed)

$2HI(g) \rightleftharpoons H_2(g) + I_2(g)$; $K_c = 0.016$

Start with conc. HI = 0.10 M, no H_2 or I_2

$$Q = \frac{(0)(0)}{(0.10)^2} = 0 < K_c\text{; some HI dissociates}$$

Start with conc. HI= conc. H_2 = conc. I_2 = 0.10 M

$$Q = \frac{(0.10)(0.10)}{(0.10)^2} = 1.0 > K_c\text{; some } H_2, I_2 \text{ react to form HI}$$

B. Extent of Reaction

1. $2\ HI(g) \rightleftharpoons H_2(g) + I_2(g)$; $K_c = 0.016$

Start with pure HI at 0.200 M

	orig.	change	equil.	
conc. HI	0.200 M	-2x	0.200 - 2x	= 0.160 M
conc. H_2	0.000 M	+x	x	= 0.020 M
conc. I_2	0.000 M	+x	x	= 0.020 M

$$\frac{x^2}{(0.200 - 2x)^2} = 0.016; \quad \frac{x}{0.200 - 2x} = 0.13; \; x = 0.020$$

2. $N_2O_4(g) \rightleftharpoons 2\ NO_2(g)$; $K_c = 0.36$

Start with pure N_2O_4 at concentration of 0.100 M

	orig.	change	equil.	
conc. N_2O_4	0.100 M	-x	0.100 - x	= 0.040 M
conc. NO_2	0.000 M	+2x	2x	= 0.120 M

$$\frac{(2x)^2}{0.100 - x} = 0.36\text{; solving, } x = 0.060$$

III Effect of Changes in Conditions on Equilibrium Position

Le Chatelier's Principle: If system at equilibrium is subjected to stress, reaction occurs in a direction so as to partially relieve the stress

A. Adding or removing a gaseous species

$$2\ HI(g) \rightleftharpoons H_2(g) + I_2(g)$$

If H_2 is added part of it reacts; [HI] greater than before equilibrium was disturbed, $[I_2]$ less. $[H_2]$ intermediate between original equilibrium value and that immediately after equilibrium was disturbed.

If H_2 is removed, some HI decomposes to bring $[H_2]$ back part way to its original equilibrium concentration. $[I_2]$ increases, [HI] decreases.

B. Increasing Volume Immediate effect is to lower concentration of molecules. To counteract this, reaction occurs which increases number of molecules in gas phase.

	Expansion	Compression
$N_2O_4(g) \rightleftharpoons 2NO_2(g)$	→	←
$N_2(g) + 3H_2(g) \rightleftharpoons 2NH_3(g)$	←	→
$H_2(g) + I_2(g) \rightleftharpoons 2HI(g)$	no effect	

LECTURE $2\frac{1}{2}$

I Effect of Change in Conditions (cont.)

A. Effect of T Increase in T is partially counteracted if an endothermic process occurs. Hence, endothermic reaction is favored by increase in T.

	ΔH	increase T	decrease T
$N_2O_4(g) \rightleftharpoons 2NO_2(g)$	+58.2 kJ	→	←
$N_2(g) + 3H_2(g) \rightleftharpoons 2NH_3(g)$	-92.4 kJ	←	→

If forward reaction is endothermic, K_c increases as T increases; if exothermic, K_c decreases as T increases.

II K_c and K_p

K_p is the equilibrium constant expression obtained using partial pressures of gases.

$$N_2(g) + 3\ H_2(g) \rightleftharpoons 2\ NH_3(g)$$

$$K_c = \frac{[NH_3]^2}{[N_2] \times [H_2]^3} \quad ; \quad K_p = \frac{[P\ NH_3]^2}{[P\ N_2] \times [P\ H_2]^3}$$

In general, $K_p = K_c(0.0821T)^{\Delta n_g}$

$N_2(g) + 3\ H_2(g) \rightleftharpoons 2\ NH_3(g)$ $\quad K_p = K_c(RT)^{-2}$

$N_2O_4(g) \rightleftharpoons 2\ NO_2(g)$ $\quad K_p = K_c(RT)$

III <u>Relation between ΔG° and K</u>

$\Delta G^\circ = -0.0191\ T \log_{10} K$

ΔG°	+50 kJ	+25 kJ	0	-25 kJ	-50 kJ
K at 25°C	2×10^{-9}	4×10^{-5}	1	2×10^{4}	6×10^{8}

For gases: $\Delta G^\circ = -0.0191\ T \log_{10} K_p$

DEMONSTRATIONS

1. Le Chatelier's Principle: Test. Dem. 19, 221; J. Chem. Educ. <u>47</u> A735 (1970)
2. Effect of T on N_2O_4 - NO_2 System: Test. Dem. 19, 131, 167

QUIZZES

<u>Quiz 1</u>

1. Consider the equilibrium

   $2\ NO(g) \rightleftharpoons N_2(g) + O_2(g)$; $\Delta H = -180.8$ kJ

   a. If we start with 1.40 mol of NO in a 4.00 L container at a certain temperature, we find that 0.20 mol of NO is left at equilibrium. Calculate K_c.

   b. Which way does the equilibrium shift if the system is compressed?

   c. Which way does the equilibrium shift if the temperature is raised?

<u>Quiz 2</u>

1. Consider the equilibrium

   $PCl_5(g) \rightleftharpoons PCl_3(g) + Cl_2(g)$; $K_c = 0.050$

   a. Suppose we start with pure PCl_5 and find that $[PCl_5] = 0.20$ M. Calculate $[PCl_3]$ and $[Cl_2]$.

   b. Suppose we add some PCl_3 to the equilibrium system in (a). How will the final equilibrium concentrations of PCl_5, PCl_3, and Cl_2 compare to their concentrations before equilibrium was disturbed?

<u>Quiz 3</u>

1. For the equililibrium

   $2\ HI(g) \rightleftharpoons H_2(g) + I_2(g)$

   $K_c = 0.016$ at 520°C

   Suppose we start with 1.00 mol of H_2 and 1.00 mol of I_2 in a 5.00 L container at 520°C. What will be the equilibrium concentrations of all three species?

Quiz 4

1. Consider the equilibrium:

   $N_2(g) + 3\ H_2(g) \rightleftharpoons 2\ NH_3(g) \quad K_c = 0.50$ at $400^\circ C$

   Suppose we start with 1.00 mol of N_2 and 2.00 mol of H_2 in a 0.500 L container. Set up an equation in terms of x (define what x is), relating the equilibrium concentrations of N_2, H_2, and NH_3. It is not necessary to solve this equation for x.

2. The heat of formation of NH_3 is -46.2 kJ. Is K_c larger or smaller than 0.50 at 200°C? Explain your reasoning.

Quiz 5

1. Consider the system

   $2\ NO(g) + O_2(g) \rightleftharpoons 2\ NO_2(g)\ ;\ \Delta H = -113$ kJ

   K_c for this reaction is 2.4 at a certain temperature. Which way will the system shift to reach equilibrium if we start with conc. NO = conc. O_2 = conc. NO_2 = 0.10 M. Show calculations.

2. Which way will the equilibrium system in (1) shift upon

   a. compression b. increasing T

   Explain your reasoning in each case.

PROBLEMS 1-30

15.1 90 min, 120 min

15.2

t(s)	0	20	40	60	80	100	
A	0.080	0.050	0.030	0.020	0.015	0.015	
B	0.100	0.070	0.050	0.040	0.035	0.035	t = 80 s
C	0.000	0.060	0.100	0.120	0.130	0.130	

15.3 a. $\dfrac{[SO_3]^2}{[SO_2]^2 \times [O_2]}$ b. $\dfrac{[H_2O]^2 \times [SO_2]^2}{[O_2]^3 \times [H_2S]^2}$ c. $\dfrac{[NH_3]^2 \times [H_2O]^4}{[NO_2]^2 \times [H_2]^7}$

15.4 a. $\dfrac{[Fe(CO)_5]}{[CO]^5}$ b. $\dfrac{[NF_3]^2 \times [HF]^6}{[F_2]^6 \times [NH_3]^2}$ c. $\dfrac{[NH_3]^2}{[H_2O]^6}$ d. $\dfrac{1}{[NH_3]^2}$

15.5 a. $2\ H_2S(g) \rightleftharpoons 2\ H_2(g) + S_2(g)$

b. $H_2(g) + I_2(g) \rightleftharpoons 2\ HI(g)$

c. $H_2(g) + I_2(s) \rightleftharpoons 2\ HI(g)$

d. $CS_2(g) + 4\ H_2(g) \rightleftharpoons CH_4(g) + 2\ H_2S(g)$

15.6 a. $K_c = (26)^{\frac{1}{2}} = 5.1$ b. $K_c = 1/(2 \times 10^{-5}) = 5 \times 10^4$

15.7 $K_c = \dfrac{[N_2] \times [O_2]}{[NO]^2} = \dfrac{(0.040)(0.040)}{(0.00035)^2} = 1.3 \times 10^4$

15.8 $[NO] = 4.00\ mol/L - 2.56\ mol/L = 1.44\ mol/L$

$$K_c = \frac{(1.28)(1.28)}{(1.44)^2} = 0.790$$

15.9 a. $[CO] = 0.78\ mol/L - \frac{1}{2}(1.36 - 0.12)mol/L = 0.16\ mol/L$

b. $[CH_3OH] = \frac{1}{2}(1.36 - 0.12)mol/L = 0.62\ mol/L$

c. $K_c = \frac{0.62}{(0.16)(0.12)^2} = 2.7 \times 10^2$

15.10 a. $\frac{(conc.\ Cl_2)(conc.\ PCl_3)}{(conc.\ PCl_5)} = \frac{(0.090)(0.18)}{0.024} = 0.68$

not at equilibrium; Q not equal to K_c

b. ← ; Q has to decrease to become equal to K_c

15.11 a. ← ; some SO_2 and O_2 have to be formed

b. $\frac{(conc.\ SO_3)^2}{(conc.\ O_2)(conc.\ SO_2)^2} = \frac{(0.15)^2}{(0.10)(0.060)^2} = 62$; system shifts to left

15.12 (a), since K_c is very small

15.13 $0.14 = \frac{[Br_2] \times [Cl_2]}{[BrCl]^2}$; $[BrCl]^2 = \frac{[Br_2] \times [Cl_2]}{0.14}$

$$= \frac{(0.0250)^2}{0.14} = 0.0045$$

$[BrCl] = 0.067\ mol/L$

15.14 $K_c = \frac{[PCl_3] \times [Cl_2]}{[PCl_5]}$; $0.050 = 3 \times [Cl_2]$; $[Cl_2] = 0.017\ mol/L$

15.15

	orig.	change	equil.
conc. ICl	0.20 M	-2x	0.20 - 2x
conc. I_2	0.00 M	+x	x
conc. Cl_2	0.00 M	+x	x

$\frac{x^2}{(0.20 - 2x)^2} = 0.11$; $\frac{x}{0.20 - 2x} = 0.33$; $x = 0.040$

$[I_2] = [Cl_2] = 0.040\ M$; $[ICl] = 0.12\ M$

15.16 a.

	orig.	change	equil.
conc. CO_2	0.00 M	+x	x
conc. H_2	0.00 M	+x	x
conc. H_2O	0.10 M	-x	0.10 - x
conc. CO	0.10 M	-x	0.10 - x

$x^2/(0.10 - x)^2 = 4.0$; $x/(0.10 - x) = 2.0$; $x = 0.067$

$[CO_2] = [H_2] = 0.067\ mol/L$; $[CO] = [H_2O] = 0.033\ mol/L$

15.16 b.

	orig.	change	equil.
conc. CO_2	0.040 M	+x	0.040 + x
conc. H_2	0.040 M	+x	0.040 + x
conc. H_2O	0.040 M	-x	0.040 - x
conc. CO	0.040 M	-x	0.040 - x

$(0.040 + x)^2/(0.040 - x)^2 = 4.0$;

$(0.040 + x)/(0.040 - x) = 2.0$; $x = 0.013$

$[CO_2] = [H_2] = 0.053$ M; $[CO] = [H_2O] = 0.027$ M

15.17

	orig.	change	equil.
conc. PCl_3	0.00 M	+x	x
conc. Cl_2	0.00 M	+x	x
conc. PCl_5	0.30 M	-x	0.30 - x

$\frac{x^2}{0.30 - x} = 0.050$; solving, $x = 0.10$

$[PCl_3] = [Cl_2] = 0.10$ mol/L; $[PCl_5] = 0.20$ mol/L

15.18

	orig.	change	equil.
conc. SO_2	0.00 M	+x	x
conc. Cl_2	0.00 M	+x	x
conc. SO_2Cl_2	0.10 M	-x	0.10 - x

$\frac{x^2}{0.10 - x} = 0.40$; solving, $x = 0.083$

$[SO_2] = [Cl_2] = 0.083$ M; $[SO_2Cl_2] = 0.017$ M

15.19 a.

	orig.	change	equil.
conc. Cl_2	0.10 M	-x	0.10 - x
conc. Cl	0.00 M	+2x	2x

$\frac{4x^2}{0.10 - x} = 0.036$; solving, $x = 0.026$ M; conc. Cl = 0.052 M

b. $[Cl]^2 = 4.2 \times 10^{-7} \times [Cl_2] = 4.2 \times 10^{-8}$; $[Cl] = 2.1 \times 10^{-4}$ M

15.20 a. $K_c = (0.10)^2/(0.10)^2 = 1.0$

b.

	orig.	change	equil.
conc. H_2	0.18 M	-x	0.18 - x
conc. H_2O	0.10 M	+x	0.10 + x

$(0.10 + x)^2/(0.18 - x)^2 = 1.0$; $\frac{0.10 + x}{0.18 - x} = 1.0$; $x = 0.04$

$[H_2O] = [H_2] = 0.14$ M

15.21 a. $\longleftarrow$ (increases number of moles

b. $\longrightarrow$

c. $\longleftarrow$ (reverse reaction endothermic)

15.21 d. $\rightarrow$

15.22 a. $\leftarrow$ b. $\leftarrow$ c. no effect

15.23 endothermic; K_c increases as T increases

15.24 a or c

15.25 $K_p = K_c(0.0821\ T)$

t(°C)	0	50	100
T (K)	273	323	373
K_c	0.0005	0.022	0.36
K_p	0.01	0.58	11

15.26 $K_p = K_c(RT)^2$; $K_c = \frac{K_p}{(RT)^2} = \frac{7.7 \times 10^{-17}}{(0.0821)^2(298)^2} = 1.3 \times 10^{-19}$

$K_p = K_c(RT)$; $K_c = \frac{K_p}{RT} = \frac{6.1 \times 10^{-13}}{(0.0821)(298)} = 2.5 \times 10^{-14}$

$K_p = K_c$; $K_c = 8.7 \times 10^2$

15.27 a. $\Delta H = \Delta H_f\ H^+(aq) + \Delta H_f\ OH^-(aq) - \Delta H_f\ H_2O(l)$

$= -229.9\ kJ + 285.8\ kJ = +55.9\ kJ$

$\Delta S° = S°\ H^+(aq) + S°\ OH^-(aq) - S°\ H_2O(l)$

$= -10.5\ J/K - 69.9\ J/K = -80.4\ J/K$

$\Delta G° = +55.9\ kJ + 298(0.0804)kJ = +79.9\ kJ$

b. $\log_{10} K = \frac{-79.9}{(2.30)(8.31 \times 10^{-3})(298)} = -14.0$; $K = 1 \times 10^{-14}$

15.28 $\Delta G° = -2.30(8.31 \times 10^{-3})(298) \log_{10}(7.7 \times 10^{-17}) = +91.8\ kJ$

$\Delta G° = -2.30(8.31 \times 10^{-3})(298) \log_{10}(6.1 \times 10^{-13}) = +69.5\ kJ$

$\Delta G° = -2.30(8.31 \times 10^{-3})(298) \log_{10}(8.7 \times 10^{2}) = -16.7\ kJ$

15.29 a. forgot to divide by volume, 5.0 L
b. upside down
c. multiplied concentrations by coefficients in equation
d. added instead of multiplying concentrations

15.30 a. impossible; if CH_4 and H_2S are formed, some CS_2 and H_2 must be consumed
b. impossible; if conc. CH_4 drops by 1.0 mol/L, that of H_2 must drop by four times as much
c. possible; Δ conc. CS_2 = -0.2 M, Δconc. H_2 = -0.8 M, Δ conc. CH_4 = +0.2 M, Δconc. H_2S = +0.4 M
d. impossible; if conc. CS_2 increases by 0.25 M, that of CH_4 must drop by same amount

CHAPTER 16

BASIC SKILLS

1. Determine the order of a reaction, given the initial rate as a function of concentration of reactants.
2. For a first order reaction, calculate
   a. the concentration of reactant after a given time, knowing its original concentration and the rate constant.
   b. the time required for the concentration to drop by a given amount, knowing the rate constant.
3. Given either the half life or the rate constant for a first order reaction, calculate the other quantity.
4. Given two of the three quantities: ΔH, E_a, E'_a (activation energy for reverse reaction), calculate the other quantity.
5. Use the Arrhenius equation (Equation 16.16) to calculate any one of the five quantities: k_1, k_2, T_2, T_1, E_a, knowing or having calculated the other quantities.
6. Determine whether a proposed mechanism for a reaction is consistent with the observed rate expression.

LECTURE NOTES

This chapter, like the preceding one, is difficult for many students. Concepts such as the rate constant, rate expression, and reaction order are abstract; students have trouble relating chemical kinetics to the real world. Perhaps most difficult of all is the relation between reaction mechanism and order (Section 16.7). Here, students must distinguish carefully between

- the equation for one step in a mechanism and the equation for the overall reaction.
- unstable intermediates and major species (reactants, products).

You should expect to spend at least 2½ lectures on this chapter (a total of 5 lectures for Chapters 15 and 16). If your schedule permits, this could profitably be stretched to 3 lectures.

LECTURE 1

I Reaction Rate
   A. Concept: $\text{rate} = \dfrac{\text{change in concentration of a species}}{\text{time during which change in conc. takes place}}$

usually expressed in terms of reactant:

$$2\ N_2O_5(g) \longrightarrow 4\ NO_2(g) + O_2(g)$$

$$\text{rate} = \frac{-\Delta \text{conc. } N_2O_5}{\Delta t} = \frac{\text{initial conc. } N_2O_5 - \text{final conc. } N_2O_5}{\text{final time} - \text{initial time}}$$

Minus sign used to make rate a positive quantity. Concentration will be expressed in moles/liter; time may be in seconds, minutes, years, - -.

B. Dependence on concentration of reactant(s)

Single reactant:

$$\text{rate} = k(\text{conc. reactant})^m \ ; \ k = \text{rate constant},\ m = \text{order}$$

Two reactants:

$$\text{rate} = k(\text{conc. reactant 1})^m(\text{conc. reactant 2})^n$$
$$\text{overall order} = m + n$$

Generally, m and n are positive integers (1, 2, 3). However they can be 0 or a fraction such as $\frac{1}{2}$.

Determination of m and k from rate-concentration data:

$$CH_3CHO(g) \longrightarrow CH_4(g) + CO(g)$$

rate	2.0 M/s	0.50 M/s	0.080 M/s
conc. CH_3CHO	1.0 M	0.50 M	0.20 M

$$\frac{\text{rate 2}}{\text{rate 1}} = \left(\frac{\text{conc. 2}}{\text{conc. 1}}\right)^m; \quad \frac{2.0}{0.50} = \left(\frac{1.0}{0.50}\right)^m; \ 4 = 2^m \ ; \ m = 2$$

hence, rate = $k(\text{conc. } CH_3CHO)^2$

$$k = \frac{\text{rate}}{(\text{conc. } CH_3CHO)^2} = \frac{2.0\ M/s}{(1.0\ M)^2} = \frac{2.0}{M \cdot s}$$

C. Relation between concentration of reactant and time

First order reaction:

$$\log_{10}\frac{X_o}{X} = \frac{kt}{2.30} \ ; \ X_o = \text{original concentration}$$
$$X = \text{concentration at time } t$$

Suppose k = 0.250/s, X_o = 1.00 M. What is the concentration of reactant after 10.0 s?

$$\log_{10}\frac{X_o}{X} = \frac{(0.250)(10.0)}{2.30} = 1.09; \quad \frac{X_o}{X} = 10^{1.09} = 12.2$$

$$X = 1.00\ M/12.2 = 0.0819\ M$$

How long does it take for concentration to drop to one half of its original value?

$$X = X_o/2;\ X_o = 2X;\ X_o/X = 2$$

$$\log_{10} 2 = kt/2.30$$

$$t = 2.30(\log_{10}2)/k = \frac{0.693}{k} = \frac{0.693}{0.250/s} = 2.77\ s$$

Note that, for a first order reaction:

- $t_{\frac{1}{2}}$ is independent of original concentration. It takes as long for concentration to drop from 1.0 M to 0.50 M as it does for concentration to drop from 2.0 M to 1.0 M.

Suppose $t_{\frac{1}{2}}$ = 1 min:

conc.	2.00 M	1.00 M	0.50 M	0.25 M
t	0	1	2	3
fraction left	1	1/2	1/4	1/8

- $t_{\frac{1}{2}}$ is inversely related to k. If $t_{\frac{1}{2}}$ is small, k is large, and vice versa.

LECTURE 2

I Activation Energy

A. In order for reaction to occur upon collision, reactant molecules must possess a certain minimum energy, called the activation energy. Otherwise, the collision is elastic; nothing happens.

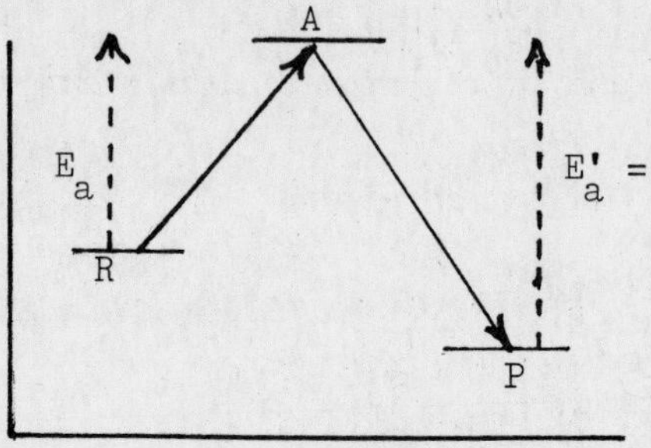

$$E_a - E'_a = \Delta H \text{ forward reaction}$$

exothermic: ΔH is −, $E_a < E'_a$; endothermic: ΔH is +, $E_a > E'_a$

B. Catalysis Catalyst increases reaction rate without being consumed in reaction. This happens because catalyst furnishes alternative path for reaction with lower activation energy.

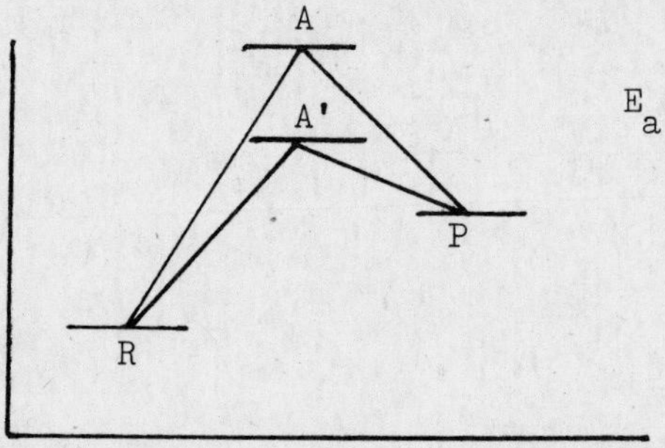

E_a is lowered, but ΔH remains the same

Example: $2\ H_2O_2(aq) \longrightarrow 2\ H_2O + O_2(g)$

Direct reaction comes about by collision between H_2O_2 molecules; E_a is high. Reaction is catalyzed by I^- ions:

$$H_2O_2(aq) + I^-(aq) \longrightarrow H_2O + IO^-(aq)$$
$$H_2O_2(aq) + IO^-(aq) \longrightarrow H_2O + O_2(g) + I^-(aq)$$
$$2\ H_2O_2(aq) \longrightarrow 2\ H_2O + O_2(g)$$

E_a much lower for two-step process

III Effect of Temperature upon Reaction Rate

In general, increase in T increases rate. Rate is approximately doubled when T increases by 10°C. Explanation:

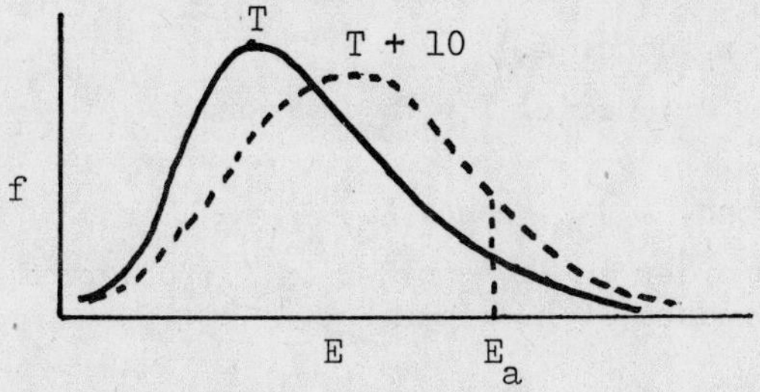

Relation between k and T:

$$\log_{10} k = A - \frac{E_a}{2.30RT} \qquad R = 8.31\ J/K \qquad E_a = \text{activation energy in joules}$$

Plot of $\log_{10}k$ vs 1/T is a straight line with a slope of $-E_a/2.30R$

$$E_a(\text{in J}) = -2.30(8.31)(\text{slope})$$

Two point equation:

$$\log_{10} \frac{k_2}{k_1} = \frac{E_a(T_2 - T_1)}{(2.30)(8.31)T_2T_1}$$

Suppose rate doubles when T increases from 25 to 35°C. Value of E_a?

$$\log_{10} \frac{k_2}{k_1} = 0.301 = \frac{E_a(10)}{(2.30)(8.31)(298)(308)}$$

$$E_a = 5.3 \times 10^4\ J = 53\ kJ$$

LECTURE $2\frac{1}{2}$

I Temperature Dependence of k (cont.)

$$\log_{10} \frac{k_2}{k_1} = \frac{E_a(T_2 - T_1)}{(2.30)(8.31)T_2T_1}$$

Suppose k = 0.0251/s at 0°C, E_a = 116 kJ. Calculate k at 100°C

$$\log_{10} \frac{k_2}{k_1} = \frac{(1.16 \times 10^5)(100)}{(2.30)(8.31)(273)(373)} = 5.96; \quad \frac{k_2}{k_1} = 9.1 \times 10^5$$

$$k_2 = (9.1 \times 10^5)(0.0251/s) = 2.3 \times 10^4/s$$

II Reaction Mechanism

Most reactions take place in a series of steps. This ordinarily results in a rather complex rate expression. To find the rate expression for a multi-step mechanism:

1. Focus on the slow step; assume rate for that step = overall rate.
2. Coefficients in rate determining step = order of reaction
3. Eliminate any unstable intermediates from rate expression

$X_2(g) \rightleftharpoons 2\,X(g)$ fast

$X(g) + A_2(g) \longrightarrow AX(g) + A(g)$ slow

$A(g) + X_2(g) \longrightarrow AX(g) + X(g)$ fast

rate = k(conc. X) x (conc. A_2)

but, $\frac{(\text{conc. }X)^2}{(\text{conc. }X_2)} = K_c$; conc. X = $K_c^{\frac{1}{2}}$ x (conc. X_2)$^{\frac{1}{2}}$

rate = k $K_c^{\frac{1}{2}}$ x (conc. A_2) x (conc. X_2)$^{\frac{1}{2}}$

Reaction is first order in A_2, $\frac{1}{2}$ order in X_2; observed rate constant is the product of k_2 times $K_c^{\frac{1}{2}}$.

DEMONSTRATIONS

1. Clock reaction: Test. Dem. 19, 85, 130, 147; J. Chem. Educ. 57 152 (1980)
2. Effect of conc. H^+ on reaction rate with Zn: Test. Dem. 84; J. Chem. Educ. 42 A607 (1965)
3. Catalysis: Test. Dem. 218; J. Chem. Educ. 55 652 (1978)
4. Effect of temperature on rate: J. Chem. Educ. 58 354 (1981)
5. Chain reaction of Cl_2 with CH_4: J. Chem. Educ. 60 597 (1983)

QUIZZES

Quiz 1

1. Given the following data for the reaction between A and B:

conc. A(M)	0.100	0.100	0.400
conc. B(M)	0.100	0.200	0.200
rate(M/min)	0.080	0.16	0.32

determine:

a. the order of reaction with respect to A
b. the order of reaction with respect to B
c. the rate constant for the reaction

Quiz 2

1. For a certain first order reaction, k = 0.052/h. Calculate:
   a. the rate when the concentration of reactant is 0.10 M.
   b. the half-life
   c. the time required for the concentration to drop to 80% of its original value.

Quiz 3

1. For a certain reaction, the activation energy is 62 kJ. Calculate:
   a. the activation energy for the reverse reaction if $\Delta H = -51$ kJ
   b. the ratio of the rate constant at 50°C to that at 0°C.
   c. the temperature at which k is 0.025/s, given that k = 0.050/s at 25°C.

Quiz 4

1. Consider the reaction mechanism:

$$A(g) + 2\ C(g) \rightleftharpoons AC_2(g) \qquad \text{fast}$$

$$AC_2(g) + A(g) \longrightarrow 2\ AC(g) \qquad \text{slow}$$

   Obtain a rate expression for the reaction of A with C to form AC (the expression should not contain the concentration of AC_2).

2. A certain reaction is zero order in reactant X and second order in reactant Y. If the concentrations of both reactants are doubled, what happens to the reaction rate?

Quiz 5

1. The rate constant, k, for a certain first order reaction is 0.12/min at 20°C.
   a. How long does it take for the concentration of reactant to drop to one fourth of its original value?
   b. What is k at 30°C if the activation energy is 29 kJ?

PROBLEMS 1-30

16.1 a. $\dfrac{-(4.0\ \text{mol/L} - 2.5\ \text{mol/L})}{10.0\ \text{s}} = 0.15\ \dfrac{\text{mol}}{\text{L·s}} \times \dfrac{60\ \text{s}}{1\ \text{min}} = 9.0\ \dfrac{\text{mol}}{\text{L·min}}$

b. 2(0.15 mol/L·s) = 0.30 mol/L·s

16.2 a. rate = $\dfrac{\Delta \text{conc. } H_2}{\Delta t}$ b. rate = $-\frac{1}{2}\dfrac{\Delta \text{conc. HI}}{\Delta t}$

16.3 a. $\dfrac{-(0.078\ \text{mol/L} - 0.126\ \text{mol/L})}{4\ \text{min}} = 0.012$ mol/L·min

b. 0.012 mol/L·min

16.4

conc. D	k	rate
0.60	0.050	0.030
0.040	7.0×10^{1}	2.8
0.50	0.17	0.085

16.5 a. rate = $k(\text{conc. } NO_2)^2$

b. $k = \dfrac{\text{rate}}{(\text{conc. } NO_2)^2} = \dfrac{2.0 \times 10^{-3}\ \text{mol/L·s}}{(0.080\ \text{mol/L})^2} = 0.31$ L/mol·s

c. liters/mole·second

d. rate = $0.31 \dfrac{\text{L}}{\text{mol·s}}(0.020\ \text{mol/L})^2 = 1.2 \times 10^{-4}$ mol/L·s

16.6 rate = $0.50 \frac{L}{mol \cdot s}$ (conc. CO)(conc. NO_2)

a. conc. CO = $\frac{rate}{k(conc.\ NO_2)} = \frac{0.10}{(0.50)(0.40)}$ = 0.50 mol/L

b. $(conc.\ CO)^2 = \frac{rate}{k} = \frac{0.10}{0.50} = 0.20$; conc. CO = 0.45 mol/L

16.7 a. rate = $k(conc.\ E)^2$

b. $k = \frac{rate}{(conc.\ E)^2} = \frac{0.080\ mol/L \cdot s}{(0.25\ mol/L)^2} = 1.3\ L/mol \cdot s$

c. $\sqrt{2} \times 0.25$ mol/L = 0.35 mol/L

16.8 a. increases rate by factor of 4 b. increases rate by factor of 2 c. increases rate by factor of 3 d. decreases rate

16.9 $\frac{rate_2}{rate_1} = \left(\frac{conc.\ A_2}{conc.\ A_1}\right)^m = 0.40^m$

a. $0.40 = 0.40^m$; m = 1 b. $1.0 = 0.40^m$; m = 0

c. $0.16 = 0.40^m$; m = 2

16.10 $\frac{rate_2}{rate_1} = \left(\frac{conc.\ A_2}{conc.\ A_1}\right)^m$; $\frac{0.016}{0.020} = \left(\frac{0.090}{0.100}\right)^m$; $0.80 = 0.90^m$; m = 2

16.11 a. compare rates when conc. D is constant

$\frac{rate_2}{rate_1} = \left(\frac{conc.\ A_2}{conc.\ A_1}\right)^m$; $\frac{1.0 \times 10^{-2}}{4.0 \times 10^{-4}} = \left(\frac{0.50}{0.10}\right)^m$

$25 = 5.0^m$; m = 2

b. compare rates when conc. A is constant

$\frac{1.0 \times 10^{-2}}{1.0 \times 10^{-2}} = \left(\frac{0.50}{0.10}\right)^n$; $1.0 = 5^n$; n = 0

c. rate = $k(conc.\ A)^2$

$k = \frac{rate}{(conc.\ A)^2} = \frac{4.0 \times 10^{-4}\ mol/L \cdot s}{(0.10\ mol/L)^2} = 4.0 \times 10^{-2} \frac{L}{mol \cdot s}$

d. rate = $4.0 \times 10^{-2} \frac{L}{mol \cdot s} (0.25\ mol/L)^2 = 2.5 \times 10^{-3} \frac{mol}{L \cdot s}$

16.12 a. compare rates when conc. H_2O_2 is constant

$\frac{6.6 \times 10^{-5}}{3.3 \times 10^{-5}} = \left(\frac{0.040}{0.020}\right)^m$; $2.0 = 2.0^m$; m = 1

b. $\frac{1.3 \times 10^{-4}}{6.6 \times 10^{-5}} = \left(\frac{0.040}{0.020}\right)^n$; $2.0 = 2.0^n$; n = 1

c. rate = k(conc. I^-)(conc. H_2O_2)

$k = \frac{3.3 \times 10^{-5}}{(0.020)(0.020)}$ L/mol·s = 0.082 L/mol·s

rate = $0.082 \frac{L}{mol \cdot s} (0.010 \frac{mol}{L})(0.030 \frac{mol}{L}) = 2.5 \times 10^{-5} \frac{mol}{L \cdot s}$

16.13

t	conc.	log conc.	1/conc.
0	0.400	-0.398	2.50
50	0.333	-0.478	3.00
100	0.286	-0.543	3.50
150	0.250	-0.602	4.00
200	0.222	-0.654	4.50

plot of 1/conc. vs t is straight line; 2nd order

16.14 a.

t	0	1	2	3	4	8	16
conc.	0.200	0.197	0.193	0.190	0.187	0.175	0.153
log conc.	-0.699	-0.706	-0.714	-0.721	-0.728	-0.757	-0.815

plot of log conc. vs t is straight line

b. $k = -2.30(\text{slope}) = -\frac{2.30(-0.116)}{16} = 0.017/\text{min}$

c. $\log \frac{X_o}{X} = \frac{0.017t}{2.30}$; $t = \frac{2.30}{0.017} \log \frac{X_o}{X} = \frac{2.30}{0.017} \log 4 = 81$ min

16.15 a. $k = 0.693/650 \text{ s} = 1.07 \times 10^{-3}/\text{s}$

b. $\log_{10} \frac{0.050}{0.0125} = \frac{(1.07 \times 10^{-3})t}{2.30} = 0.602$; $t = 1.30 \times 10^{3}$ s

c. $\log \frac{0.050}{X} = \frac{(1.07 \times 10^{-3})(4.90 \times 10^{3})}{2.30} = 2.28$

$\frac{0.050}{X} = 190$; $X = 2.6 \times 10^{-4}$ M

16.16 a. $\log_{10} \frac{0.200}{X} = \frac{(2.50 \times 10^{-3})(60)}{2.30} = 0.0652$

$0.200/X = 1.16$; $X = 0.172$ mol/L

$\log_{10} \frac{0.200}{X} = \frac{(2.50 \times 10^{-3})(24)(60)}{2.30} = 1.57$

$0.200/X = 37$; $X = 0.0054$ mol/L

b. $t_{\frac{1}{2}} = \frac{0.693}{k} = \frac{0.693}{2.50 \times 10^{-3}} = 277$ min

16.17 $\log_{10} \frac{1.00}{0.700} = \frac{k(13.2)}{2.30} = 0.155$; $k = 2.70 \times 10^{-2}/\text{min}$

$t_{\frac{1}{2}} = \frac{0.693}{k} = \frac{0.693}{0.0270} = 25.7$ min

16.18 a. System 3 b. -20 kJ c. System 2

16.19

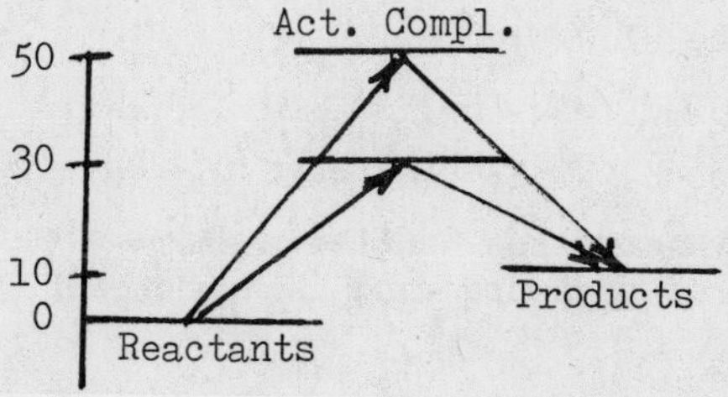

16.20 $\Delta H = +52\text{ kJ} - 62\text{ kJ} = -10\text{ kJ}$

$E_a' = 163\text{ kJ} - (-10\text{ kJ}) = 173\text{ kJ}$

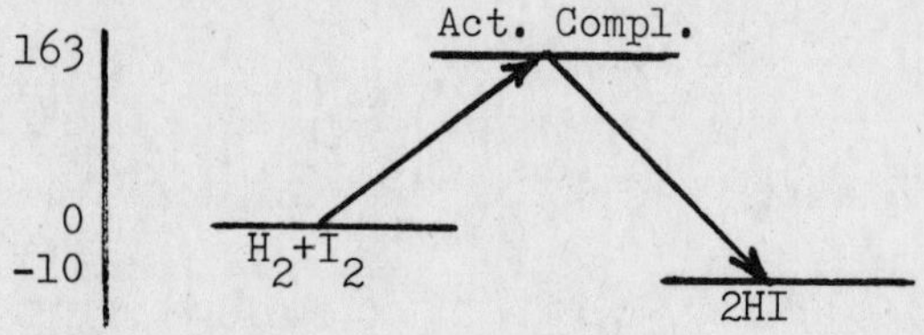

16.21

log k	-1.98	-1.00	-0.22	0.46
1/T	0.00143	0.00133	0.00125	0.00118

slope $\approx -9.8 \times 10^3$; $E_a = (-2.30)(8.31)(-9.8 \times 10^3)\text{J}$
$= 1.9 \times 10^2\text{ kJ}$

16.22 $\log_{10} \frac{43}{1.4} = \frac{E_a(50)}{(2.30(8.31)(400)(450)}$; $E_a = 1.0 \times 10^2\text{ kJ}$

16.23 $\log_{10} \frac{k_2}{k_1} = \frac{(8.2 \times 10^4)(200)}{(2.30)(8.31)(300)(500)} = 5.72$; $\frac{k_2}{k_1} = 5.3 \times 10^5$

$k_2 = (5.3 \times 10^5)(1.2 \times 10^{-2}\text{ L/mol·s}) = 6.4 \times 10^3\text{ L/mol·s}$

16.24 $\log_{10} \frac{k_2}{k_1} = \frac{6.5 \times 10^4(10)}{(2.30)(8.31)(308)(298)} = 0.370$; $\frac{k_2}{k_1} = 2.34$

$k_1 = k_2/2.34 = 0.427\ k_2$; reduced by 57.3%

16.25 a. $f = e^0 = 1.00$ b. $f = e^{-10000/2493} = e^{-4.01} = 0.0181$

c. $f = e^{-8.02} = 3.29 \times 10^{-4}$

16.26 rate = k(conc. Br)(conc. H_2)

$\frac{(\text{conc. Br})^2}{\text{conc. }Br_2} = K_c$; conc. Br $= K_c^{\frac{1}{2}}(\text{conc. }Br_2)^{\frac{1}{2}}$

rate = k(conc. H_2)(conc. $Br_2)^{\frac{1}{2}} \times K_c^{\frac{1}{2}}$

16.27 a. rate = k(conc. CO)(conc. NO_2); not consistent

b. rate = k(conc. N_2O_4)(conc. $CO)^2$

= kK_c(conc. $NO_2)^2$(conc. $CO)^2$; not consistent

c. rate = k(conc. $NO_2)^2$; OK

d. rate = k(conc. $NO_2)^2$; OK

16.28 step 2 must be the slow step

rate = k(conc. N_2O_2)(c onc. H_2)

conc. $N_2O_2 = K_c$(conc. $NO)^2$; rate = kK_c(conc. $NO)^2$(conc. H_2)

16.29 a. rate ordinarily decreases as time passes

b. rate increases as pressure (and hence concentration) increases

c. rate increases

16.30 a. fewer molecules have enough energy to react when they collide
b. would be true only for first order reaction
c. flame supplies activation energy for spontaneous reaction
d. does not change ΔH or ΔG

CHAPTER 17

BASIC SKILLS

1. Discuss the methods of preparing NH_3, HNO_3, and H_2SO_4 from components of the atmosphere; indicate how equiibrium and rate principles are applied in these preparations.
2. Explain what is meant by relative humidity; by the greenhouse effect.
3. Describe how thermodynamic and kinetic principles apply to the formation and decomposition of species in the upper atmosphere, including ozone.
4. Discuss the formation, effects, and methods of removal of the major chemical air pollutants (SO_2, SO_3, CO, NO, NO_2).

LECTURE NOTES

This chapter, like Chapter 13, is descriptive. It illustrates how the principles of chemical thermodynamics, equilibrium, and kinetics apply to processes involving components of the atmosphere. We encourage you to cover this chapter for a couple of reasons. For one thing, students find it interesting; many of the topics covered are ones that they have heard about on TV and in the newspapers. Moreover, after three tough, principles-oriented chapters, it's time for review.

The amount of time devoted to this chapter depends upon what topics you wish to cover. We suggest two lectures but this could be reduced to $1\frac{1}{2}$ or even to one lecture with selective omissions.

LECTURE 1

I Components of the Atmosphere

Substance	N_2	O_2	Ar	H_2O	CO_2
Mole Fraction	0.781	0.210	0.0093	varies	0.00034

Concentrations of minor components are often expressed in parts per million:

$$\text{ppm} = 10^6\ X; \quad \text{ppm } CO_2 = 340$$

II Preparation of Industrial Chemicals

A. NH_3 $\quad N_2(g) + 3\ H_2(g) \rightleftharpoons 2\ NH_3(g)\ ;\ \Delta H = -92.4\text{ kJ}$

increase in pressure shifts equilibrium to right (4 mol gas $\rightarrow$ 2 mol gas). Also increases rate (higher concentration)

increase in temperature shifts equilibrium to left (reverse reaction endothermic). Also increases rate

Haber process: P = 200-600 atm, T = 400-450°C, solid catalyst used to speed up reaction

B. HNO_3 Made in three steps from NH_3 by Ostwald process

$$4\ NH_3(g) + 5\ O_2(g) \rightarrow 4\ NO(g) + 6\ H_2O(g)$$

$$2\ NO(g) + O_2(g) \rightarrow 2\ NO_2(g)$$

$$3\ NO_2(g) + H_2O(l) \rightarrow NO(g) + 2\ HNO_3(aq)$$

Note that rate of second step actually decreases when T increases. This comes about because it occurs by two-step mechanism:

(1) $NO(g) + NO(g) \rightleftharpoons N_2O_2(g)$ fast

(2) $N_2O_2(g) + O_2(g) \rightarrow 2\ NO_2(g)$ slow

rate = $k_2K_1(\text{conc. } O_2)(\text{conc. } NO)^2$

as T increases, K_1 decreases faster than k_2 increases

C. H_2SO_4 Made by three-step process from sulfur

$$S(s) + O_2(g) \rightarrow SO_2(g)$$

$$2\ SO_2(g) + O_2(g) \rightleftharpoons 2\ SO_3(g)$$

$$SO_3(g) + H_2O \rightarrow H_2SO_4(aq)$$

Second step is most difficult one; use Pt or V_2O_5 catalyst

Thermodynamics of second step:

$\Delta H = 2\ \Delta H_f\ SO_3(g) - 2\ \Delta H_f\ SO_2(g) = -198.2$ kJ

$S^o = 2\ S^o\ SO_3(g) - 2\ S^o\ SO_2(g) - S^o\ O_2(g) = -189.6$ J/K

$\Delta G^o = -198.2 + 0.1896\ T$; spontaneous below about 1050 K

III <u>Water and Carbon Dioxide</u>

A. <u>Relative humidity</u>

$$R.H. = 100 \times \frac{P\ H_2O}{P^oH_2O}$$

$P\ H_2O$ = pressure water vapor in air

P^oH_2O = equilibrium vapor pressure H_2O

Suppose R.H. is 85.0% on day when t = 25°C (P^o = 23.8 mm Hg)

$$P\ H_2O = 0.850(23.8 \text{ mm Hg}) = 20.2 \text{ mm Hg}$$

Cool to 20°C, where P^o = 17.5 mm Hg

$$R.H. = \frac{20.2}{17.5} \times 100 > 100\%; \text{ condensation}$$

Dew point = t at which condensation occurs on cooling (about 22°C here)

B. <u>Greenhouse Effect</u>

T of earth is determined by an equilibrium between:

- energy from sun (ultraviolet, visible, infrared)
- energy radiated by earth (infrared)

Water vapor and carbon dioxide in atmosphere absorb IR radiation, prevent it from escaping into space. In absence of this effect, average t would be -25°C instead of 15°C.

CO_2 concentration is increasing because of combustion of fossil fuels (forms CO_2) and reduction in vegetation (photosynthesis consumes CO_2). Could increase to 400 ppm by 2000 A.D., causing 1°C increase in t.

LECTURE 2

I Upper Atmosphere

A. Components present: atoms, cations

$$O_2(g) \longrightarrow 2\,O(g) \qquad \Delta E = 494 \text{ kJ/mol}; \ \lambda = 242 \text{ nm}$$

$$O_2(g) \longrightarrow O_2^+(g) + e^- \quad \Delta E = 1109 \text{ kJ/mol}; \ \lambda = 108 \text{ nm}$$

$$\lambda = \frac{1.196 \times 10^5}{\Delta E}$$

UV radiation available only at high altitudes; O_3 in air absorbs UV, prevents it from reaching earth

B. Ozone

Mechanism of formation: $O_2(g) \longrightarrow 2\,O(g)$

$$O_2(g) + O(g) \longrightarrow O_3(g)$$

Mechanism of decomposition: $O_3(g) + O\,(g) \longrightarrow 2\,O_2(g)$

Equilibrium reached to produce about 10 ppm O_3 at 30 km. Can be disturbed by presence of Cl atoms:

$$\begin{array}{l} Cl(g) + O_3(g) \longrightarrow ClO(g) + O_2(g) \\ ClO(g) + O(g) \longrightarrow Cl(g) + O_2(g) \\ \hline O_3(g) + O(g) \longrightarrow 2\,O_2(g) \end{array}$$

Activation energy less than for direct decomposition; Cl atoms come from decomposition of aerosol propellants such as $CFCl_3$ and CF_2Cl_2.

II Air Pollution

A. Oxides of Sulfur (SO_2, SO_3)

Main source is burning of sulfur-containing coal. SO_2 is slowly converted to SO_3 and then to H_2SO_4, which is the major source of acid rain in the eastern U.S. and Canada. Sulfur dioxide can be removed from stack gases by adding $CaCO_3$:

$$CaCO_3(s) + SO_2(g) + \tfrac{1}{2}\,O_2(g) \longrightarrow CO_2(g) + CaSO_4(s)$$

B. Carbon Monoxide

Formed mostly by incomplete combustion of hydrocarbons:

$$C_8H_{18}(l) + \frac{17}{2}\,O_2(g) \longrightarrow 8\,CO(g) + 9\,H_2O(l)$$

CO forms complex with hemoglobin more stable than that with O_2

$$Hem{\cdot}O_2(aq) + CO(g) \rightleftharpoons O_2(g) + Hem{\cdot}CO(aq); \quad K = 210$$

Symptoms of CO poisoning show up when 10% of hemoglobin is in the form of the CO complex. Ratio of $[CO]/[O_2]$ at that point?

$$210 = \frac{[O_2] \times [Hem{\cdot}CO]}{[CO] \times [Hem{\cdot}O_2]}; \quad \frac{[CO]}{[O_2]} = \frac{[Hem{\cdot}CO]}{210 \times [Hem{\cdot}O_2]}$$

$$= \frac{10}{210 \times 90} = 5.3 \times 10^{-4}$$

$X\ O_2 = 0.21$; $X\ CO = 5.3 \times 10^{-4} \times 0.21 = 1.1 \times 10^{-4} = 110$ ppm

C. Oxides of Nitrogen (NO, NO_2)

$$N_2(g) + O_2(g) \rightleftharpoons 2\ NO(g); \quad \Delta H = +180.8 \text{ kJ}$$

NO formed in high temperature combustion of fuels (K_c increases with T). NO slowly converted to NO_2 in air.

Removed by catalytic converter in automobile (Pt-Rh catalyst)

$$CO(g) + NO(g) \longrightarrow CO_2(g) + \tfrac{1}{2} N_2(g)$$

DEMONSTRATIONS

1. Properties of liquid air: Test. Dem. 24, 92
2. Ostwald process: Test. Dem. 38, 99, 169
3. Contact process for H_2SO_4: Test. Dem. 42
4. Preparation of ozone: Test. Dem. 57, 127

QUIZZES

Quiz 1

1. Consider the equilibrium:

   $$N_2(g) + 3\ H_2(g) \rightleftharpoons 2\ NH_3(g)$$

   1.20 mol N_2 and 2.60 mol H_2 are placed in a 3.00 L container. When equilibrium is reached, 1.00 mol of NH_3 is present. Calculate K_c.
2. Explain briefly what is meant by the greenhouse effect.

Quiz 2

1. Consider the oxidation of carbon monoxide:

   $$CO(g) + \tfrac{1}{2} O_2(g) \longrightarrow CO_2(g)$$

   Using data in Tables 14.1 and 14.3, calculate the temperature range over which this reaction is spontaneous at 1 atm.
2. Explain why CO is a dangerous air pollutant.

Quiz 3

1. For the upper atmosphere reaction

   $O(g) + H_2O(g) \longrightarrow 2\ OH(g)$

   the activation energy is 77.4 kJ. Calculate the ratio of the rate constant at 25°C to that at -25°C.

2. Explain what is meant by "acid rain" and what causes it.

Quiz 4

1. Suppose the mechanism for the reaction of NO with O_2 is:

   $NO(g) + NO(g) \rightleftharpoons N_2O_2(g)$ fast

   $N_2O_2(g) + O_2(g) \longrightarrow NO_2(g) + NO_2(g)$ slow

   Derive the rate expression in terms of the concentrations of NO and O_2.

2. Explain how the air pollutant NO is formed and how it is removed from automobile exhaust.

Quiz 5

1. Consider the equilibrium:

   $N_2(g) + O_2(g) \rightleftharpoons 2\ NO(g)$

   At 2000°C, K_c for this reaction is 0.10. If the original concentrations of N_2 and O_2 are 0.20 M, what is the equilibrium concentration of NO at 2000°C?

2. Explain what is meant by the "ozone layer" and what effect it has on our health.

PROBLEMS 1-30

17.1 a. $N_2(g) + 3\ H_2(g) \longrightarrow 2\ NH_3(g)$

b. $S(s) + O_2(g) \longrightarrow SO_2(g)$

$SO_2(g) + \frac{1}{2}\ O_2(g) \longrightarrow SO_3(g)$

$SO_3(g) + H_2O \longrightarrow H_2SO_4(aq)$

c. $2\ NH_3(g) + H_2SO_4(l) \longrightarrow (NH_4)_2SO_4(s)$

17.2 increase in P increases both rate and yield
increase in T increases rate, decreases yield

17.3 a. $P = (R.H./100)P^o = 0.64(31.8\text{ mm Hg}) = 20\text{ mm Hg}$

b. $R.H. = 100\ P/P^o = 100 \times 20/23.8 = 84\%$

17.4 a. $P = \dfrac{nRT}{V} = \dfrac{(0.540/18.02\text{ mol})(0.0821\text{ L·atm/mol·K})(303\text{ K})}{30.0\text{ L}}$

$= 0.0248\text{ atm} \times \dfrac{760\text{ mm Hg}}{1\text{ atm}} = 18.8\text{ mm Hg}$

b. $R.H. = 100 \times 18.8/31.8 = 59.1\%$

17.5 $\lambda_{max} = \dfrac{1.196 \times 10^5}{331}\text{ nm} = 361\text{ nm}$

17.6 $\Delta E = \dfrac{1.196 \times 10^5}{242} = 494$ kJ/mol = B.E.

17.7 a. false; price of H_2 has increased
b. true
c. false; absorbs IR radiation given off by earth

17.8 a. could deplete ozone layer
b. promotes smog formation
c. reduces visibility, blackens buildings
d. possible lead poisoning

17.9 a. absorption of IR radiation by CO_2 and H_2O in air
b. particles far apart, moving rapidly
c. produced by catalytic oxidation of SO_2

17.10 NO and hydrocarbons are produced in early morning, as automobile traffic builds up. Later, NO is converted to NO_2 and O_3 is produced.

17.11 a. $\Delta G^\circ = \Delta H = +78.5$ kJ

b. $\Delta G^\circ = \Delta H - T\Delta S^\circ = +78.5 \text{ kJ} - 300(0.0176)\text{kJ} = +73.2$ kJ

c. $\Delta G^\circ = +78.5 \text{ kJ} - 1000(0.0176)\text{kJ} = +60.9$ kJ

17.12 a. ΔS° should be negative (decrease in number moles gas)

$\Delta S^\circ = 2\ S^\circ\ NH_3(g) - S^\circ\ N_2(g) - 3\ S^\circ\ H_2(g)$

$= 2(192.5 \text{ J/K}) - 191.5 \text{ J/K} - 3(130.6 \text{ J/K}) = -198.3$ J/K

b. ΔS° should be negative (decrease in number moles gas)

$\Delta S^\circ = S^\circ\ NH_4Cl(s) - S^\circ\ NH_3(g) - S^\circ\ HCl(g)$

$= 94.6 \text{ J/K} - 192.5 \text{ J/K} - 186.7 \text{ J/K} = -284.6$ J/K

c. ΔS° should be negative (decrease in number moles gas)

$\Delta S^\circ = S^\circ\ SO_3(g) - S^\circ\ SO_2(g) - \frac{1}{2}\ S^\circ\ O_2(g)$

$= 256.2 \text{ J/K} - 248.5 \text{ J/K} - 102.5 \text{ J/K} = -94.8$ J/K

d. ΔS° should be positive; increase in number moles gas

$\Delta S^\circ = 2\ S^\circ\ NO(g) + S^\circ\ O_2(g) - 2\ S^\circ\ NO_2(g)$

$= 2(210.6 \text{ J/K}) + 205.0 \text{ J/K} - 481.0 \text{ J/K} = +145.2$ J/K

17.13 a. $\Delta H = \Delta H_f\ NH_4NO_3(s) - \Delta H_f\ NH_3(g) - \Delta H_f\ HNO_3(l)$

$= -365.1 \text{ kJ} + 46.2 \text{ kJ} + 173.2 \text{ kJ} = -145.7$ kJ

b. $\Delta S^\circ = S^\circ\ NH_4NO_3(s) - S^\circ\ NH_3(g) - S^\circ\ HNO_3(l)$

$= 151.0 \text{ J/K} - 192.5 \text{ J/K} - 155.6 \text{ J/K} = -197.1$ J/K

c. $\Delta G^\circ = -145.7 \text{ kJ} + 298(0.1971)\text{kJ} = -87.0$ kJ

17.14 a. $\Delta H = \Delta H_f CO_2(g) + 2\ \Delta H_f H_2O(g) - \Delta H_f CH_4(g) - 4\Delta H_f NO(g)$

$= -393.5 \text{ kJ} - 483.6 \text{ kJ} + 74.8 \text{ kJ} - 361.6 \text{ kJ} = -1163.9$ kJ

b. $\Delta S^\circ = 2S^\circ N_2(g) + S^\circ CO_2(g) + 2S^\circ H_2O(g) - S^\circ CH_4(g) - 4S^\circ NO(g)$

$= 2(191.5 \text{ J/K}) + 213.6 \text{ J/K} + 2(188.7 \text{ J/K}) - 186.2$ J/K

$- 4(210.6 \text{ J/K}) = -54.6$ J/K

17.14 c. $\Delta G^\circ = 0$ when $T = \frac{1163.9}{0.0546} = 2.13 \times 10^4$ K

reaction is spontaneous at all reasonable temperatures

17.15 $\Delta H = \Delta H_f\ MgSO_4(s) - \Delta H_f\ MgO(s) - \Delta H_f\ SO_2(g)$

$= -1278.2$ kJ + 601.8 kJ + 296.1 kJ = -380.3 kJ

$\Delta S^\circ = S^\circ\ MgSO_4(s) - S^\circ\ MgO(s) - S^\circ\ SO_2(g) - \frac{1}{2} S^\circ\ O_2(g)$

= 91.6 J/K − 26.8 J/K − 248.5 J/K − 102.5 J/K = −286.2 J/K

$\Delta G^\circ = 0$ when $T = \frac{380.3}{0.2862} = 1329$ K = 1056°C

spontaneous below 1056°C

17.16 a. increase in T: ← increase in P has no effect
b. increase in T: → increase in P: ←
c. increase in T: ← increase in P: →
d. increase in T: → increase in P has no effect

17.17 a. $\frac{[\text{Hem}\cdot\text{CO}]}{[\text{Hem}\cdot\text{O}_2]} \times \frac{[\text{O}_2]}{[\text{CO}]} = 210$; $\frac{[\text{Hem}\cdot\text{CO}]}{[\text{Hem}\cdot\text{O}_2]} = 210 \times \frac{[\text{CO}]}{[\text{O}_2]} = 210$

b. $\frac{[\text{Hem}\cdot\text{CO}]}{[\text{Hem}\cdot\text{O}_2]} = 210 \times 0.15 = 32$

17.18 a. 0.47 mol NH_3, 0.53/4 = 0.13 mol N_2, 0.40 mol H_2

b. $V = \frac{(1.00\text{ mol})(0.0821\text{ L}\cdot\text{atm/mol}\cdot\text{K})(673\text{ K})}{300\text{ atm}} = 0.184$ L

c. $[NH_3]$ = 0.47 mol/0.184 L = 2.6 mol/L

$[N_2]$ = 0.13 mol/0.184 L = 0.71 mol/L

$[H_2]$ = 0.40 mol/0.184 L = 2.2 mol/L

d. $K_c = \frac{[NH_3]^2}{[N_2] \times [H_2]^3} = \frac{(2.6)^2}{(0.71)(2.2)^3} = 0.89$

17.19 a. $K_p = K_c(RT)^{-\frac{1}{2}} = \frac{20}{(0.0821 \times 973)^{\frac{1}{2}}} = 2.2$

b. $\Delta G^\circ = -(2.30)(8.31 \times 10^{-3})(973) \log_{10} 2.2 = -6.4$ kJ

17.20 a. $\log_{10} K_p = \frac{-206.7}{(2.30)(8.31 \times 10^{-3})(298)} = -36.3$; $K_p = 5 \times 10^{-37}$

b. $K_p = K_c(RT)^2$

$K_c = K_p/(RT)^2 = \frac{5 \times 10^{-37}}{(0.0821 \times 298)^2} = 8 \times 10^{-40}$

17.21 a. $k = \frac{\text{rate}}{(\text{conc. NO})(\text{conc. } N_2O)} = \frac{1.92 \times 10^{-15}\text{ mol/L}\cdot\text{s}}{(5.0 \times 10^{-8}\text{ mol/L})^2}$

$= 0.77$ L/mol·s

b. $\text{conc. } N_2O = \frac{\text{rate}}{k(\text{conc. NO})} = \frac{7.4 \times 10^{-20}\text{ mol/L}\cdot\text{s}}{(0.77\text{ L/mol}\cdot\text{s})(2.0 \times 10^{-8}\text{ mol/L})}$

$= 4.8 \times 10^{-12}$ mol/L

17.22 rate = $1.2 \times 10^7 \ \frac{L}{mol \cdot s}(2 \times 10^{-8} \ mol/L)^2 = 5 \times 10^{-9}$ mol/L·s

rapid; $\frac{1}{4}$ of concentration used up in one second

17.23 rate = $5.0 \times 10^6 \ \frac{L}{mol \cdot s} \times 3.0 \times 10^{-8} \ \frac{mol}{L} \times 1.2 \times 10^{-14} \ \frac{mol}{L}$

$= 1.8 \times 10^{-15}$ mol/L·s

17.24 catalyzed reaction: rate = $5.4 \times 10^9 \frac{L}{mol \cdot s}$ x conc. NO_2 x conc0

direct reaction: rate = $5.0 \times 10^6 \ \frac{L}{mol \cdot s}$ x conc. O_3 x conc. O

ratio = $1.1 \times 10^3 \times \frac{\text{conc. } NO_2}{\text{conc. } O_3} = \frac{1.1 \times 10^3}{2000} = 0.55$

17.25 $t_{\frac{1}{2}} = \frac{0.693 \ s}{3 \times 10^{-26}} \times \frac{1 \ hr}{3.6 \times 10^3 \ s} \times \frac{1 \ d}{24 \ h} \times \frac{1 \ yr}{365 \ d} = 7 \times 10^{17}$ yr

not likely

17.26 compare 1st and 2nd points:

$\frac{3.2 \times 10^{-9}}{1.6 \times 10^{-9}} = \left(\frac{4.0 \times 10^{-6}}{2.0 \times 10^{-6}}\right)^m$; $2.0 = 2.0^m$; m = 1

compare 2nd and 3rd points:

$\frac{1.6 \times 10^{-9}}{0.90 \times 10^{-9}} = \left(\frac{4.0 \times 10^{-8}}{3.0 \times 10^{-8}}\right)^m$; $1.8 = 1.3^n$; n = 2

rate = k(conc. CO)(conc. NO)^2

$k = \frac{\text{rate}}{(\text{conc. CO})(\text{conc. NO})^2} = \frac{3.2 \times 10^{-9} \ mol/L \cdot min}{(4.0 \times 10^{-6} \ M)(4.0 \times 10^{-8} \ M)^2}$

$= 5.0 \times 10^{11} \ L^2/mol^2 \cdot min$

17.27 $\Delta H = \Delta H_f \ O(g) - \Delta H_f \ H(g) - \Delta H_f \ OH(g)$

$= 248 \ kJ - 218 \ kJ - 40 \ kJ = -10 \ kJ$

$E_a' = E_a - \Delta H = 39$ kJ

17.28 $\log_{10} \frac{2.8 \times 10^{-3}}{k_1} = \frac{7.7 \times 10^4(300 - 248)}{(2.30)(8.31)(300)(248)} = 2.8$

$2.8 \times 10^{-3}/k_1 = 650$; $k_1 = 4.3 \times 10^{-6}$ L/mol·s

17.29 rate = k(conc. N_2O_2)(conc. O_2) ; $\frac{\text{conc. } N_2O_2}{(\text{conc. NO})^2} = K_c$

rate = kK_c(conc. NO)^2(conc. O_2)

17.30 a. $2 \ NO_2(g) + F_2(g) \longrightarrow 2 \ NO_2F_2(g)$

b. rate = k(conc. N_2O_4)(conc. F_2)

$= kK_c$(conc. NO_2)^2(conc. F_2)

CHAPTER 18

BASIC SKILLS

1. Using the solubility rules, predict whether a precipitate will form when two electrolyte solutions are mixed and write a net ionic equation for any reaction that occurs.
2. Using the chemical equation for a precipitation reaction, relate the amounts of two different reactants.
3. Given the formula of a slightly soluble ionic compound, write its K_{sp} expression.
4. Use the value of K_{sp} to:
   a. determine the concentration of an ion in solution, given that of the other ion in equilibrium with it.
   b. decide whether or not a precipitate will form when two solutions are mixed.
   c. calculate the solubility of an electrolyte in pure water or in a solution containing a common ion.
5. Given the solubility of an electrolyte in pure water, calculate K_{sp}.
6. Explain how hard water can be softened, using a zeolite.

LECTURE NOTES

Students generally find this material straightforward. They have fewer difficulties with solubility product calculations than with any other type of equilibrium constant. Two lectures should suffice for this chapter.

LECTURE 1

I Precipitation Reactions

A. Solubility Rules (Table 18.1, p. 548)

Use in predicting results of precipitation reactions:

1. Mix solutions of $Ba(NO_3)_2$ + Na_2CO_3. What happens?

   Ions present: Ba^{2+}, NO_3^-; Na^+, CO_3^{2-}

   Possible products: $BaCO_3$, $NaNO_3$

   According to solubility rules, $BaCO_3$ is insoluble:

   $$Ba^{2+}(aq) + CO_3^{2-}(aq) \longrightarrow BaCO_3(s)$$

2. Mix solutions of $BaCl_2$, NaOH

Ions present: Ba^{2+}, Cl^-; Na^+, OH^-

Possible precipitates: $Ba(OH)_2$, NaCl

both are soluble; no reaction

B. Stoichiometry

1. $Cu^{2+}(aq) + 2\ OH^-(aq) \longrightarrow Cu(OH)_2(s)$

What volume of 0.200 M $CuSO_4$ solution is required to react with 50.0 mL of 0.100 M NaOH?

$$\text{no. moles } OH^- = 0.0500\ L \times \frac{0.100\ \text{mol NaOH}}{1\ L} \times \frac{1\ \text{mol } OH^-}{1\ \text{mol NaOH}}$$

$$= 0.00500\ \text{mol } OH^-$$

$$\text{no. moles } Cu^{2+} = 0.00500\ \text{mol } OH^- \times \frac{1\ \text{mol } Cu^{2+}}{2\ \text{mol } OH^-} = 0.00250\ \text{mol } Cu^{2+}$$

$$\text{volume} = 0.00250\ \text{mol } Cu^{2+} \times \frac{1\ \text{mol } CuSO_4}{1\ \text{mol } Cu^{2+}} \times \frac{1\ L}{0.200\ \text{mol } CuSO_4}$$

$$= 0.0125\ L = 12.5\ mL$$

2. $2\ Ag^+(aq) + CrO_4^{2-}(aq) \longrightarrow Ag_2CrO_4(s)$

Find that 22.0 mL of 0.100 M $AgNO_3$ is required to titrate 30.0 mL of solution containing CrO_4^{2-}. Concentration CrO_4^{2-}?

$$\text{no. moles } Ag^+ = 0.0220\ L \times \frac{0.100\ \text{mol } AgNO_3}{1\ L} \times \frac{1\ \text{mol } Ag^+}{1\ \text{mol } AgNO_3}$$

$$= 0.00220\ \text{mol } Ag^+$$

$$\text{no. moles } CrO_4^{2-} = 0.00220\ \text{mol } Ag^+ \times \frac{1\ \text{mol } CrO_4^{2-}}{2\ \text{mol } Ag^+}$$

$$= 0.00110\ \text{mol } CrO_4^{2-}$$

$$\text{conc. } CrO_4^{2-} = \frac{0.00110\ \text{mol}}{0.0300\ L} = 0.0367\ M$$

II Solubility Equilibrium; K_{sp}

A. Expression for K_{sp}

$AgCl(s) \rightleftharpoons Ag^+(aq) + Cl^-(aq)$

$K_{sp}\ AgCl = [Ag^+] \times [Cl^-] = 1.6 \times 10^{-10}$

$PbCl_2(s) \rightleftharpoons Pb^{2+}(aq) + 2\ Cl^-(aq)$

$K_{sp}\ PbCl_2 = [Pb^{2+}] \times [Cl^-]^2 = 1.7 \times 10^{-5}$

K_{sp} can be calculated from measured solubility:

Solubility $PbCl_2$ in pure water = 0.016 mol/L

$[Pb^{2+}] = 1.6 \times 10^{-2} M$; $[Cl^-] = 3.2 \times 10^{-2}\ M$

$K_{sp}\ PbCl_2 = (1.6 \times 10^{-2})(3.2 \times 10^{-2})^2 = 1.7 \times 10^{-5}$

LECTURE 2

I Uses of K_{sp}

A. Calculation of concentration of one ion, knowing that of other

What is $[Pb^{2+}]$ in a solution in equilibrium with $PbCl_2$ ($K_{sp} = 1.7 \times 10^{-5}$) if $[Cl^-] = 0.020$ M?

$$[Pb^{2+}] = \frac{1.7 \times 10^{-5}}{(2.0 \times 10^{-2})^2} = 0.042 \text{ M}$$

B. Determination of whether precipitate will form

Compare original concentration product, P, to K_{sp}

If $P < K_{sp}$, no precipitate (equilibrium not established)

If $P > K_{sp}$, precipitate forms until P becomes equal to K_{sp}

1. Suppose enough Ag^+ is added to a solution 0.001 M in CrO_4^{2-} to make conc. $Ag^+ = 0.001$ M. Assuming conc. CrO_4^{2-} is unchanged, will precipitate of Ag_2CrO_4 form ($K_{sp} = 2 \times 10^{-12}$)?

   $P = (\text{orig. conc. } Ag^+)^2(\text{conc. } CrO_4^{2-}) = (1 \times 10^{-3})^2(1 \times 10^{-3})$

   $= 1 \times 10^{-9} > K_{sp}$; precipitate forms

2. Add 100 mL of 0.001 M $AgNO_3$ to 100 mL of 0.001 M Na_2CrO_4. Will precipitate form?

   $$\text{conc. } Ag^+ = \text{conc. } CrO_4^{2-} = 0.001 \text{ M} \times \frac{0.100 \text{ L}}{0.200 \text{ L}} = 5 \times 10^{-4} \text{ M}$$

   $P = (5 \times 10^{-4})^2(5 \times 10^{-4}) = 1 \times 10^{-10} > K_{sp}$; ppt. forms

C. Determination of solubility

1. Pure water

   $$AgCl(s) \rightleftharpoons Ag^+(aq) + Cl^-(aq);\ s = \text{solubility in mol/L}$$
   $$\quad s \qquad\qquad s \qquad\quad s$$

   $K_{sp} = s^2$; $s = (K_{sp})^{\frac{1}{2}} = (1.6 \times 10^{-10})^{\frac{1}{2}} = 1.3 \times 10^{-5}$ M

   $$PbCl_2(s) \rightleftharpoons Pb^{2+}(aq) + 2\ Cl^-(aq)$$
   $$\quad s \qquad\qquad s \qquad\qquad 2s$$

   $K_{sp} = 4s^3$; $s = (K_{sp}/4)^{1/3} = (1.7 \times 10^{-5}/4)^{1/3}$

   $= 1.6 \times 10^{-2}$ M

2. In solution containing a common ion

   Solubility of $PbCl_2$ in 0.100 M HCl?

   $$PbCl_2(s) \rightleftharpoons Pb^{2+}(aq) + 2\ Cl^-(aq)$$
   $$\quad s \qquad\qquad s \qquad\qquad s + 0.10 \approx 0.10$$

   $s(0.10)^2 = 1.7 \times 10^{-5}$; $s = 1.7 \times 10^{-3}$ M

   solubility much less than in pure water

II Water Softening

"Hard" water caused by Ca^{2+}, Mg^{2+}, Fe^{3+}. Evaporation leaves insoluble compounds such as $CaCO_3$, $CaSO_4$, $Mg(OH)_2$, $Fe(OH)_3$. Also gives precipitates with soaps, which are sodium salts of organic acids.

To soften water, pass through column of zeolite, containing Na^+ ions and Z^- ions (network covalent anion).

$$Ca^{2+}(aq) + 2\ NaZ(s) \longrightarrow CaZ_2(s) + 2\ Na^+(aq)$$

Replaces Ca^{2+} ions by Na^+ ions. To reverse process, flush with concentrated solution of NaCl.

DEMONSTRATIONS

1. Preparation of $Hg(SCN)_2$, Pharaoh's Serpents: Test. Dem. 29, 94
2. Precipitation of PbI_2; Test. Dem. 45
3. Estimation of K_{sp}: Test. Dem. 131, 173
4. Equilibrium between AgCl and Ag_2CrO_4: J. Chem. Educ. 54 618 (1977)
5. Common ion effect: Test. Dem. 19, 86
6. Softening of water by ion exchange: Test. Dem. 219; J. Chem. Educ. 53 302 (1976)

QUIZZES

Quiz 1

1. Write balanced net ionic equations for any precipitation reactions that occur when 0.10 M solutions of the following are mixed.

   a. $CaCl_2$ and K_2CO_3 b. $NiSO_4$ and NaOH c. KCl and $NaNO_3$

2. Show by calculation whether or not a precipitate will form when 500 mL of 0.010 M $BaCl_2$ is mixed with 500 mL of 0.010 M NaF (K_{sp} BaF_2 = 2×10^{-6}).

Quiz 2

1. Write balanced net ionic equations for:

   a. the formation of a precipitate when solutions of $SbCl_3$ and Na_2S are mixed.

   b. the removal of Mg^{2+} ions using the zeolite NaZ.

2. K_{sp} of PbI_2 is 1×10^{-8}. Calculate the solubility of PbI_2 in:

   a. pure water b. 0.2 M NaI

Quiz 3

1. Write balanced net ionic equations for any precipitation reactions that occur when the following solutions are mixed.

   a. $FeCl_3$, $Ba(OH)_2$ b. Na_2SO_4, $Pb(NO_3)_2$ c. $ZnSO_4$, BaS

2. Complete the following table for Ag_2CrO_4 ($K_{sp} = 2 \times 10^{-12}$)

$[Ag^+]$	$[CrO_4^{2-}]$
______	2×10^{-6} M
2×10^{-6} M	------

Quiz 4

1. 250 mL of 0.100 M $CaCl_2$ solution is treated with 0.125 M $AgNO_3$ solution until precipitation is complete.
   a. Write a balanced net ionic equation for the reaction that occurs.
   b. How many moles of precipitate are formed?
   c. What volume of $AgNO_3$ solution is required?
   d. What is the concentration of Ca^{2+} in the final solution?

Quiz 5

1. Write a balanced net ionic equation for the reaction that occurs when:
   a. solutions of $CuSO_4$ and KOH are mixed.
   b. solutions of $Pb(NO_3)_2$ and NaCl are mixed.
   c. a zeolite which has been used to remove Ca^{2+} ions from hard water is treated with a concentrated NaCl solution.
2. The concentration of Ca^{2+} in a certain water supply is 2.0 x 10^{-4} M. If 250 L of this water is treated with a zeolite to remove all the Ca^{2+}, what volume of 1.0 M NaCl solution is required to regenerate all the NaZ?

PROBLEMS 1-30

18.1 K_2CO_3 and $NiSO_4$ are soluble

18.2 a. Add NaOH; filter off the $Fe(OH)_2$ precipitated

b. Add $BaCl_2$ to dilute H_2SO_4; filter off $BaSO_4$

c. Precipitate AgCl by adding HCl or NaCl

18.3 a. $Cu^{2+}(aq) + 2\ OH^-(aq) \longrightarrow Cu(OH)_2(s)$

b. $Zn^{2+}(aq) + S^{2-}(aq) \longrightarrow ZnS(s)$

$Ba^{2+}(aq) + SO_4^{2-}(aq) \longrightarrow BaSO_4(s)$

18.4 a. $Hg_2^{2+}(aq) + 2\ Cl^-(aq) \longrightarrow Hg_2Cl_2(s)$

b. no precipitate; both KOH and $NaNO_3$ are soluble

c. $Fe^{3+}(aq) + 3\ OH^-(aq) \longrightarrow Fe(OH)_3(s)$

$Ba^{2+}(aq) + SO_4^{2-}(aq) \longrightarrow BaSO_4(s)$

d. $2\ As^{3+}(aq) + 3\ S^{2-}(aq) \longrightarrow As_2S_3(s)$

e. $Ni^{2+}(aq) + CO_3^{2-}(aq) \longrightarrow NiCO_3(s)$

18.5 a. $Ag^+(aq) + Cl^-(aq) \longrightarrow AgCl(s)$

b. $Pb^{2+}(aq) + S^{2-}(aq) \longrightarrow PbS(s)$

c. no reaction

d. $Fe^{3+}(aq) + 3\ OH^-(aq) \longrightarrow Fe(OH)_3(s)$

18.6 a. no. moles $Ag^+ = 0.105 \frac{mol}{L} \times 0.0100\ L = 0.00105\ mol\ Ag^+$

no. moles $Cl^- = 0.00105$

no. moles $ScCl_3 = 0.00105\ mol\ Cl^- \times \frac{1\ mol\ ScCl_3}{3\ mol\ Cl^-}$

$= 0.000350\ mol\ ScCl_3$

volume $= 0.000350\ mol\ ScCl_3 \times \frac{1\ L}{0.0250\ mol} = 0.0140\ L$ (14.0 mL)

b. no. moles AgCl = no. moles $Ag^+ = 0.00105$

mass AgCl $= 0.00105\ mol\ AgCl \times \frac{143.32\ g\ AgCl}{1\ mol\ AgCl} = 0.150\ g\ AgCl$

18.7 a. no. moles $S^{2-} = 0.0814 \frac{mol}{L} \times 0.0500\ L = 0.00407\ mol$

no. moles $Ag^+ = 0.00407\ mol\ S^{2-} \times \frac{2\ mol\ Ag^+}{1\ mol\ S^{2-}}$

$= 0.00814\ mol\ Ag^+$

volume $= 0.00814\ mol\ Ag^+ \times \frac{1\ L}{0.100\ mol\ Ag^+} = 0.0814\ L$ (81.4 mL)

b. no. moles $Cl^- = 0.197 \frac{mol}{L} \times 0.0626\ L = 0.0123\ mol\ Cl^-$

no. moles Ag^+ = no. moles $Cl^- = 0.0123\ mol\ Ag^+$

volume $= 0.0123\ mol\ Ag^+ \times \frac{1\ L}{0.100\ mol\ Ag^+} = 0.123\ L$ (123 mL)

c. no. moles $CrO_4^{2-} = 0.240 \frac{mol}{L} \times 0.0127\ L = 0.00305\ mol$

no. moles $Ag^+ = 0.00305\ mol\ CrO_4^{2-} \times \frac{2\ mol\ Ag^+}{1\ mol\ CrO_4^{2-}}$

$= 0.00610\ mol\ Ag^+$

volume $= 0.00610\ mol\ Ag^+ \times \frac{1\ L}{0.100\ mol\ Ag^+} = 0.0610\ L$ (61.0 mL)

Net ionic equations:

$2\ Ag^+(aq) + S^{2-}(aq) \longrightarrow Ag_2S(s)$

$Ag^+(aq) + Cl^-(aq) \longrightarrow AgCl(s)$

$2\ Ag^+(aq) + CrO_4^{2-}(aq) \longrightarrow Ag_2CrO_4(s)$

18.8 a. no. moles $Hg_2^{2+} = 0.0350 \frac{mol}{L} \times 0.0350\ L = 0.00122\ mol\ Hg_2^{2+}$

no. moles $I^- = 0.100 \frac{mol}{L} \times 0.0400\ L = 0.00400\ mol\ I^-$

b. $Hg_2^{2+}(aq) + 2\ I^-(aq) \longrightarrow Hg_2I_2(s)$

if all Hg_2^{2+} consumed: 0.00122 mol Hg_2I_2

if all I^- is consumed: 0.00200 mol Hg_2I_2

yield = 0.00122 mol Hg_2I_2

c. moles I^- left = 0.00400 - 0.00244 = 0.00156 mol I^-

18.9 no. moles SCN^- = $0.0502 \frac{mol}{L} \times 0.0402\ L = 0.00202$ mol SCN^-

no. moles Ag^+ = 0.00202

mass Ag = $0.00202\ mol \times \frac{107.87\ g}{1\ mol} = 0.218$ g

% Ag = $\frac{0.218\ g}{0.249\ g} \times 100 = 87.6$

18.10 no. moles Ag^+ = $0.120 \frac{mol}{L} \times 0.0156\ L = 0.00187$ mol

no. moles Cl^- = 0.00187 mol

mass Cl^- = $0.00187\ mol \times \frac{35.45\ g}{1\ mol} = 0.0663$ g

% Cl^- = $\frac{0.0663\ g}{0.400\ g} \times 100 = 16.6$

18.11 molar mass $Mg_2P_2O_7$ = (48.60 + 61.94 + 112.00)g/mol
= 222.54 g/mol

mass P = 0.3233 g $Mg_2P_2O_7 \times \frac{61.94\ g\ P}{222.54\ g\ Mg_2P_2O_7}$ = 0.08998 g P

% P = $\frac{0.08988\ g}{1.250\ g} \times 100 = 7.190$

18.12 a. $TlI(s) \rightleftharpoons Tl^+(aq) + I^-(aq)$; $K_{sp} = [Tl^+] \times [I^-]$

b. $Eu(OH)_3(s) \rightleftharpoons Eu^{3+}(aq) + 3\ OH^-(aq)$; $K_{sp} = [Eu^{3+}] \times [OH^-]^3$

c. $Pb_3(PO_4)_2(s) \rightleftharpoons 3\ Pb^{2+}(aq) + 2\ PO_4{}^{3-}(aq)$

$K_{sp} = [Pb^{2+}]^3 \times [PO_4{}^{3-}]^2$

d. $Zn(OH)_2(s) \rightleftharpoons Zn^{2+}(aq) + 2\ OH^-(aq)$; $K_{sp} = [Zn^{2+}] \times [OH^-]^2$

18.13 a. $Cu_2P_2O_7(s) \rightleftharpoons 2\ Cu^{2+}(aq) + P_2O_7{}^{4-}(aq)$

b. $Ni_3(AsO_4)_2(s) \rightleftharpoons 3\ Ni^{2+}(aq) + 2\ AsO_4{}^{3-}(aq)$

c. $Fe(OH)_3(s) \rightleftharpoons Fe^{3+}(aq) + 3\ OH^-(aq)$

d. $Mg(NbO_3)_2(s) \rightleftharpoons Mg^{2+}(aq) + 2\ NbO_3{}^-(aq)$

18.14

$[Mg^{2+}]$	0.1 M	0.005 M	0.01 M	0.0002 M
$[F^-]$	0.001 M	0.004 M	0.003 M	0.02 M

18.15 a. $[Cu^{2+}] = \frac{K_{sp}\ CuS}{[S^{2-}]} = \frac{1 \times 10^{-35}}{1 \times 10^{-3}} = 1 \times 10^{-32}$ M

b. $[Zn^{2+}] = \frac{K_{sp}\ ZnS}{[S^{2-}]} = \frac{1 \times 10^{-20}}{1 \times 10^{-3}} = 1 \times 10^{-17}$ M

c. $[Ag^+]^2 = \frac{K_{sp}\ Ag_2S}{[S^{2-}]} = \frac{1 \times 10^{-49}}{1 \times 10^{-3}} = 1 \times 10^{-46}$

$[Ag^+] = 1 \times 10^{-23}$ M

18.16 a. $[Cl^-] = \frac{K_{sp}\ AgCl}{[Ag^+]} = \frac{1.6 \times 10^{-10}}{1.0 \times 10^{-4}} = 1.6 \times 10^{-6}$ M

18.16 b. $[Ag^+] = \frac{K_{sp}\ AgCl}{[Cl^-]} = \frac{1.6 \times 10^{-10}}{2.0 \times 10^{-2}} = 8.0 \times 10^{-9}$ M

$\%\ Ag^+$ left $= \frac{8.0 \times 10^{-9}}{1.0 \times 10^{-4}} \times 100 = 0.0080\ \%$

18.17 0.010 M: $P = (4.1 \times 10^{-6})(1.0 \times 10^{-2}) = 4.1 \times 10^{-8}$

$P > K_{sp}$; precipitate forms

0.0010 M: $P = (4.1 \times 10^{-6})(1.0 \times 10^{-3}) = 4.1 \times 10^{-9}$

$P < K_{sp}$; no precipitate

18.18 $P\ BaCrO_4 = (1.0 \times 10^{-2})(1.0 \times 10^{-4}) = 1.0 \times 10^{-6}$

$P > K_{sp}\ BaCrO_4$; precipitate

$P\ SrCrO_4 = (1.0 \times 10^{-2})(1.0 \times 10^{-4}) = 1.0 \times 10^{-6}$

$P < K_{sp}\ SrCrO_4$; no precipitate

$BaCrO_4$ precipitates

18.19 a. no. moles $Ag^+ = 0.10\ L \times 0.0045\ \frac{mol}{L} = 4.5 \times 10^{-5}$ mol

conc. $Ag^+ = \frac{4.5 \times 10^{-5}\ mol}{0.35\ L} = 1.3 \times 10^{-4}$ M

b. no. moles $CrO_4^{2-} = 0.25\ L \times 0.00075\ \frac{mol}{L} = 1.9 \times 10^{-4}$ mol

conc. $CrO_4^{2-} = \frac{1.9 \times 10^{-4}\ mol}{0.35\ L} = 5.4 \times 10^{-4}$ M

c. $P\ Ag_2CrO_4 = (1.3 \times 10^{-4})^2(5.4 \times 10^{-4}) = 9.1 \times 10^{-12}$

$P > K_{sp}$; precipitate forms

18.20 no. moles $Pb^{2+} = 0.20\ L \times 0.10\ mol/L = 2.0 \times 10^{-2}$ mol

conc. $Pb^{2+} = \frac{2.0 \times 10^{-2}\ mol}{0.50\ L} = 4.0 \times 10^{-2}$ M

no. moles $Br^- = 0.30\ L \times 0.020\ mol/L = 6.0 \times 10^{-3}$ mol

conc. $Br^- = \frac{6.0 \times 10^{-3}\ mol}{0.50\ L} = 1.2 \times 10^{-2}$ M

$P\ PbBr_2 = (4.0 \times 10^{-2})(1.2 \times 10^{-2})^2 = 5.8 \times 10^{-6}$

$P > K_{sp}\ PbBr_2 = 5 \times 10^{-6}$; precipitate forms (just barely)

18.21 $[Cl^-] = [Tl^+] = 1.4 \times 10^{-2}$ M

$K_{sp}\ TlCl = (1.4 \times 10^{-2})^2 = 2.0 \times 10^{-4}$

18.22 a. $4s^3 = 2 \times 10^{-12}$; $s = 8 \times 10^{-5}$ mol/L

b. $s^2 = 2 \times 10^{-8}$; $s = 1.4 \times 10^{-4}\ \frac{mol}{L} \times \frac{84.31\ g}{1\ mol} = 0.01$ g/L

18.23 $s^2 = 1.4 \times 10^{-9}$; $s = 3.7 \times 10^{-5}$ mol/L

mass $BaSO_4 = 0.150\ L \times 3.7 \times 10^{-5}\ \frac{mol}{L} \times \frac{233.39\ g}{1\ mol} = 0.0013$ g

18.24 a. $Cu_2P_2O_7(s) \rightleftharpoons 2\ Cu^{2+}(aq) + P_2O_7^{4-}(aq)$

s 2s s $K_{sp} = (2s)^2 s = 4s^3$

b. $Ni_3(AsO_4)_2(s) \rightleftharpoons 3\ Ni^{2+}(aq) + 2\ AsO_4^{3-}(aq)$

s 3s 2s

$K_{sp} = (3s)^3(2s)^2 = 108\ s^5$

c. $Fe(OH)_3(s) \rightleftharpoons Fe^{3+}(aq) + 3\ OH^-(aq)$

s s 3s $K_{sp} = s(3s)^3 = 27\ s^4$

d. $Mg(NbO_3)_2(s) \rightleftharpoons Mg^{2+}(aq) + 2\ NbO_3^-(aq)$

s s 2s $K_{sp} = s(2s)^2 = 4s^3$

18.25 a. ⟵ b. ⟶ c. ⟶ d. ⟵

18.26 a. $4s^3 = K_{sp}$; $s^3 = \frac{1 \times 10^{-8}}{4} = 2.5 \times 10^{-9}$; $s = 1 \times 10^{-3}$ M

b. $[Pb^{2+}] = s = \frac{1 \times 10^{-8}}{[I^-]^2} = \frac{1 \times 10^{-8}}{(1 \times 10^{-2})^2} = 1 \times 10^{-4}$ M

c. $[I^-]^2 = \frac{1 \times 10^{-8}}{[Pb^{2+}]} = \frac{1 \times 10^{-8}}{2 \times 10^{-2}} = 0.5 \times 10^{-6}$; $[I^-] = 7 \times 10^{-4}$ M

$s = \frac{[I^-]}{2} \approx 4 \times 10^{-4}$ M

18.27 Na^+ ions replace Ca^{2+} - - - ions in the zeolite

18.28 moles $Ca^{2+} = 65 \times 10^{-3}$ g $Ca^{2+} \times \frac{10^3 \text{ L}}{1 \text{ m}^3} \times \frac{1.20 \text{ m}^3}{1 \text{ d}} \times 7$ d

$\times \frac{1 \text{ mol } Ca^{2+}}{40.08 \text{ g } Ca^{2+}} = 13.6$ mol Ca^{2+}

mass NaCl = 13.6 mol $Ca^{2+} \times \frac{2 \text{ mol NaCl}}{1 \text{ mol } Ca^{2+}} \times \frac{58.44 \text{ g NaCl}}{1 \text{ mol NaCl}} \times \frac{1 \text{ kg}}{10^3 \text{ g}}$

= 1.6 kg

18.29 a. deposits $CaCO_3$

b. Ca^{2+} salts of soaps

c. solubility = $(K_{sp})^{\frac{1}{2}} = 3.7 \times 10^{-5}$ M

18.30 a. not unless solution was obtained by shaking with pure AgCl

b. cannot compare directly, since $s^2 = K_{sp}$ in one case and $4s^3 = K_{sp}$ in other

c. no common ion

CHAPTER 19

BASIC SKILLS

1. Given one of the three quantities: $[H^+]$, $[OH^-]$, pH, calculate the other two quantities.
2. Predict whether a given acid or base is strong or weak.
3. Write a net ionic equation to explain why a molecule, cation, or anion gives an acidic or basic solution.
4. Predict whether a given ion or ionic compound will give an acidic, basic, or neutral water solution.
5. Write net ionic equations to describe the reaction of
   a. a strong acid with a strong base.
   b. a weak acid with a strong base.
   c. a strong acid with a weak base.
6. Choose an apprpriate indicator for a given acid-base titration.
7. Use titration data to calculate the concentration of an acidic or basic species in solution or its percent in a solid mixture.
8. Given the equation for an acid-base reaction, select the Bronsted acid and Bronsted base; the Lewis acid and Lewis base; the conjugate acid-base pairs.

LECTURE NOTES

This is a chapter that should be covered slowly and carefully. Students have trouble with it because there are so many different concepts. They have particular difficulty

- predicting whether a given species (in particular, a salt) will give an acidic, basic, or neutral solution.

- writing a chemical equation to show why a species is acidic or basic.

- writing equations for acid-base reactions.

To master these skills, students must know which acids and bases are strong.

You will probably want to spend at least $2\frac{1}{2}$ lectures, more likely 3, on this material.

LECTURE 1

I Acidic and Basic Water Solutions

A. The H^+ ion (or H_3O^+ ion) is characteristic of acidic water solutions; OH^- ion gives basic solutions their characteristic properties. There is an equilibrium between these two ions in water or in any aqueous solution:

$$H_2O \rightleftharpoons H^+(aq) + OH^-(aq)$$

$$K_w = [H^+] \times [OH^-] = 1.0 \times 10^{-14} \text{ at } 25°C$$

1. Pure water: $[H^+] = [OH^-] = 1.0 \times 10^{-7}$ M; neutral solution
2. Acidic solution: $[H^+] > 1.0 \times 10^{-7}M > [OH^-]$
3. Basic solution: $[OH^-] > 1.0 \times 10^{-7}$ M $> [H^+]$

In seawater, $[H^+] = 5 \times 10^{-9}$ M; $[OH^-] = ?$

$$[OH^-] = \frac{1.0 \times 10^{-14}}{5.0 \times 10^{-9}} = 2 \times 10^{-6} \text{ M; basic}$$

b. $pH = -\log_{10} H^+$

Neutral solution: pH = 7.0
Acidic solution: pH < 7.0
Basic solution : pH > 7.0

Suppose $[H^+] = 2.4 \times 10^{-6}$ M ; $pH = -\log_{10}(2.4 \times 10^{-6})$
$= -(0.38 - 6.00) = 5.62$

Suppose pH = 8.68, $[H^+]$ = ? Use 10^x or INV and LOG keys
$[H^+] = 10^{-8.68} = 2.1 \times 10^{-9}$ M

II Strong and Weak Acids

A. Strong acids are completely dissociated to form H^+ ions in water:

$$HX(aq) \longrightarrow H^+(aq) + X^-(aq)$$

There are only a few common strong acids; learn their formulas

HCl, HBr, HI; HNO_3, $HClO_4$, H_2SO_4

B. Weak acids are partially dissociated to H^+ ions

$$HX(aq) \rightleftharpoons H^+(aq) + X^-(aq)$$

Usually, conc. HX molecules >> conc. H^+ ions. However, enough H^+ ions are formed to make the solution acidic

1. Molecular weak acids (many thousands)

$$HF(aq) \rightleftharpoons H^+(aq) + F^-(aq)$$

Acid-Base indicators are molecular weak acids:

$$\underset{\text{blue}}{HIn(aq)} \rightleftharpoons H^+(aq) + \underset{\text{yellow}}{In^-(aq)}$$

At high $[H^+]$, mostly in form of HIn (blue). At low $[H^+]$, In^- dominates, solution is yellow.

2. Anions containing an ionizable H atom:

$H_2SO_4(aq) \longrightarrow H^+(aq) + HSO_4^-(aq)$; H_2SO_4 is strong acid
$HSO_4^-(aq) \rightleftharpoons H^+(aq) + SO_4^{2-}(aq)$; HSO_4^- is weak acid

3. Cations. All cations except those of Group 1 metals, Ca^{2+}, Sr^{2+}, Ba^{2+}.

$NH_4^+(aq) \rightleftharpoons H^+(aq) + NH_3(aq)$

$Zn(H_2O)_4^{2+}(aq) \rightleftharpoons H^+(aq) + Zn(H_2O)_3(OH)^+(aq)$

This reaction explains why solutions of zinc salts such as $ZnCl_2$ are acidic.

LECTURE 2

I Strong and Weak Bases

A. Strong bases - completely ionized to form OH^- ion in solution. Hydroxides of Group 1 metals:

$NaOH(s) \rightarrow Na^+(aq) + OH^-(aq)$

Heavier Group 2 metals:

$M(OH)_2(s) \rightarrow M^{2+}(aq) + 2\ OH^-(aq)$; M = Ca, Sr, Ba

B. Weak bases

1. Molecular:

$NH_3(aq) + H_2O \rightleftharpoons NH_4^+(aq) + OH^-(aq)$

2. Anions derived from weak acids:

$F^-(aq) + H_2O \rightleftharpoons HF(aq) + OH^-(aq)$

$CO_3^{2-}(aq) + H_2O \rightleftharpoons HCO_3^-(aq) + OH^-(aq)$

II Acid-Base Properties Salt Solutions

Must consider effect of cation and anion separately. Then combine these effects to give overall result for salt.

A. Cations

1. Neutral - derived from strong bases: Li^+, Na^+, K^+; Ca^{2+}, Sr^{2+}, Ba^{2+}
2. Acidic - all other cations, including those of transition metals

B. Anions

1. Neutral - derived from strong acids: Cl^-, Br^-, I^-; NO_3^-, ClO_4^-, SO_4^{2-}
2. Acidic: HSO_4^-, $H_2PO_4^-$
3. Basic: all other anions. In general, an anion derived from a weak acid is expected to be basic.

C. Overall result

Salt	Cation	Anion	
$NaNO_3$	Na^+(N)	NO_3^-(N)	neutral
KF	K^+(N)	F^-(B)	basic
$FeCl_3$	Fe^{3+}(A)	Cl^-(N)	acidic
$CaSO_4$	Ca^{2+}(N)	SO_4^{2-}(N)	neutral

III Acid-Base Reactions

A. Types of reactions

1. Strong acid + strong base: $HNO_3 + Ca(OH)_2$

principal species: H^+, NO_3^- ions; Ca^{2+}, OH^- ions

$H^+(aq) + OH^-(aq) \longrightarrow H_2O$

2. Weak acid + strong base: HF + KOH

principal species: HF molecules; K^+, OH^- ions

$HF(aq) + OH^-(aq) \longrightarrow F^-(aq) + H_2O$

3. Strong acid + weak base: $HClO_4 + NH_3$

principal species: H^+, ClO_4^- ions; NH_3 molecules

$H^+(aq) + NH_3(aq) \longrightarrow NH_4^+(aq)$

$H_2SO_4 + Na_2CO_3$:

$H^+(aq) + CO_3^{2-}(aq) \longrightarrow HCO_3^-(aq)$

$H^+(aq) + HCO_3^-(aq) \longrightarrow H_2CO_3(aq) \longrightarrow CO_2(g) + H_2O$

LECTURE 3

I Acid-Base Reactions (cont.)

A. Titrations - measure volume of one solution required to react with known volume of other solution. If one concentration is known, other concentration can be calculated.

1. 26.2 mL of 0.120 M HCl required to react with 20.0 mL of $Ba(OH)_2$ solution. Molarity of $Ba(OH)_2$?

$$H^+(aq) + OH^-(aq) \longrightarrow H_2O$$

$$\text{moles } H^+ = 0.0262\ L \times \frac{0.120\ \text{mol HCl}}{1\ L} \times \frac{1\ \text{mol } H^+}{1\ \text{mol HCl}} = 0.00314\ \text{mol } H^+$$

$$\text{moles } Ba(OH)_2 = 0.00314\ \text{mol } H^+ \times \frac{1\ \text{mol } OH^-}{1\ \text{mol } H^+} \times \frac{1\ \text{mol } Ba(OH)_2}{2\ \text{mol } OH^-}$$

$$= 0.00157\ \text{mol } Ba(OH)_2$$

$M = 0.00157\ \text{mol}/0.0200\ L = 0.0785\ \text{mol}/L$

B. Choice of indicator in acid-base titration

1. Strong acid - weak base. Solution at equivalence point is weakly acidic. Choose indicator which turns color below pH 7. Methyl red changes at pH 5.

2. Weak acid- strong base. Solution at equivalence point is weakly basic. Choose indicator which changes color above pH 7. Phenolphthalein changes at pH 9.

3. Strong acid - strong base. Solution at equivalence point is neutral. However, pH changes so rapidly near the end point that any indicator will work.

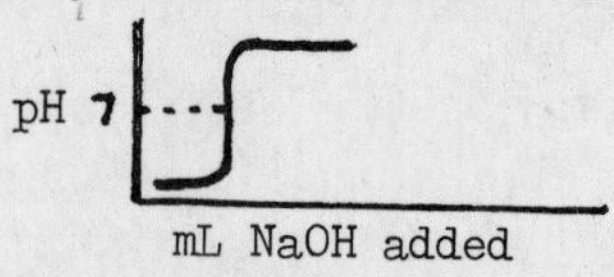

II Other Models of Acids and Bases

A. Bronsted-Lowry

An acid is a proton donor; a base is a proton acceptor

$HF(aq) + OH^-(aq) \longrightarrow F^-(aq) + H_2O$
acid base

$NH_3(aq) + H_2O \longrightarrow NH_4^+(aq) + OH^-(aq)$
base acid

$HCl(aq) + H_2O \longrightarrow H_3O^+(aq) + Cl^-(aq)$
acid base

H_3O^+ is referred to as the conjugate acid of H_2O

OH^- is the conjugate base of H_2O

b. Lewis

Acid accepts electron pair, base donates electron pair

$H^+(aq) + H_2O \longrightarrow H_3O^+(aq)$
acid base

$Zn^{2+}(aq) + 4\ H_2O \longrightarrow Zn(H_2O)_4^{2+}(aq)$
acid base

DEMONSTRATIONS

1. Indicator colors: Test. Dem. 12, 61, 147; J. Chem. Educ. 61 172 (1984)
2. HCl, NH_3 fountains: Test. Dem. 14
3. Grape juice as an indicator: Test. Dem. 204
4. pH of phosphate salts: Test. Dem. 40
5. Acid-base properties of salts: Test. Dem. 62
6. Strong and weak acid titrations: J. Chem. Educ. 56 194 (1979)

QUIZZES

Quiz 1

1. The pH of a solution is 2.68. Calculate $[H^+]$ and $[OH^-]$.
2. Write balanced net ionic equations for:
   a. the reaction between solutions of $HC_2H_3O_2$ and $Ca(OH)_2$.
   b. the reaction of NaCN with water to give a basic solution.
   c. the reaction that explains why a solution of zinc chloride is acidic.

Quiz 2

1. It is found that 16.0 mL of 0.105 M $Ba(OH)_2$ is required to react with 22.0 mL of HCl. What is the molarity of the HCl?
2. Classify each of the following salts as acidic, basic or neutral.

   a. NaCl   b. $Ca(C_2H_3O_2)_2$ c. $Ni(NO_3)_2$

Quiz 3

1. Write balanced net ionic equations to explain why:
   a. a water solution of Na_2CO_3 is basic.
   b. a water solution of $ZnSO_4$ is acidic.
   c. a reaction occurs when solutions of HCl and $Ca(OH)_2$ are mixed.

Quiz 4

1. Classify solutions of each of the following as acidic, basic, or neutral.

   a. H_2SO_4 b. NH_3 c. $Ca(OH)_2$ d. Na_2CO_3 e. $Al(NO_3)_3$
2. For each solution in (1) which is acidic or basic, write a net ionic equation to explain how it got that way.

Quiz 5

1. A 1.000 g mixture of NaOH and NaCl is titrated with 50.0 mL of 0.100 M HCl. What is the percent NaOH in the mixture?
2. Calculate $[H^+]$ and the pH of 0.100 M $Ba(OH)_2$.

PROBLEMS 1-30

19.1 a. $[OH^-] = \dfrac{1.0 \times 10^{-14}}{2.0 \times 10^{-4}} = 5.0 \times 10^{-11}$ M

b. $[OH^-] = \dfrac{1.0 \times 10^{-14}}{6.0} = 1.7 \times 10^{-15}$ M

c. $[OH^-] = \dfrac{1.0 \times 10^{-14}}{3.2 \times 10^{-9}} = 3.1 \times 10^{-6}$ M

d. $[OH^-] = \dfrac{1.0 \times 10^{-14}}{5 \times 10^{-6}} = 2 \times 10^{-9}$ M

19.2 a. $pH = -\log_{10}(1 \times 10^{-4}) = 4.0$ b. $pH = -\log_{10}(12) = -1.08$

c. $pH = -\log_{10}(2.7 \times 10^{-6}) = 5.57$

d. $pH = -\log_{10}(7.6 \times 10^{-9}) = 8.12$

a, b, c are acidic; d is basic

19.3 a. $[H^+] = 1 \times 10^{-6}$ M; $[OH^-] = 1 \times 10^{-8}$ M; acidic

b. $[H^+] = 6 \times 10^{-8}$ M; $[OH^-] = 2 \times 10^{-7}$ M: basic

c. $[H^+] = 1.0$ M; $[OH^-] = 1.0 \times 10^{-14}$ M; acidic

d. $[H^+] = 6.0$ M; $[OH^-] = 1.7 \times 10^{-15}$ M; acidic

19.4 pH solution(2) = 6.7; (1) more acidic, (2) has a higher pH

19.5 $\frac{[H^+] \text{ solution 1}}{[H^+] \text{ solution 2}} = \frac{2 \times 10^{-4}}{2 \times 10^{-5}} = 10$; $\frac{[OH^-] \text{solution 1}}{[OH^-] \text{solution 2}} = 0.1$

19.6 $[H^+] = 3.2 \times 10^{-9}$ M; $[OH^-] = \frac{1.0 \times 10^{-14}}{3.2 \times 10^{-9}} = 3.1 \times 10^{-6}$ M

19.7 a. $HNO_3(aq) \rightarrow H^+(aq) + NO_3^-(aq)$ strong
b. $HNO_2(aq) \rightleftharpoons H^+(aq) + NO_2^-(aq)$ weak
c. $HBr(aq) \rightarrow H^+(aq) + Br^-(aq)$ strong
d. $HC_2H_3O_2(aq) \rightleftharpoons H^+(aq) + C_2H_3O_2^-(aq)$ weak
e. $H_3PO_4(aq) \rightleftharpoons H^+(aq) + H_2PO_4^-(aq)$ weak

19.8 a. $Al(H_2O)_6^{3+}(aq) \rightleftharpoons H^+(aq) + Al(H_2O)_5(OH)^{2+}(aq)$
b. $Zn(H_2O)_3(OH)^+(aq) \rightleftharpoons H^+(aq) + Zn(H_2O)_2(OH)_2(aq)$
c. $H_2PO_4^-(aq) \rightleftharpoons H^+(aq) + HPO_4^{2-}(aq)$
d. $Mn(H_2O)_6^{2+}(aq) \rightleftharpoons H^+(aq) + Mn(H_2O)_5(OH)^+(aq)$

19.9 a. $[H^+] = 0.40$ M, $[OH^-] = \frac{1.0 \times 10^{-14}}{0.40} = 2.5 \times 10^{-14}$ M

$pH = -\log_{10}(0.40) = 0.40$

b. $[OH^-] = 0.33$ M, $[H^+] = \frac{1.0 \times 10^{-14}}{0.33} = 3.0 \times 10^{-14}$ M; pH = 13.52

c. $[OH^-] = \frac{12.0/56.11 \text{ mol}}{0.200 \text{ L}} = 1.07$ M;

$[H^+] = \frac{1.0 \times 10^{-14}}{1.07} = 9.3 \times 10^{-15}$ M ; pH = 14.03

d. $[H^+] = \frac{12 \times 0.020}{0.200} = 1.2$ M; $[OH^-] = \frac{1.0 \times 10^{-14}}{1.2} = 8.3 \times 10^{-15}$ M

pH = -0.08

19.10 a. false (much less than 0.10 M) b. true c. true
d. false; more than 1

19.11 LiOH and $Ca(OH)_2$ are strong; $CH_3CH_2NH_2$ and F^- are weak

19.12 a. $LiOH(s) \rightarrow Li^+(aq) + OH^-(aq)$
b. $CH_3CH_2NH_2(aq) + H_2O \rightleftharpoons CH_3CH_2NH_3^+(aq) + OH^-(aq)$
c. $Ca(OH)_2(s) \rightarrow Ca^{2+}(aq) + 2\ OH^-(aq)$
d. $F^-(aq) + H_2O \rightleftharpoons HF(aq) + OH^-(aq)$

19.13 a. $BO_3^{3-}(aq) + H_2O \rightleftharpoons HBO_3^{2-}(aq) + OH^-(aq)$
b. $CH_3NH_2(aq) + H_2O \rightleftharpoons CH_3NH_3^+(aq) + OH^-(aq)$
c. $CN^-(aq) + H_2O \rightleftharpoons HCN(aq) + OH^-(aq)$

19.14 a. acidic b. basic c. neutral d. acidic e. basic

19.15 a. $NH_4^+(aq) \rightleftharpoons H^+(aq) + NH_3(aq)$
b. $CN^-(aq) + H_2O \rightleftharpoons HCN(aq) + OH^-(aq)$

19.15 c. no reaction

d. $Al(H_2O)_6^{3+}(aq) \rightleftharpoons H^+(aq) + Al(H_2O)_5(OH)^{2+}(aq)$

e. $NO_2^-(aq) + H_2O \rightleftharpoons HNO_2(aq) + OH^-(aq)$

19.16 a. neutral

b. basic; $CO_3^{2-}(aq) + H_2O \rightleftharpoons HCO_3^-(aq) + OH^-(aq)$

c. acidic; $Al(H_2O)_6^{3+}(aq) \rightleftharpoons H^+(aq) + Al(H_2O)_5(OH)^{2+}(aq)$

d. acidic; $HSO_4^-(aq) \rightleftharpoons H^+(aq) + SO_4^{2-}(aq)$

19.17 a. KCN, KF, K_2CO_3, $KC_2H_3O_2$, - -

b. KCl, KBr, KI, KNO_3, - -

c. NaBr, KBr, $CaBr_2$, $BaBr_2$, - -

d. $AlBr_3$, $CrBr_3$, $FeBr_3$, $CuBr_2$, - -

19.18 a. $HC_3H_5O_2(aq) + OH^-(aq) \rightarrow C_3H_5O_2^-(aq) + H_2O$

b. $CH_3NH_2(aq) + H^+(aq) \rightarrow CH_3NH_3^+(aq)$

c. $H^+(aq) + OH^-(aq) \rightarrow H_2O$

19.19 a. $F^-(aq) + H^+(aq) \rightarrow HF(aq)$

b. $OH^-(aq) + H^+(aq) \rightarrow H_2O$

c. $NH_3(aq) + H^+(aq) \rightarrow NH_4^+(aq)$

19.20 acidic: 19.18b, 19.19a, 19.19c
neutral: 19.18c, 19.19b
basic: 19.18a

19.21 no. moles HCl = $0.108 \frac{mol}{L} \times 2.26 \times 10^{-2}$ L = 2.44×10^{-3} mol

no. moles KOH = no. moles HCl = 2.44×10^{-3} mol

M KOH = $\frac{2.44 \times 10^{-3} \text{ mol}}{2.00 \times 10^{-2} \text{ L}}$ = 0.122 M

19.22 a. no. moles HF = $0.280 \frac{mol}{L} \times 0.0126$ L = 0.00353 mol

no. moles NaOH = no. moles HF = 0.00353 mol

volume = $\frac{0.00353 \text{ mol}}{0.208 \text{ mol/L}}$ = 0.0170 L = 17.0 mL

b. no. moles $HClO_4$ = $0.144 \frac{mol}{L} \times 0.0150$ L = 0.00216 mol

no. moles NaOH = no. moles $HClO_4$ = 0.00216 mol

volume = $\frac{0.00216 \text{ mol}}{0.208 \text{ mol/L}}$ = 0.0104 L = 10.4 mL

c. no. moles $HC_2H_3O_2$ = 7.50 g x 0.997 x $\frac{1 \text{ mol}}{60.05 \text{ g}}$ = 0.125 mol

no. moles NaOH = no. moles $HC_2H_3O_2$ = 0.125 mol

volume = $\frac{0.125 \text{ mol}}{0.208 \text{ mol/L}}$ = 0.601 L = 601 mL

19.23 no. moles NaOH = $0.125 \frac{mol}{L} \times 0.0222\ L = 0.00278$ mol

no. moles $C_6H_8O_6$ = no. moles NaOH = 0.00278 mol

mass $C_6H_8O_6 = 0.00278\ mol \times \frac{176.12\ g}{1\ mol} = 0.490$ g

% Vitamin C = $\frac{0.490}{0.508} \times 100 = 96.5$

19.24 a. any indicator OK
b. phenolphthalein
c. methyl red
d. phenolphthalein

19.25 a. $[OH^-] = 0.2000$ M; $[H^+] = 5.0 \times 10^{-14}$ M; pH = 13.30

b. $[OH^-] = \frac{2.500 \times 10^{-2}\ L \times 0.2000\ mol/L}{7.500 \times 10^{-2}\ L} = 0.0667$ M

$[H^+] = 1.5 \times 10^{-13}$ M; pH = 12.82

c. $[OH^-] = \frac{1.0 \times 10^{-5}\ L \times 0.2000\ mol/L}{9.999 \times 10^{-2}\ L} = 2.0 \times 10^{-5}$ M

$[H^+] = 5.0 \times 10^{-10}$ M; pH = 9.30

d. pH = 7.00

e. $[H^+] = \frac{1.0 \times 10^{-4}\ L \times 0.2000\ mol/L}{1.001 \times 10^{-1}\ L} = 2.0 \times 10^{-4}$ M

pH = 3.70

f. $[H^+] = \frac{5.000 \times 10^{-2}\ L \times 0.2000\ mol/L}{1.500 \times 10^{-1}\ L} = 0.0667$ M; pH = 1.18

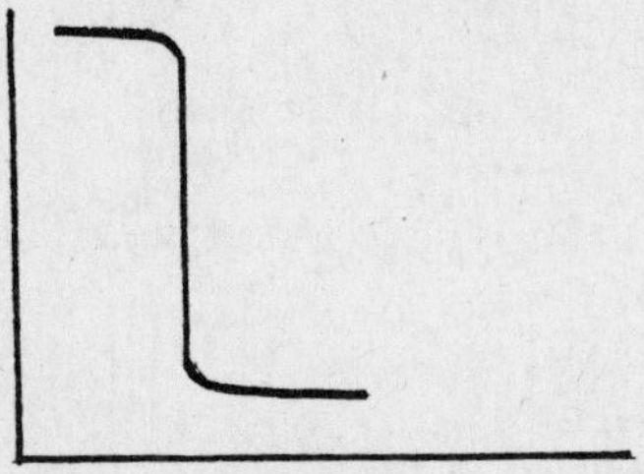

19.26 a. H_2SO_4 b. HCO_3^- c. NO_2^- d. NH_3

19.27

	acid	base	acid-base pairs
a.	HNO_2	H_2O	HNO_2, NO_2^-; H_3O^+, H_2O
b.	H_2O	S^{2-}	H_2O, OH^-; HS^-, S^{2-}
c.	$HC_2H_3O_2$	CN^-	$HC_2H_3O_2$, $C_2H_3O_2^-$; HCN, CN^-

19.28

	B-L acid	B-L base	Lewis acid	Lewis base
a.	H_2O	H_2O	- - - -	H_2O
b.	- - -	- - -	Cu^{2+}	- - -
c.	HCO_3^-	HCO_3^-	- - - -	HCO_3^-

19.29 a. no. moles H^+ originally = $0.150 \frac{mol}{L} \times 0.02500\ L$

$= 0.00375\ mol$

no. moles OH^- originally = $0.100 \frac{mol}{L} \times 0.01500\ L$

$= 0.00150\ mol$

no. moles H^+ left = 0.00375 mol - 0.00150 mol = 0.00225 mol

$$[H^+] = \frac{2.25 \times 10^{-3}\ mol}{4.00 \times 10^{-2}\ L} = 0.0562\ M$$

b. $pH = -\log_{10}(0.0562) = 1.250$

c. $$[OH^-] = \frac{1.0 \times 10^{-14}}{0.0562} = 1.8 \times 10^{-13}\ M$$

d. $$\text{conc. } K^+ = \frac{0.100 \frac{mol}{L} \times 1.500 \times 10^{-2}\ L}{4.00 \times 10^{-2}\ L} = 0.0375\ M$$

19.30 a. HCN, NH_4^+

b. Na_2CO_3, $NaHCO_3$

c. methyl red

d. CO_3^{2-}, HCO_3^-

CHAPTER 20

BASIC SKILLS

1. Calculate K_a for a weak acid HB, given the $[H^+]$ or pH of a solution prepared by dissolving HB in water to a known initial concentration.
2. Given the initial concentration of a weak acid and the value of K_a, calculate $[H^+]$.
3. Given the composition of a buffer system, determine its pH before and after the addition of known amounts of strong acid or base.
4. Relate the value of K_b for a weak base to K_a for its conjugate acid.
5. Given the original concentration of a weak base and the value of K_b, calculate $[OH^-]$.
6. Use the reciprocal rule and the rule of multiple equilibria to calculate values of equilibrium constants.

LECTURE NOTES

The most difficult topic in this chapter is that of buffers. Students have a great deal of trouble calculating the pH of a buffer after addition of strong acid or base. Such a calculation requires that they apply two different concepts: stoichiometry and equilibrium principles. Otherwise, students usually find this material relatively straightforward. With a little practice, they become adept at applying the reciprocal rule and the rule of multiple equilibria.

Two lectures should suffice for this chapter.

LECTURE 1

I K_a for a Weak Acid

A. $HF(aq) \rightleftharpoons H^+(aq) + F^-(aq)$; $K_a = \frac{[H^+] \times [F^-]}{[HF]} = 7.0 \times 10^{-4}$

$HC_2H_3O_2(aq) \rightleftharpoons H^+(aq) + C_2H_3O_2^-(aq)$; $K_a = \frac{[H^+] \times [C_2H_3O_2^-]}{[HC_2H_3O_2]}$

$= 1.8 \times 10^{-5}$

$HC_2H_3O_2$ is weaker than HF; at a given concentration, $[H^+]$ is lower, pH is higher.

B. Calculation of K_a

pH of 0.100 M $HC_2H_3O_2$ solution is 2.87; Ka = ?

$[H^+] = [C_2H_3O_2^-] = 1.3 \times 10^{-3}$ M

$HC_2H_3O_2 = 0.100$ M $- 0.0013$ M $= 0.099$ M

$$K_a = \frac{(1.3 \times 10^{-3})^2}{0.099} = 1.7 \times 10^{-5}$$

II <u>Calculation of $[H^+]$ in Solution Prepared by Dissolving Weak Acid in Water.</u>

A. Calculate $[H^+]$ in 0.200 M $HC_2H_3O_2$ solution

Let $[H^+] = x$; $[C_2H_3O_2^-] = x$; $[HC_2H_3O_2] = 0.200 - x$

$$\frac{x^2}{0.200 - x} = 1.8 \times 10^{-5}$$

To solve, assume $0.200 - x = 0.200$; $x^2 = 0.200(1.8 \times 10^{-5})$
$= 3.6 \times 10^{-6}$
$x = 1.9 \times 10^{-3}$ M

K_a seldom known to better than ± 5%, so approximation is OK if x ≤ 5% of original concentration of weak acid, i.e., % ionization ≤ 5%

in this case: $\%\text{ ion.} = \frac{1.9 \times 10^{-3}}{0.200} \times 100 = 1.0\%$

B. Calculate $[H^+]$ in 0.100 M HF

$$\frac{x^2}{0.100 - x} = 7.0 \times 10^{-4}$$

Set $0.100 - x = 0.100$, solve; $x = 8.4 \times 10^{-3} >$ 5% of 0.100

Make second approximation:

$$\frac{x^2}{0.100 - 0.008} = 7.0 \times 10^{-4}; \; x = 8.0 \times 10^{-3} \text{ M}$$

III <u>Buffer Systems</u>

A. Prepared by adding both the weak acid HB and its conjugate base B^- to water.

$$[H^+] = K_a \times \frac{[HB]}{[B^-]}$$

Calculate $[H^+]$ in solution prepared by adding 0.200 mol $HC_2H_3O_2$, 0.200 mol $C_2H_3O_2^-$ to one liter. Neglect the small amount of acetic acid that dissociates.

$$[H^+] = 1.8 \times 10^{-5} \times \frac{0.200}{0.200} = 1.8 \times 10^{-5} \text{ M} \; ; \; pH = 4.74$$

B. Note that since HB and B^- are present in roughly equal amounts, $[H^+]$ of buffer is roughly equal to K_a of weak acid.

To establish a buffer of pH 7, choose a system such as $H_2PO_4^-$ - HPO_4^{2-}, where K_a is close to 10^{-7} (actually, 6.2×10^{-8}).

C. Effect of adding strong acid or base to buffer

$HB(aq) + OH^-(aq) \longrightarrow H_2O + B^-(aq)$

$B^-(aq) + H^+(aq) \longrightarrow HB(aq)$

As a result of these reactions, H^+ or OH^- ions are consumed so do not drastically effect pH of water.

Calculate pH of $HC_2H_3O_2$ - $C_2H_3O_2^-$ buffer referred to above after addition of 0.020 mol NaOH.

	orig. conc.	change	final
HAc	0.200 M	-0.020 M	0.180 M
Ac^-	0.200 M	+0.020 M	0.220 M

$$[H^+] = 1.8 \times 10^{-5} \times \frac{0.180}{0.220} = 1.5 \times 10^{-5}$$

pH = 4.82 (slightly lower)

LECTURE 2

I K_b of Weak Base

A. $B^-(aq) + H_2O \rightleftharpoons HB(aq) + OH^-(aq)$; $K_b = \frac{[HB] \times [OH^-]}{[B^-]}$

NH_3: $K_b = 1.8 \times 10^{-5}$

$C_2H_3O_2^-$: $K_b = 5.6 \times 10^{-10}$; weaker base than NH_3

B. Relation between K_b and K_a of conjugate weak acid

$$K_b \times K_a = \frac{[HB] \times [OH^-]}{[B^-]} \times \frac{[H^+] \times [B^-]}{[HB]} = [H^+] \times [OH^-] = 1.0 \times 10^{-14}$$

	K_a		K_b
HF	7.0×10^{-4}	F^-	1.4×10^{-11}
$HC_2H_3O_2$	1.8×10^{-5}	$C_2H_3O_2^-$	5.6×10^{-10}
NH_4^+	5.6×10^{-10}	NH_3	1.8×10^{-5}

strength of base is inversely related to that of conjugate weak acid.

C. Calculation of $[OH^-]$ in solution of weak base

pH of 0.10 M NaF solution?

$$\frac{[HF] \times [OH^-]}{[F^-]} = 1.4 \times 10^{-11} \quad ; \quad \frac{[OH^-]^2}{0.10} = 1.4 \times 10^{-11}$$

$[OH^-] = 1.2 \times 10^{-6}$; pH = 8.08 (slightly basic)

II Relations between Equilibrium Constants

Often the equilibrium constant required for a given reaction cannot be found directly. Instead, it must be calculated from other equilibrium constants.

A. Reciprocal Rule

$A \rightleftharpoons B \quad K = [B]/[A]$

$B \rightleftharpoons A \quad K' = [A]/[B] = 1/K$

$H^+(aq) + OH^-(aq) \rightleftharpoons H_2O$; $K = 1/K_w = 1.0 \times 10^{14}$

$H^+(aq) + NH_3(aq) \rightleftharpoons NH_4^+(aq)$; $K = 1/K_a\ NH_4^+ = 1.8 \times 10^9$

B. Rule of Multiple Equilibria

If Equation 1 + Equation 2 = Equation 3

$$K_1 \times K_2 = K_3$$

$H_2S(aq) \rightleftharpoons 2\ H^+(aq) + S^{2-}(aq)$; $K = ?$

$H_2S(aq) \rightleftharpoons H^+(aq) + HS^-(aq)$; $K_1 = 1 \times 10^{-7}$

$HS^-(aq) \rightleftharpoons H^+(aq) + S^{2-}(aq)$; $K_2 = 1 \times 10^{-15}$

$H_2S(aq) \rightleftharpoons 2\ H^+(aq) + S^{2-}(aq)$; $K_3 = 1 \times 10^{-22}$

$$\frac{[H^+]^2 \times [S^{2-}]}{[H_2S]} = 1 \times 10^{-22}$$

DEMONSTRATIONS

1. Buffer action: Test. Dem. 62, 128, 155
2. Blood buffer: J. Chem. Educ. 60 493 (1983)

QUIZZES

Quiz 1

1. For the weak acid HClO, $K_a = 3.2 \times 10^{-8}$
   a. Calculate the pH of 0.20 M HClO.
   b. Calculate K_b of the ClO^- ion.
   c. Calculate K for: $H^+(aq) + ClO^-(aq) \rightleftharpoons HClO(aq)$

Quiz 2

1. For the weak acid H_2CO_3, $K_a = 4.2 \times 10^{-7}$. A buffer is prepared by adding 0.100 mol of H_2CO_3 and 0.200 mol of HCO_3^- to a liter of water.
   a. Calculate the pH of the buffer.
   b. Calculate the pH of the buffer after 0.050 mol H^+ is added.

Quiz 3

1. A 0.100 M solution of the weak acid HX has a pH of 4.50.
   a. Calculate K_a of HX.
   b. Calculate the pH of a 0.200 M solution of HX.

Quiz 4

1. For the weak base NO_2^-, $K_b = 2.2 \times 10^{-11}$
   a. Calculate the pH of a 0.10 M $NaNO_2$ solution
   b. Calculate K_a of HNO_2

Quiz 5

1. K_a of H_2S is 1×10^{-7}; K_a of HS^- is 1×10^{-15}.
   a. Calculate K_b of HS^-.
   b. Calculate K for: $H_2S(aq) \rightleftharpoons 2\,H^+(aq) + S^{2-}(aq)$
   c. Calculate $[H^+]$ in a solution 0.10 M in H_2S and 1×10^{-10} M in S^{2-}.

PROBLEMS 1-30

20.1 a. $HNO_2(aq) \rightleftharpoons H^+(aq) + NO_2^-(aq)$ $\quad K_a = \dfrac{[H^+] \times [NO_2^-]}{[HNO_2]}$

b. $H_2PO_4^-(aq) \rightleftharpoons H^+(aq) + HPO_4^{2-}(aq)$ $\quad K_a = \dfrac{[H^+] \times [HPO_4^{2-}]}{[H_2PO_4^-]}$

c. $H_2SO_3(aq) \rightleftharpoons H^+(aq) + HSO_3^-(aq)$ $\quad K_a = \dfrac{[H^+] \times [HSO_3^-]}{[H_2SO_3]}$

20.2 a. $pK_a = -\log_{10}(4.5 \times 10^{-4}) = 3.35$

b. $pK_a = -\log_{10}(4.2 \times 10^{-7}) = 6.38$

c. $pK_a = -\log_{10}(6.2 \times 10^{-8}) = 7.21$

20.3 a. D > C > B > A b. A

20.4 $[H^+] = [N_3^-] = 0.0014$ M; $[HN_3] = 0.100 - 0.0014 = 0.099$ M

$K_a = \dfrac{[H^+] \times [N_3^-]}{[HN_3]} = \dfrac{(0.0014)^2}{0.099} = 2.0 \times 10^{-5}$

20.5 $[H^+] = [CHO_2^-] = 7.2 \times 10^{-3}$ M; $[HCHO_2] = 0.150/0.500 - 0.007$
$= 0.293$ M

$K_a = \dfrac{[H^+] \times [CHO_2^-]}{[HCHO_2]} = \dfrac{(7.2 \times 10^{-3})^2}{0.293} = 1.8 \times 10^{-4}$

20.6 $[H^+] = [B^-] = 0.10 \text{ M} \times 0.00025 = 2.5 \times 10^{-5}$ M

$K_a = \dfrac{[H^+] \times [B^-]}{[HB]} = \dfrac{(2.5 \times 10^{-5})^2}{0.10} = 6.2 \times 10^{-9}$

20.7 a. $\dfrac{[H^+]^2}{1.0} \approx 2.1 \times 10^{-9}$; $[H^+]^2 = 2.1 \times 10^{-9}$; $[H^+] = 4.6 \times 10^{-5}$M

b. $\dfrac{[H^+]^2}{0.20} \approx 2.1 \times 10^{-9}$; $[H^+]^2 = 4.2 \times 10^{-10}$; $[H^+] = 2.0 \times 10^{-5}$ M

20.8 a. orig. conc. acetic acid = 3.75 mol/2.50 L = 1.50 M

$\dfrac{[H^+]^2}{1.50} \approx 1.8 \times 10^{-5}$; $[H^+]^2 = 2.7 \times 10^{-5}$; $[H^+] = 5.2 \times 10^{-3}$ M

pH = 2.28

20.8 b. orig. conc. acetic acid = 0.700 mol/2.50 L = 0.280 M

$$\frac{[H^+]^2}{0.280} \approx 1.8 \times 10^{-5};\ [H^+]^2 = 5.0 \times 10^{-6};\ [H^+] = 2.2 \times 10^{-3}\ M$$

pH = 2.66

20.9 a. $$\frac{[H^+]^2}{1.5} \approx 1.4 \times 10^{-4};\ [H^+]^2 = 2.1 \times 10^{-4};\ [H^+] = 1.4 \times 10^{-2}\ M$$

b. $$[OH^-] = \frac{1.0 \times 10^{-14}}{1.4 \times 10^{-2}} = 7.1 \times 10^{-13}\ M$$

c. pH = 1.85

d. $$\%\ \text{dissociation} = \frac{[H^+]}{1.5} \times 100 = \frac{1.4 \times 10^{-2}}{1.5} \times 100 = 0.93\%$$

20.10 a. $$\frac{[H^+]^2}{0.10} \approx 6.6 \times 10^{-5};\ [H^+]^2 = 6.6 \times 10^{-6};\ [H^+] = 2.6 \times 10^{-3}\ M$$

b. $$[OH^-] = \frac{1.0 \times 10^{-14}}{2.6 \times 10^{-3}} = 3.8 \times 10^{-12}\ M$$

c. pH = 2.59

d. $$\%\ \text{diss.} = \frac{2.6 \times 10^{-3}}{0.10} \times 100 = 2.6\%$$

20.11 a. $$\frac{x^2}{0.120 - x} = 0.0014$$

$$x^2 + 0.0014x - 0.000168 = 0$$

$$x = \frac{-0.0014 \pm (2.0 \times 10^{-6} + 6.72 \times 10^{-4})^{\frac{1}{2}}}{2}$$

$$= \frac{-0.0014 \pm 0.0260}{2} = 0.012\ M$$

b. $$\frac{x^2}{0.120 - x} = 0.0014;\quad x^2 \approx 0.120(0.0014);\ x = 0.013\ M$$

$$\frac{x^2}{0.107} \approx 0.0014;\quad x^2 = 0.107(0.0014);\ x = 0.012\ M$$

20.12 a. $$[H^+] = K_a \times \frac{HNO_2}{NO_2^-} = 4.5 \times 10^{-4}\frac{(0.10)}{(0.20)} = 2.2 \times 10^{-4}$$

pH = 3.66

b. $$[H^+] = 4.5 \times 10^{-4}\frac{(0.10)}{(0.10)} = 4.5 \times 10^{-4};\ pH = 3.35$$

c. $$[H^+] = 4.5 \times 10^{-4}\frac{(0.10)}{(0.050)} = 9.0 \times 10^{-4};\ pH = 3.05$$

d. $$[H^+] = 4.5 \times 10^{-4}\frac{(0.10)}{(0.010)} = 4.5 \times 10^{-3};\ pH = 2.35$$

20.13 $$[H^+] = K_a\ H_2PO_4^- \times \frac{[H_2PO_4^-]}{[HPO_4^{2-}]} = 6.2 \times 10^{-8} \times \frac{0.025}{0.025} = 6.2 \times 10^{-8}$$

pH = 7.21

20.14 a. $K_a \approx 4 \times 10^{-4}$; $HNO_2 - NO_2^-$

b. $K_a \approx 5 \times 10^{-10}$; $NH_4^+ - NH_3$ or $H_3BO_3 - H_2BO_3^-$

c. $K_a \approx 4 \times 10^{-7}$; $H_2CO_3 - HCO_3^-$

20.15 a. $\frac{[HF]}{[F^-]} = \frac{[H^+]}{K_a\ HF} = \frac{1.0 \times 10^{-3}}{7.0 \times 10^{-4}} = 1.4$

b. $1.4 \times 0.100 = 0.14$ mol HF

c. $\frac{0.100}{1.4} = 0.071$ mol NaF

20.16 a. $[H^+] = K_a\ HC_2H_3O_2 \times \frac{[HC_2H_3O_2]}{[C_2H_3O_2^-]} = 1.8 \times 10^{-5} \times \frac{0.040}{0.050}$

$= 1.4 \times 10^{-5}$

pH = 4.85

b. $[HC_2H_3O_2] = 0.040\ M - 0.010\ M = 0.030\ M$

$[C_2H_3O_2^-] = 0.050\ M + 0.010\ M = 0.060\ M$

$[H^+] = 1.8 \times 10^{-5} \times \frac{0.030}{0.060} = 9.0 \times 10^{-6}$; pH = 5.05

c. $[HC_2H_3O_2] = 0.040\ M + 0.010\ M = 0.050\ M$

$[C_2H_3O_2^-] = 0.050\ M - 0.010\ M = 0.040\ M$

$[H^+] = 1.8 \times 10^{-5} \times \frac{0.050}{0.040} = 2.2 \times 10^{-5}$; pH = 4.66

20.17 a. $\frac{[H_2CO_3]}{[HCO_3^-]} = \frac{[H^+]}{K_a} = \frac{4.0 \times 10^{-8}}{4.2 \times 10^{-7}} = 0.095$

b. Take $[H_2CO_3] = 0.095$ M, $[HCO_3^-] = 1.00$ M

10% conversion: $[HCO_3^-] = 0.90(1.00) = 0.90$ M

$[H_2CO_3] = 0.095\ M + 0.10\ M = 0.195\ M$

$[H^+] = K_a \times \frac{[H_2CO_3]}{[HCO_3^-]} = 4.2 \times 10^{-7} \times \frac{0.195}{0.90} = 9.1 \times 10^{-8}$ M

pH = 7.04

c. 10% conversion: $[H_2CO_3] = 0.90(0.095) = 0.086$ M

$[HCO_3^-] = 1.00 + 0.009 = 1.01$ M

$[H^+] = K_a \times \frac{[H_2CO_3]}{[HCO_3^-]} = 4.2 \times 10^{-7} \times \frac{0.086}{1.01} = 3.6 \times 10^{-8}$

pH = 7.44

20.18 a, b (gives 0.10 mol $HC_2H_3O_2$, 0.10 mol $C_2H_3O_2^-$)

20.19 $\frac{[OH^-]^2}{0.20} \approx 1.8 \times 10^{-5}$; $[OH^-]^2 = 3.6 \times 10^{-6}$

$[OH^-] = 1.9 \times 10^{-3}$ M; $[H^+] = \frac{1.0 \times 10^{-14}}{1.9 \times 10^{-3}} = 5.3 \times 10^{-12}$ M

pH = 11.28

20.20 a. $K_b = \frac{1.0 \times 10^{-14}}{1.9 \times 10^{-5}} = 5.3 \times 10^{-10}$

b. $\frac{[OH^-]^2}{0.050} \approx 5.3 \times 10^{-10}$; $[OH^-]^2 = 2.6 \times 10^{-11}$

$[OH^-] = 5.1 \times 10^{-6}$ M; $[H^+] = \frac{1.0 \times 10^{-14}}{5.1 \times 10^{-6}} = 2.0 \times 10^{-9}$ M   pH = 8.70

20.21 a. $C_6H_7NH^+$; K_a larger

b. K_b $C_6H_7N = \frac{1.0 \times 10^{-14}}{2.5 \times 10^{-5}} = 4.0 \times 10^{-10}$

K_b $C_5H_5N = \frac{1.0 \times 10^{-14}}{2.4 \times 10^{-7}} = 4.2 \times 10^{-8}$

c. C_5H_5N

20.22 $\frac{[OH^-]^2}{0.25} \approx 2.5 \times 10^{-5}$; $[OH^-]^2 = 6.2 \times 10^{-6}$; $[OH^-] = 2.5 \times 10^{-3}$M

$[H^+] = \frac{1.0 \times 10^{-14}}{2.5 \times 10^{-3}} = 4.0 \times 10^{-12}$ M; pH = 11.40

20.23 $\frac{x^2}{0.12 - x} = 5.9 \times 10^{-3}$; $x^2 \approx 0.12(5.9 \times 10^{-3})$; x = 0.027

$\frac{x^2}{0.093} = 5.9 \times 10^{-3}$; $x^2 \approx 0.093(5.9 \times 10^{-3})$; x = 0.023

$[H^+] = \frac{1.0 \times 10^{-14}}{0.023} = 4.3 \times 10^{-13}$ M; pH = 12.37

20.24 a. $K = 1/K_a$ HClO $= 1/(3.2 \times 10^{-8}) = 3.1 \times 10^7$

b. $K = 1/K_b$ $HPO_4^{2-} = 1/(1.6 \times 10^{-7}) = 6.2 \times 10^6$

c. $K = 1/K_a$ $H_3PO_4 = 1/(7.5 \times 10^{-3}) = 1.3 \times 10^2$

20.25 a. $HNO_2(aq) \rightleftharpoons H^+(aq) + NO_2^-(aq)$ $K_1 = K_a$ HNO_2

$F^-(aq) + H^+(aq) \rightleftharpoons HF(aq)$ $K_2 = 1/K_a$ HF

$HNO_2(aq) + F^-(aq) \rightleftharpoons NO_2^-(aq) + HF(aq)$

$K = \frac{K_a \; HNO_2}{K_a \; HF} = \frac{4.5 \times 10^{-4}}{7.0 \times 10^{-4}} = 0.64$

b. $K = 1/K_b$ $NO_2^- = 1/(2.2 \times 10^{-11}) = 4.5 \times 10^{10}$

c. $K = 1/K_a$ $HNO_2 = 1/(4.5 \times 10^{-4}) = 2.2 \times 10^3$

20.26 a. $K = 1/K_a\ HC_4H_2O_4^- = 1/(6.0 \times 10^{-7}) = 1.7 \times 10^6$

b. $K = \dfrac{1}{K_a\ H_2C_4H_2O_4 \times K_a\ HC_4H_2O_4^-} = \dfrac{1}{(1.2 \times 10^{-2})(6.0 \times 10^{-7})}$

$= 1.4 \times 10^8$

c. $K = \dfrac{K_a\ H_2C_4H_2O_4}{K_w} = \dfrac{1.2 \times 10^{-2}}{1.0 \times 10^{-14}} = 1.2 \times 10^{12}$

20.27 a. The benzoate ion is the anion of a weak acid, so produces OH^- ions in water.

b. HNO_2 is a weak acid, not completely ionized.

c. Reacts with both H^+ and OH^- ions

20.28 $HCl < HC_2H_3O_2 < NH_4Cl < NaC_2H_3O_2 < NH_3 < NaOH$

20.29 a. main species: $HC_2H_3O_2$

$\dfrac{[H^+]^2}{0.20} = 1.8 \times 10^{-5}$; $[H^+]^2 = 3.6 \times 10^{-6}$

$[H^+] = 1.9 \times 10^{-3}$ M; pH = 2.72

b. main species: K^+, CN^-

$\dfrac{[OH^-]^2}{0.30} = 2.5 \times 10^{-5}$; $[OH^-]^2 = 7.5 \times 10^{-6}$; $[OH^-] = 2.7 \times 10^{-3}$

$[H^+] = \dfrac{1.0 \times 10^{-14}}{2.7 \times 10^{-3}} = 3.7 \times 10^{-12}$ M; pH = 11.43

c. main species: K^+, OH^-

$[OH^-] = 0.40$ M; $[H^+] = 2.5 \times 10^{-14}$ M; pH = 13.60

20.30 a. $\dfrac{[H^+]^2}{0.20} \approx 3.5 \times 10^{-5}$; $[H^+]^2 = 7.0 \times 10^{-6}$

$[H^+] = 2.6 \times 10^{-3}$ M; pH = 2.59

b. $K_b = \dfrac{1.0 \times 10^{-14}}{3.5 \times 10^{-5}} = 2.9 \times 10^{-10} \approx \dfrac{[OH^-]^2}{0.20}$

$[OH^-]^2 = 5.8 \times 10^{-11}$; $[OH^-] = 7.6 \times 10^{-6}$ M

$[H^+] = 1.3 \times 10^{-9}$ M; pH = 8.89

c. $[H^+] = K_a = 3.5 \times 10^{-5}$ M; pH = 4.46

CHAPTER 21

BASIC SKILLS

1. Given the charge and composition of a complex ion, determine the charge and coordination number of the central metal cation.
2. Given the composition of a complex, sketch its geometry, including any geometric isomers.
3. Write the orbital diagram (valence bond model) of a transition metal ion and complexes derived from it.
4. For any octahedral complex, draw diagrams for the "high spin" and "low spin" forms (crystal field model).
5. Given the formation constant for a complex ion, relate the concentrations of free cation, complex ion, and ligand.

LECTURE NOTES

One thing to keep in mind throughout this chapter is the amount of "jargon" involved. Such terms as "chelating agent", "coordination number", "square planar", and "ligand" are unfamiliar to students. They also have trouble visualizing the structure of a complex given only its formula, e.g., $Co(en)_2(NH_3)_2^{3+}$.

The two main topics of this chapter are:

- geometry of complexes, including isomerism. Here, it helps to emphasize the geometric relations in an octahedron, a figure which is new to many students.

- electronic structure of complexes. Here, it seems unnecessary to cover both the valence bond and crystal field models. In the lecture outline below, we deal only with the valence bond model. If you choose the crystal field approach (and limit it to octahedral complexes, as the text does), somewhat less time may be required. Either way, two lectures should be quite adequate for this chapter.

<u>LECTURE 1</u>

I <u>Complex Ions and Coordination Compounds</u>

A. A complex ion consists of a central metal cation (usually derived from a transition metal) joined by coordinate covalent bonds to two or more molecules or anions called ligands.

Complex ion	Cation	Ligands	Coordination no.
$Ag(NH_3)_2^+$	Ag^+	2 NH_3 molecules	2
$Cu(H_2O)_4^{2+}$	Cu^{2+}	4 H_2O molecules	4
$Fe(CN)_6^{3-}$	Fe^{3+}	6 CN^- ions	6

$$\underset{H'}{\overset{H}{}}\,H-N:Ag:N-H\underset{H}{\overset{H}{}}$$

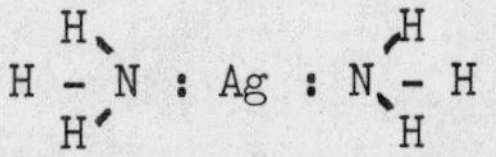

coordinate covalent bonds between Ag and N atoms

Coordination compound contains complex ion. Examples:

$$[Cu(H_2O)_4]SO_4, \quad [Ag(NH_3)_2]NO_3, \quad K_3[Fe(CN)_6]$$

Charge of central cation in $Zn(H_2O)_3(OH)^+$? +2

$$[Co(NH_3)_6]Cl_3 \quad ? \quad +3$$

B. Nature of ligands

Ordinarily contains at least one unshared pair of electrons

$$:N\underset{H}{\overset{H}{-}}H \qquad H\overset{\ddot{O}}{\cdots}H \qquad (:\ddot{O}-H)^- \qquad (:\ddot{C}\ddot{l}:)^-$$

If the ligand contains two or more unshared pairs on different, nonadjacent atoms, it can act as a chelating agent, forming more than one bond with the central metal ion.

$$\left[\begin{array}{c} :\ddot{O}: \\ :\ddot{O}: \end{array} C - C \begin{array}{c} :\ddot{O}: \\ :\ddot{O}: \end{array} \right]^{2-} \qquad H - \underset{H}{\overset{\cdot\cdot}{N}} - CH_2 - CH_2 - \underset{H}{\overset{\cdot\cdot}{N}} - H$$

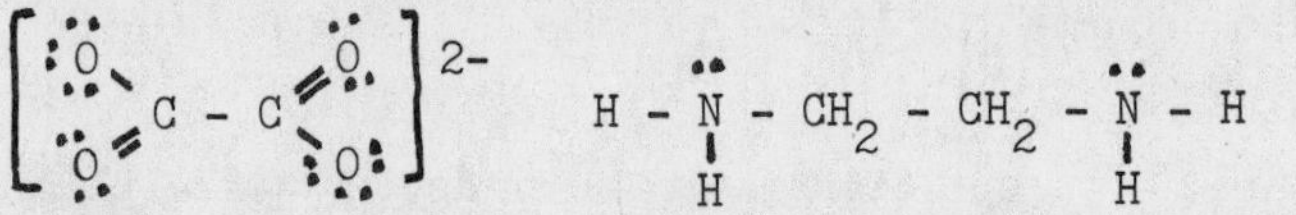

oxalate ion (ox)   ethylenediamine (en)

both form bidentate chelates:

$Co(en)_3^{3+}$ coord. no. $Co^{3+} = 6$

$Pt(ox)_2^{2-}$ coord. no. $Pt^{2+} = 4$

II Geometry of Complex Ions

A. Coordination no. = 2; linear $(NH_3 - Ag - NH_3)^+$

180° bond angle

B. Coordination no. = 4

1. Tetrahedral: $Zn(NH_3)_4^{2+}$

2. Square planar: $Cu(H_2O)_4^{2+}$

$$\left[\begin{array}{ccc} H_2O & & H_2O \\ & Cu & \\ H_2O & & H_2O \end{array} \right]^{2+}$$

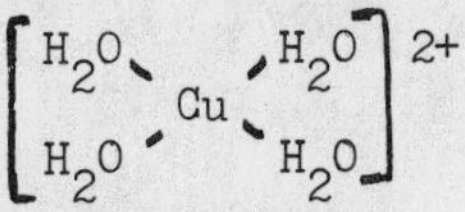

can show isomerism:

$$\left[\begin{array}{ccc} Br & & Br \\ & Ni & \\ Cl & & Cl \end{array} \right]^{2-} \qquad \left[\begin{array}{ccc} Cl & & Br \\ & Ni & \\ Br & & Cl \end{array} \right]^{2-}$$

cis isomer   trans isomer
(like groups close)   (like groups far apart)

C. Coordination number = 6; octahedral

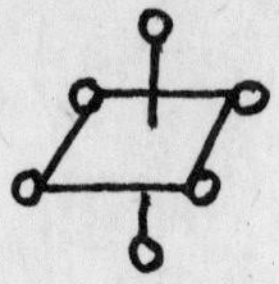

any position is trans to one position and cis to the other four

$Co(NH_3)_4Cl_2^+$ two isomers

$Co(NH_3)_3Cl_3$ two isomers

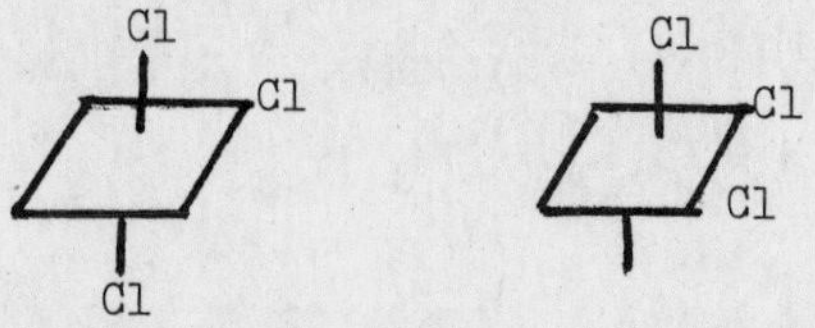

LECTURE 2

I Electronic Structure of Complex Ions (valence bond model)
Electron pairs contributed by ligands enter hybrid orbitals of central metal ion.

A. Orbitals occupied by ligand pairs

Coord. no. = 2: sp

Coord. no. = 4: sp^3 (tetrahedral) or dsp^2 (square planar)

Coord. no. = 6: d^2sp^3

B. Orbital diagrams
Start by determining coordination number. Then locate ligand pairs (must know whether complex is tetrahedral or square planar if coordination number is 4). Finally, locate electrons of central metal cation, remembering that:

- there are no outer s electrons in transition metal ions
- metal ion electrons are distributed by Hund's rule

Example: $Ni(CN)_4^{2-}$ square planar (8 e^- beyond Ar in Ni^{2+})

3d 4s 4p

$_{18}Ar$ (↑↓)(↑↓)(↑↓)(↑↓)(↑↓) (↑↓) (↑↓)(↑↓)( )

$Cr(H_2O)_6^{3+}$:

3d 4s 4p

$_{18}Ar$ (↑)(↑)(↑)(↑↓)(↑↓) (↑↓) (↑↓)(↑↓)(↑↓)

II Labile and Inert Complexes

Most complexes are labile; exchange ligands virtually instantaneously.

$$Ni(H_2O)_6^{2+}(aq) + 6\ NH_3(aq) \longrightarrow Ni(NH_3)_6^{2+}(aq) + 6\ H_2O$$

color changes from green to blue as soon as NH_3 is added

A few complexes are inert; ligand exchange is slow. Most complexes of Co^{3+}, Cr^{3+}, Pt^{4+} are of this type.

III Equilibria Involved in Complex Ion Formation

A. K_f

$$Ag^+(aq) + 2\ NH_3(aq) \rightleftharpoons Ag(NH_3)_2^+(aq);\ K_f = 2 \times 10^7$$

$$Ag^+(aq) + 2\ CN^-(aq) \rightleftharpoons Ag(CN)_2^-(aq)\ ;\ K_f = 1 \times 10^{21}$$

$Ag(CN)_2^-$ more stable than $Ag(NH_3)_2^+$

B. Application

1. Calculate ratio of $[Ag^+]/[Ag(NH_3)_2^+]$ in 0.1 M NH_3

$$\frac{[Ag(NH_3)_2^+]}{[Ag^+] \times [NH_3]^2} = 2 \times 10^7$$

$$\frac{Ag^+}{Ag(NH_3)_2^+} = \frac{1}{(2 \times 10^7) \times [NH_3]^2} = \frac{1}{2 \times 10^5} = 5 \times 10^{-6}$$

2. Calculate K for:

$$AgCl(s) + 2\ NH_3(aq) \rightleftharpoons Ag(NH_3)_2^+(aq) + Cl^-(aq)$$

$$K = K_{sp}\ AgCl \times K_f\ Ag(NH_3)_2^+$$

$$= (1.6 \times 10^{-10})(2 \times 10^7) = 3 \times 10^{-3}$$

Solubility of AgCl in 6 M NH_3?

$$\frac{s^2}{6^2} = 3 \times 10^{-3};\quad s^2 = 0.11;\quad s \approx 0.3\ M$$

DEMONSTRATIONS

1. Formation of $Cu(NH_3)_4^{2+}$ ion: Test. Dem. 22
2. EDTA complexes: Test. Dem. 132
3. Paramagnetism of complex ions: Test. Dem. 215; J. Chem. Educ. 54 431 (1977)
4. Complexes of Co^{2+}: Test. Dem. 222; J. Chem. Educ. 57 453 (1980)
5. Complex ions and precipitates of Ag^+: J. Chem. Educ. 57 813 (1980)
6. Complex ions and precipitates of Ni^{2+}: J. Chem. Educ. 57 900 (1980)

QUIZZES

Quiz 1

1. Show the geometry, including isomers, of

   a. $AgCl_2^-$ b. $Zn(en)_2^{2+}$ (tetrahedral) c. $Co(en)_2Cl_2^+$

2. Give the formula of a simple salt (e.g., NaCl) which would be similar to each of the following in conductivity and colligative properties.

   a. $K_2[Zn(OH)_4]$ b. $[Co(NH_3)_3(en)Cl]SO_4$ c. $[Pt(NH_3)_4]Cl_2$

Quiz 2

1. Draw orbital diagrams for each of the following:

   a. Fe^{3+} b. Zn^{2+} c. $Fe(CN)_6^{3-}$ d. $Zn(CN)_4^{2-}$ (tetrahedral)

2. Give the charge and coordination number of the central metal ion in:

   a. $[Co(NH_3)_4(en)]Cl_2$ b. $[Pt(en)(H_2O)Cl_3]^+$ c. $Al(OH)_6^{3-}$

Quiz 3

1. The formation constant for $Al(OH)_4^-$ is 1×10^{33}. At what pH is the ratio $[Al^{3+}]/[Al(OH)_4^-] = 1 \times 10^{-13}$?

2. Show the geometric isomers of $[Pt(NH_3)(H_2O)Cl_2]$ (square planar)

Quiz 4

1. Draw orbital diagrams for

   a. Cr^{3+} b. Co^{2+} c. $Cr(H_2O)_4(en)^{3+}$ d. CoF_6^{4-}

2. Given that K_{sp} for $Al(OH)_3$ is 5×10^{-33} and K_f $Al(OH)_4^-$ is 1×10^{33}, calculate K for

   $Al(OH)_3(s) + OH^-(aq) \rightleftharpoons Al(OH)_4^-(aq)$

Quiz 5

1. Draw all the isomers of the following complexes

   a. $Cr(NH_3)_3(NO_3)_3$ b. $Co(en)_2Br_2^+$ c. $Ni(NH_3)_5(H_2O)^{2+}$

2. What is the coordination number and charge of the central metal ion in each of the complexes in Question 1?

PROBLEMS 1-30

21.1 a. 2 NH_3 molecules, 2 H_2O molecules, 2 Br^- ions b. +3

c. $[Cr(NH_3)_2(H_2O)_2Br_2]SO_4$

21.2 a. $Pt(NH_3)_2Cl_2$

b. $Pt(en)(NO_2)_2$

c. $Pt(NH_3)_2ClBr$

21.3 a. 0 b. 2.00 c. 4.00 d. 3.00

21.4 a. 4 b. 6 c. 6 d. 6

21.5 a. $Ag(en)^+$ b. $Fe(H_2O)_6^{2+}$ c. $Zn(CN)_4^{2-}$ d. $Fe(C_2O_4)_3^{3-}$

21.6 a. monodentate
b. tetradentate (unshared pairs on N, O atoms)
c. tridentate (unshared pairs on N atoms)
d. monodentate

21.7 a. 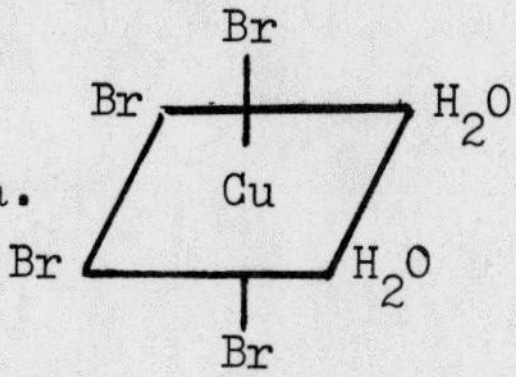

b. 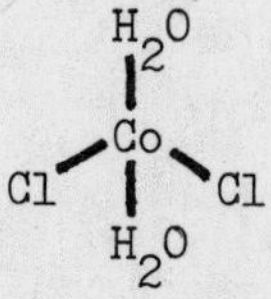

c. 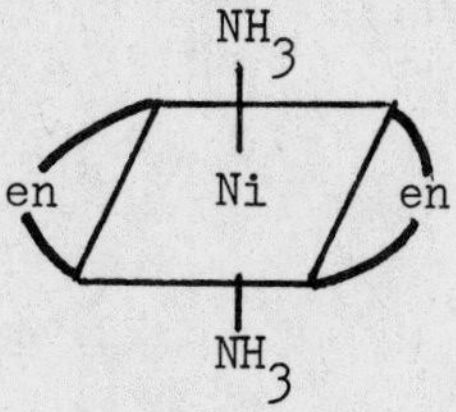

d. 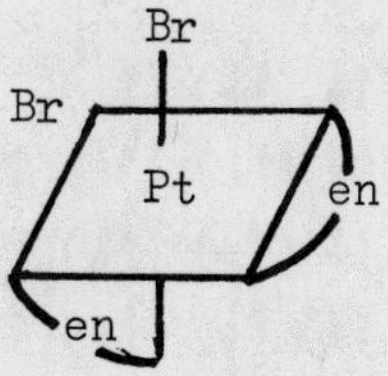

e. 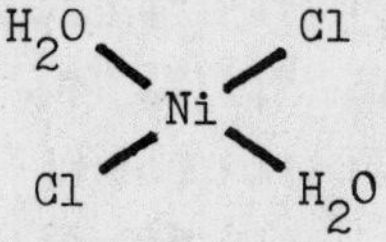

21.8 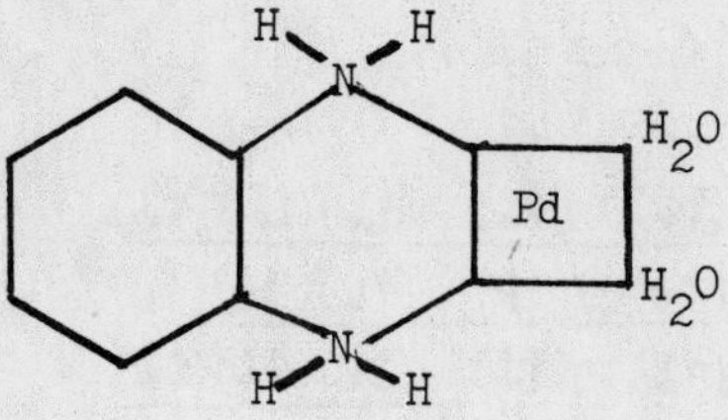

21.9 b. 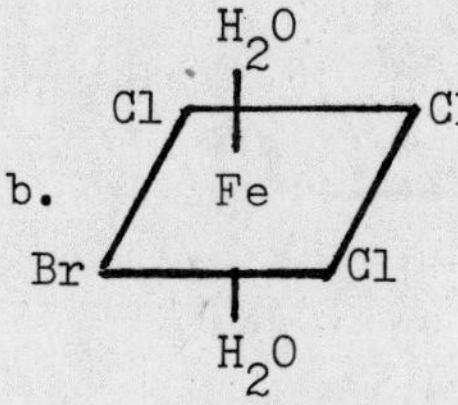, 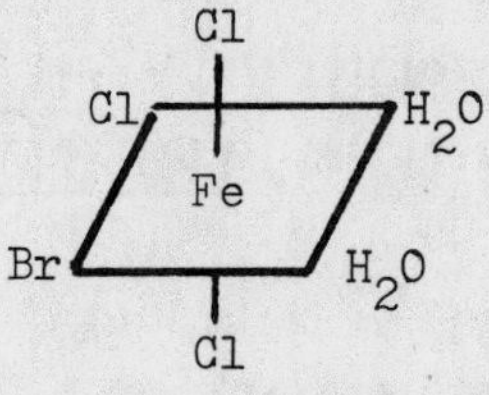, 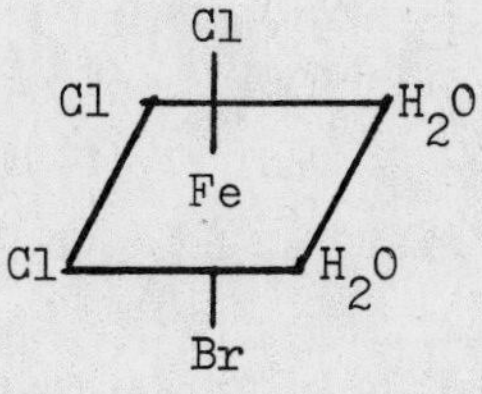

c. 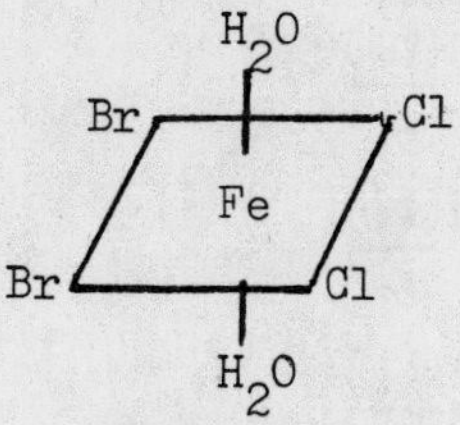, 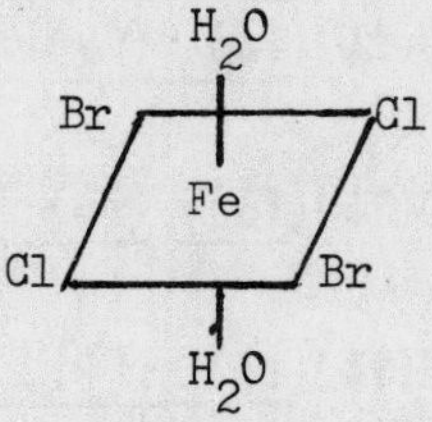, 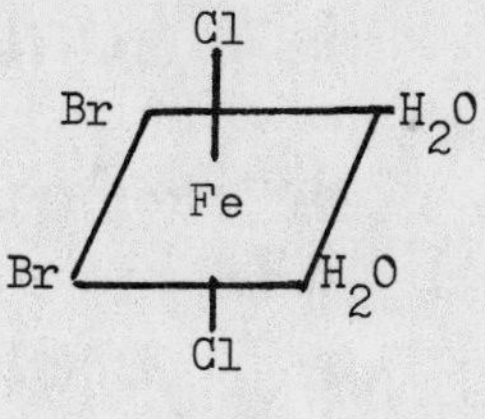

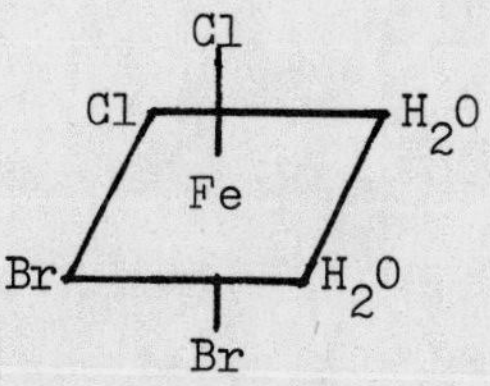, 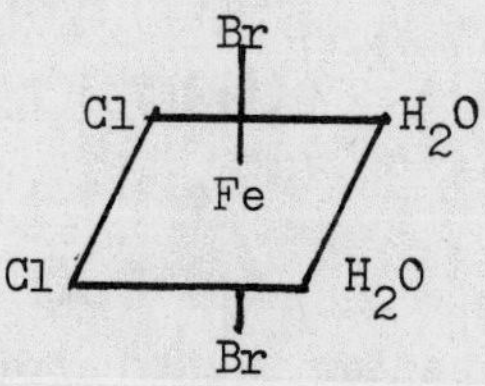

21.10 $Co(en)_3^{3+}$, $Co(en)_2Cl_2^{+}$ (2 isomers), $Co(en)Cl_4^{-}$,

$CoCl_6^{3-}$; total of 5 complexes

21.11 a. $1s^2 2s^2 2p^6 3s^2 3p^6 3d^1$

b. $1s^2 2s^2 2p^6 3s^2 3p^6 3d^3$

c. $1s^2 2s^2 2p^6 3s^2 3p^6 3d^7$

d. $1s^2 2s^2 2p^6 3s^2 3p^6 3d^{10}$

e. $1s^2 2s^2 2p^6 3s^2 3p^6 4s^2 3d^{10} 4p^6 4d^4$

21.12

	1s	2s	2p	3s	3p	3d
a.	(↑↓)	(↑↓)	(↑↓)(↑↓)(↑↓)	(↑↓)	(↑↓)(↑↓)(↑↓)	(↑)( )( )( )( )
b.	(↑↓)	(↑↓)	(↑↓)(↑↓)(↑↓)	(↑↓)	(↑↓)(↑↓)(↑↓)	(↑)(↑)(↑)( )( )
c.	(↑↓)	(↑↓)	(↑↓)(↑↓)(↑↓)	(↑↓)	(↑↓)(↑↓)(↑↓)	(↑↓)(↑↓)(↑)(↑)(↑)
d.	(↑↓)	(↑↓)	(↑↓)(↑↓)(↑↓)	(↑↓)	(↑↓)(↑↓)(↑↓)	(↑↓)(↑↓)(↑↓)(↑↓)(↑↓)

e. all orbitals filled through 4p; 4 unpaired electrons in 4d

1 unpaired e^- in a, 3 in b, 3 in c, 0 in d

21.13

	3d	4s	4p
a.	(↑↓)(↑↓)(↑)(↑↓)(↑↓)	(↑↓)	(↑↓)(↑↓)(↑↓)
b.	(↑↓)(↑↓)(↑↓)(↑↓)(↑↓)	(↑↓)	(↑↓)(↑↓)( )
c.	(↑↓)(↑↓)(↑↓)(↑↓)(↑↓)	(↑↓)	(↑↓)(↑↓)(↑↓)
d.	(↑)(↑)(↑)(↑↓)(↑↓)	(↑↓)	(↑↓)(↑↓)(↑↓)

21.14

	3d	4s	4p	
$CuCl_2^{-}$	(↑↓)(↑↓)(↑↓)(↑↓)(↑↓)	(↑↓)	(↑↓)( )( )	0 unpaired e^-
$CuCl_4^{2-}$	(↑↓)(↑↓)(↑↓)(↑↓)(↑↓)	(↑↓)	(↑↓)(↑↓)(↑)	1 unpaired e^-

21.15 a. 2 b. 2 c. 0

21.16

	3d	4s	4p
Cu^{2+}	(↑↓)(↑↓)(↑↓)(↑↓)(↑↓)	(↑↓)	(↑↓)(↑↓)(↑)
Ni^{2+}	(↑↓)(↑↓)(↑↓)(↑↓)(↑↓)	(↑↓)	(↑↓)(↑↓)( )

	4d	5s	5p
Pd^{2+}	(↑↓)(↑↓)(↑↓)(↑↓)(↑↓)	(↑↓)	(↑↓)(↑↓)( )

	5d	6s	6p
Pt^{2+}	(↑↓)(↑↓)(↑↓)(↑↓)(↑↓)	(↑↓)	(↑↓)(↑↓)( )

21.17 a.
( )( ) , (↑)(↑)
(↑↓)(↑↓)(↑↓) , (↑↓)(↑)(↑)

b.
( )( ) , (↑)(↑)
(↑↓)(↑↓)(↑) , (↑)(↑)(↑)

21.18 Electrons must enter two lower d orbitals, unpaired

21.19 ( )( ) ($\uparrow$)($\uparrow$)

($\uparrow\downarrow$)($\uparrow\downarrow$)($\uparrow\downarrow$) ($\uparrow\downarrow$)($\uparrow$)($\uparrow$)

$Co(NH_3)_6^{3+}$ CoF_6^{3-}

21.20 a. ( )( ) ($\uparrow\downarrow$)($\uparrow$)($\uparrow$) 2 unpaired e^-

b. ( )( ) ($\uparrow\downarrow$)($\uparrow\downarrow$)($\uparrow\downarrow$) 0 unpaired e^-

c. ( )( ) ($\uparrow\downarrow$)($\uparrow\downarrow$)($\uparrow\downarrow$) 0 unpaired e^-

d. ( )( ) ($\uparrow$)($\uparrow$)( ) 2 unpaired e^-

e. ( )( ) ($\uparrow\downarrow$)($\uparrow$)($\uparrow$) 2 unpaired e^-

21.21 $\lambda = \frac{1.196 \times 10^5}{300} = 399$ nm

21.22 Study color (absorption spectrum) as function of time

21.23 $\log_{10}\frac{X_o}{X} = \frac{(2.2 \times 10^{-6})(3600)(24)}{2.30} = 0.083$; $X_o/X = 1.21$

$X = 0.83\ X_o$; fraction reacted = 1 - 0.83 = 0.17

21.24 $\frac{[Mn(C_2O_4)_2^{2-}]}{[Mn^{2+}]} = 6.3 \times 10^5 \times [C_2O_4^{2-}]^2$

a. $6.3 \times 10^5\ (0.10)^2 = 6.3 \times 10^3$

b. $6.3 \times 10^5(1.0 \times 10^{-4})^2 = 6.3 \times 10^{-3}$

21.25 $[Ag(NH_3)_2^+] \approx 0.10$ M

$\frac{[Ag(NH_3)_2^+]}{[Ag^+]} = 2 \times 10^7 \times [NH_3]^2$; $[Ag^+] = \frac{[Ag(NH_3)_2^+]}{(2 \times 10^7) \times [NH_3]^2}$

$= 1 \times 10^{-9}$ M

21.26 $CuCl_2^-(aq) \rightleftharpoons Cu^+(aq) + 2\ Cl^-(aq)$; $K = 1/K_f\ CuCl_2^-$

$Cu^+(aq) + 2\ NH_3(aq) \rightleftharpoons Cu(NH_3)_2^+(aq)$; $K = K_f\ Cu(NH_3)_2^+$

$CuCl_2^-(aq) + 2\ NH_3(aq) \rightleftharpoons Cu(NH_3)_2^+(aq) + 2\ Cl^-(aq)$

$$K = \frac{K_f\ Cu(NH_3)_2^+}{K_f\ CuCl_2^-} = \frac{5 \times 10^{10}}{1 \times 10^5} = 5 \times 10^5$$

21.27 a. two pairs of unshared electrons
b. forms stable complex, $Ag(NH_3)_2^+$
c. all four positions are equivalent; each position is equidistant from all others

21.28 a. kinetically unstable; dissociates rapidly
b. thermodynamically stable; large value of K_f
c. ligands at corners of square, metal ion in center
d. complex with maximum number of unpaired electrons
e. complex with minimum number of unpaired electrons

21.29 a. true; structure is $[Pt(NH_3)_4Cl_2]Cl$

b. true

c. false; A is more stable

21.30 a. in 100 g of compound:

$$\text{no. moles Co} = 22.0 \text{ g Co} \times \frac{1 \text{ mol Co}}{58.93 \text{ g Co}} = 0.373 \text{ mol Co}$$

$$\text{no. moles N} = 31.4 \text{ g N} \times \frac{1 \text{ mol N}}{14.01 \text{ g N}} = 2.24 \text{ mol N}$$

$$\text{no. moles H} = 6.78 \text{ g H} \times \frac{1 \text{ mol H}}{1.008 \text{ g H}} = 6.73 \text{ mol H}$$

$$\text{no. moles Cl} = 39.8 \text{ g Cl} \times \frac{1 \text{ mol Cl}}{35.45 \text{ g Cl}} = 1.12 \text{ mol Cl}$$

formula: $[Co(NH_3)_6]Cl_3$

b. $[Co(NH_3)_6]Cl_3(s) \longrightarrow Co(NH_3)_6^{3+}(aq) + 3\ Cl^-(aq)$

CHAPTER 22

BASIC SKILLS

1. Classify cations as being in Group I, II, III, or IV of the qualitative analysis scheme.
2. Describe how successive groups are separated from one another in cation analysis.
3. Given the results of tests carried out with a Group I unknown, state whether each ion is present, absent, or in doubt.
4. Write balanced net ionic equations for the following types of reactions carried out in qualitative analysis:
   - formation and decomposition of complex ions
   - formation and dissolving of precipitates
5. Apply equilibrium principles (Ch. 18-21) to problems in qualitative analysis.

LECTURE NOTES

This chapter offers a way to review chemical equations for solution reactions and solution equilibria. It is placed here in the text, after complex ions, so that the chemistry of qualitative analysis can be covered intelligibly. Note that the chemistry of only one group (Group I) is covered in detail.

Two lectures should be sufficient for this chapter.

<u>LECTURE 1</u>

I <u>Cation Group Separations</u>

A. Group I (Ag^+, Pb^{2+}, Hg_2^{2+}). Form insoluble chlorides, so are precipitated by HCl.

B. Group II (Cu^{2+}, Bi^{3+}, Hg^{2+}, Cd^{2+}, Sn^{4+}, Sb^{3+}) Form very insoluble sulfides. Hence, these cations are precipitated by H_2S in acidic solution (pH = 0.5), where $[S^{2-}]$ is very low.

C. Group III (Al^{3+}, Cr^{3+}, Co^{2+}, Fe^{2+}, Mn^{2+}, Zn^{2+}, Ni^{2+}) Sulfides of these ions are more soluble than those of Group II. However, precipitates do form at pH 9. Al^{3+} and Cr^{3+} come down as hydroxides, other cations as sulfides.

D. Group IV (Mg^{2+}, Ca^{2+}, Ba^{2+}, Na^+, K^+, NH_4^+) The alkaline earth cations can be precipitated as carbonates. Na^+ and K^+ give distinctive flame tests; the NH_4^+ ion can be detected by heating with strong base:

$$NH_4^+(aq) + OH^-(aq) \longrightarrow NH_3(g) + H_2O$$

II <u>Group I Analysis</u>

Flow sheet:

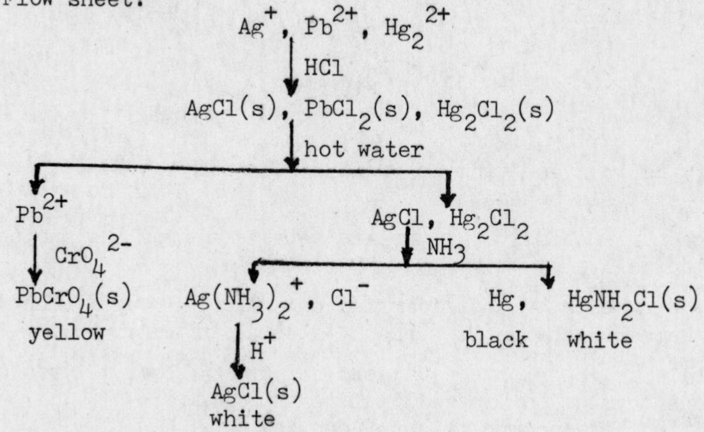

III <u>Reactions in Qualitative Analysis</u>

A. Complex ion formation

$$Ni^{2+}(aq) + 6\ NH_3(aq) \longrightarrow Ni(NH_3)_6^{2+}(aq)$$
$$Zn^{2+}(aq) + 4\ OH^-(aq) \longrightarrow Zn(OH)_4^{2-}(aq)$$

Note list of complex ions in Table 22.2

B. Complex ion decomposition

Complexes containing OH^- ions or NH_3 molecules as ligands can be decomposed by adding strong acid (H^+ ions). In this way, OH^- ions $\rightarrow$ H_2O, NH_3 molecules $\rightarrow$ NH_4^+ ions.

$$Zn(OH)_4^{2-}(aq) + 4\ H^+(aq) \longrightarrow Zn^{2+}(aq) + 4\ H_2O$$
$$Ni(NH_3)_6^{2+}(aq) + 6\ H^+(aq) \longrightarrow Ni^{2+}(aq) + 6\ NH_4^+(aq)$$

Note confirmatory test for Ag^+:

$$Ag(NH_3)_2^+(aq) + Cl^-(aq) + 2H^+(aq) \longrightarrow AgCl(s) + 2\ NH_4^+(aq)$$

LECTURE 2

I <u>Reactions in Qualitative Analysis</u> (cont.)

A. Precipitate formation

1. Direct addition of anion, as in precipitation of Group I

$$Ag^+(aq) + Cl^-(aq) \longrightarrow AgCl(s)$$
$$Hg_2^{2+}(aq) + 2\ Cl^-(aq) \longrightarrow Hg_2Cl_2(s)$$

2. Addition of H_2S (source of S^{2-} ions)

$$Cu^{2+}(aq) + H_2S(aq) \longrightarrow CuS(s) + 2\ H^+(aq)$$
$$2Bi^{3+}(aq) + 3\ H_2S(aq) \longrightarrow Bi_2S_3(s) + 6\ H^+(aq)$$

3. Addition of NH_3 (source of OH^- ions)

$Pb^{2+}(aq) + 2\ NH_3(aq) + 2\ H_2O \longrightarrow Pb(OH)_2(s) + 2\ NH_4^+(aq)$

$Al^{3+}(aq) + 3\ NH_3(aq) + 3\ H_2O \longrightarrow Al(OH)_3(s) + 3\ NH_4^+(aq)$

very useful where cation forms hydroxo complex; not so useful if it forms complex with ammonia

B. Dissolving precipitates

1. Strong acid will often dissolve solid containing basic anion, particularly OH^- or CO_3^{2-}:

$Fe(OH)_3(s) + 3\ H^+(aq) \longrightarrow Fe^{3+}(aq) + 3\ H_2O$

$CaCO_3(s) + 2\ H^+(aq) \longrightarrow Ca^{2+}(aq) + H_2CO_3(aq)$

2. NH_3 or NaOH may be used if cation forms complex with NH_3 or OH^-:

$Zn(OH)_2(s) + 2\ OH^-(aq) \longrightarrow Zn(OH)_4^{2-}(aq)$

$Zn(OH)_2(s) + 4\ NH_3(aq) \longrightarrow Zn(NH_3)_4^{2+}(aq) + 2\ OH^-(aq)$

II <u>Equilibria in Qualitative Analysis</u>

A. K_{sp} of $PbCl_2 = 1.3 \times 10^{-4}$ at 100°C. Calculate solubility.

$4s^3 = 1.3 \times 10^{-4}$

$s = 0.032$ M

This is large enough to give a positive test with CrO_4^{2-}.

B. Precipitate $PbCl_2$ in Group I with 6 M HCl. Calculate conc. Pb^{2+} remaining, taking K_{sp} $PbCl_2 = 1.7 \times 10^{-5}$ at 25°C.

$[Pb^{2+}] = \dfrac{1.7 \times 10^{-5}}{6^2} = 4.7 \times 10^{-7}$

Will PbS ($K_{sp} = 1 \times 10^{-27}$) precipitate in Group II where $[S^{2-}] = 1 \times 10^{-20}$ M?

$(4.7 \times 10^{-7})(1 \times 10^{-20}) = 5 \times 10^{-27} > K_{sp}$ PbS

PbS is likely to come down in Group II if Pb^{2+} is present

C. Separation of Group II, III cations

$[S^{2-}]$ in Group II analysis $= 1 \times 10^{-20}$ M

CuS: conc. Cu^{2+} x conc. $S^{2-} = 1 \times 10^{-21} > K_{sp}$ CuS $= 10^{-35}$

ZnS: conc. Zn^{2+} x conc. $S^{2-} = 1 \times 10^{-21} < K_{sp}$ ZnS $= 10^{-20}$

CuS comes down in Group II, ZnS doesn't

D. Dissolving of hydroxides in acid

$Mg(OH)_2(s) \rightleftharpoons Mg^{2+}(aq) + 2\ OH^-(aq) \qquad K_1$

$2\ H^+(aq) + 2\ OH^-(aq) \rightleftharpoons 2\ H_2O \qquad K_2$

$Mg(OH)_2(s) + 2H^+(aq) \rightleftharpoons Mg^{2+}(aq) + 2H_2O \qquad K_3$

$K_1 = K_{sp}\ Mg(OH)_2 = 1 \times 10^{-11}$

$K_2 = 1/K_w^2 = 1/(1 \times 10^{-14})^2 = 1 \times 10^{28}$

$K_3 = (1 \times 10^{-11})(1 \times 10^{28}) = 1 \times 10^{17}$

Since K_3 is so large, $Mg(OH)_2$ (and indeed all other insoluble hydroxides), dissolves readily in strong acid

DEMONSTRATIONS

1. Test for NH_4^+ ion: Test. Dem. 27

QUIZZES

Quiz 1

1. Give the formula of the precipitate formed in group separations of the following cations:

   a. Ag^+ b. Cu^{2+} c. Ca^{2+} d. Zn^{2+}

2. Write a balanced equation for the reaction by which:
   a. a precipitate of $Fe(OH)_3$ dissolves in strong acid.
   b. a precipitate of $Al(OH)_3$ dissolves in strong base.
   c. a precipitate forms when NH_3 is added to a solution containing Fe^{3+} ions.

Quiz 2

1. The chloride precipitate from a Group I unknown is completely insoluble in both hot water and ammonia. Indicate whether each of the Group I cations is present, absent, or in doubt.

2. Write a balanced equation for the reaction by which:
   a. Sb^{3+} forms a precipitate with H_2S.
   b. Sb^{3+} forms a precipitate with NH_3.
   c. the precipitate formed in (a) dissolves in NaOH.

Quiz 3

1. An NH_4^+ - NH_3 buffer is used in qualitative analysis to maintain a pH of 9.00. Taking K_a $NH_4^+ = 5.6 \times 10^{-10}$, calculate the ratio $[NH_4^+]/[NH_3]$ in the buffer.

2. What reagent is used in qualitative analysis to:
   a. precipitate Group I?
   b. precipitate Ni^{2+}?
   c. test for NH_4^+?

Quiz 4

1. Indicate the cation analysis group for each of the following:

   a. Pb^{2+} b. Na^+ c. Co^{2+} d. Sn^{4+}

2. Write a balanced equation for the reaction by which:

   a. Pb^{2+} forms a precipitate with ammonia.
   b. the precipitate in (a) dissolves in strong acid.
   c. the precipitate in (a) dissolves in strong base.

Quiz 5

1. Given: $H_2S(aq) \rightleftharpoons 2\ H^+(aq) + S^{2-}(aq)$; $K = 1 \times 10^{-20}$

   calculate:

   a. $[S^{2-}]$ in Group II analysis, where $[H_2S] = 0.1$ M and pH = 0.5

   b. $[S^{2-}]$ in Group III analysis, where $[H_2S] = 0.1$ M and pH = 9.0

2. Describe what happens in each step of a Group I analysis with an unknown that contains only Ag^+.

PROBLEMS 1-30

22.1

Pb^{2+}	I	Cl^-	$PbCl_2$
Cu^{2+}	II	H_2S	CuS
Cr^{3+}	III	OH^-	$Cr(OH)_3$
Ca^{2+}	IV	CO_3^{2-}	$CaCO_3$

22.2 a. form insoluble chlorides

b. do not form insoluble chlorides or sulfides

22.3 Na^+, K^+ and NH_4^+ are often present in reagents used to test for other cations.

22.4 $H_2S(aq) \rightleftharpoons 2\ H^+(aq) + S^{2-}(aq)$

Position of equilibrium and hence concentration of S^{2-} depends upon the concentration of H^+. The $[H^+]$ is kept high to avoid precipitating Group III cations.

22.5 a. Ag^+ b. Sn^{4+}, Sb^{3+} c. H_2S

22.6 a. would give positive test for Na^+, since reagents used to test for Groups I-III contain Na^+.

b. would precipitate both CuS and NiS

c. would form Sn^{4+}

22.7 a. 4 b. $1s^2 2s^2 2p^6$ c. tetrahedral

22.8 Forms white precipitate with HCl. When heated with water, part of precipitate dissolves. Treatment of hot water solution with K_2CrO_4 gives yellow precipitate. When the remaining precipitate is treated with NH_3, it dissolves. Treatment of the solution with HNO_3 gives a white precipitate.

22.9 Add NH_3. If the precipitate is $PbCl_2$, nothing happens. If it is AgCl, the precipitate dissolves. If it is Hg_2Cl_2, the precipitate turns black.

22.10 a. no precipitate forms b. no effect c. AgCl does not dissolve

22.11 Pb^{2+} absent, Ag^+ present, Hg_2^{2+} absent

22.12 a. Test water solubility: AgCl is insoluble, $AgNO_3$ soluble.

b. Test solubility in hot water; $PbCl_2$ dissolves, AgCl doesn't

c. Add NH_3; Hg_2^{2+} forms precipitate, Ag^+ doesn't.

22.12 d. Add SO_4^{2-}; Pb^{2+} gives precipitate, Hg_2^{2+} does not.

22.13 $AgNO_3$ is present (dissolves in water, forms ppt. with HCl)
Hg_2Cl_2 is present (turns black with NH_3)
AgCl is absent (would dissolve in NH_3, reprecipitate with acid)
$PbSO_4$ is in doubt

22.14 a. $Ag^+(aq) + 2\ NH_3(aq) \longrightarrow Ag(NH_3)_2^+(aq)$

b. $Pb^{2+}(aq) + 3\ OH^-(aq) \longrightarrow Pb(OH)_3^-(aq)$

c. $Ni^{2+}(aq) + 6\ NH_3(aq) \longrightarrow Ni(NH_3)_6^{2+}(aq)$

22.15 a. $Al(OH)_4^-(aq) + 4\ H^+(aq) \longrightarrow Al^{3+}(aq) + 4\ H_2O$

b. $Ag(NH_3)_2^+(aq) + 2\ H^+(aq) \longrightarrow Ag^+(aq) + 2\ NH_4^+(aq)$

c. no reaction

d. $Zn(OH)_4^{2-}(aq) + 4\ H^+(aq) \longrightarrow Zn^{2+}(aq) + 4\ H_2O$

22.16 a. $Bi^{3+}(aq) + 3\ OH^-(aq) \longrightarrow Bi(OH)_3(s)$

b. $2\ Bi^{3+}(aq) + 3\ S^{2-}(aq) \longrightarrow Bi_2S_3(s)$

c. $Bi^{3+}(aq) + 3\ NH_3(aq) + 3\ H_2O \longrightarrow Bi(OH)_3(s) + 3\ NH_4^+(aq)$

d. $2\ Bi^{3+}(aq) + 3\ H_2S(aq) \longrightarrow Bi_2S_3(s) + 6\ H^+(aq)$

22.17 a. $Ni(OH)_2(s) + 2\ H^+(aq) \longrightarrow Ni^{2+}(aq) + 2\ H_2O$

b. $BaCO_3(s) + 2\ H^+(aq) \longrightarrow Ba^{2+}(aq) + H_2CO_3(aq)$

c. $Al(OH)_3(s) + 3\ H^+(aq) \longrightarrow Al^{3+}(aq) + 3\ H_2O$

d. $Sb(OH)_4^-(aq) + 4\ H^+(aq) \longrightarrow Sb^{3+}(aq) + 4\ H_2O$

22.18 a. $Zn^{2+}(aq) + 4\ NH_3(aq) \longrightarrow Zn(NH_3)_4^{2+}(aq)$

b. $Zn(OH)_2(aq) + 4\ NH_3(aq) \longrightarrow Zn(NH_3)_4^{2+}(aq) + 2\ OH^-(aq)$

c. $Al^{3+}(aq) + 3\ NH_3(aq) + 3\ H_2O \longrightarrow Al(OH)_3(s) + 3\ NH_4^+(aq)$

22.19 a. $Al^{3+}(aq) + 4\ OH^-(aq) \longrightarrow Al(OH)_4^-(aq)$

b. $Al(OH)_3(s) + OH^-(aq) \longrightarrow Al(OH)_4^-(aq)$

c. $Fe^{3+}(aq) + 3\ OH^-(aq) \longrightarrow Fe(OH)_3(s)$

22.20 a. $CaCO_3(s) + 2\ H^+(aq) \longrightarrow Ca^{2+}(aq) + H_2CO_3(aq)$

b. $Mg(OH)_2(s) + 2\ H^+(aq) \longrightarrow Mg^{2+}(aq) + 2\ H_2O$

c. $ZnS(s) + 2\ H^+(aq) \longrightarrow Zn^{2+}(aq) + H_2S(aq)$

22.21 $AgCl(s) + 2\ NH_3(aq) \longrightarrow Ag(NH_3)_2^+(aq) + Cl^-(aq)$

$AgCl(s) + 2\ S_2O_3^{2-}(aq) \longrightarrow Ag(S_2O_3)_2^{3-}(aq) + Cl^-(aq)$

$AgCl(s) + 2\ CN^-(aq) \longrightarrow Ag(CN)_2^-(aq) + Cl^-(aq)$

22.22 a. $Pb^{2+}(aq) + 2\ Cl^-(aq) \longrightarrow PbCl_2(s)$

b. $PbCl_2(s) \longrightarrow Pb^{2+}(aq) + 2\ Cl^-(aq)$

$\underline{Pb^{2+}(aq) + CrO_4^{2-}(aq) \longrightarrow PbCrO_4(s)}$

$PbCl_2(s) + CrO_4^{2-}(aq) \longrightarrow PbCrO_4(s) + 2\ Cl^-(aq)$

22.23 a. $[H^+] = 1.0 \times 10^{-9}$ M

b. $\frac{[NH_4^+]}{[NH_3]} = \frac{[H^+]}{K_a} = \frac{1.0 \times 10^{-9}}{5.6 \times 10^{-10}} = 1.8$

c. 5.0 mL x 1.8 = 9.0 mL

22.24 $H_2S(aq) \rightleftharpoons H^+(aq) + HS^-(aq)$

$\frac{[H^+]^2}{0.10} \approx 1 \times 10^{-7}$; $[H^+]^2 = 1 \times 10^{-8}$; $[H^+] = 1 \times 10^{-4}$ M

pH = 4.0

22.25 a. $[Pb^{2+}] = \frac{K_{sp}\ PbCl_2}{[Cl^-]^2} = \frac{1.7 \times 10^{-5}}{(0.50)^2} = 6.8 \times 10^{-5}$ M

b. (conc. Pb^{2+}) x (conc. S^{2-}) = $(6.8 \times 10^{-5})(1 \times 10^{-20})$

$= 7 \times 10^{-25} > K_{sp}$ PbS

precipitate should form

22.26 a. conc. Cl^- = $\frac{0.10 \text{ mol/L} \times 0.010 \text{ L}}{0.030 \text{ L}}$ = 0.033 mol/L

conc. Pb^{2+} = $\frac{0.010 \text{ mol/L} \times 0.020 \text{ L}}{0.030 \text{ L}}$ = 0.0067 mol/L

b. (conc. Pb^{2+})(conc. Cl^-)^2 = $(6.7 \times 10^{-3})(3.3 \times 10^{-2})^2$

$= 7.3 \times 10^{-6} < K_{sp}$

no precipitate forms

22.27 a. $[S^{2-}] = 1 \times 10^{-20} \times \frac{H_2S}{[H^+]^2} = \frac{(1 \times 10^{-20})(1 \times 10^{-1})}{(1 \times 10^{-2})^2}$

$= 1 \times 10^{-17}$ M

b. conc. Cd^{2+} x conc. S^{2-} = $(1 \times 10^{-1})(1 \times 10^{-17})$

$= 1 \times 10^{-18} > K_{sp}$; CdS precipitates

c. conc. Zn^{2+} x conc. S^{2-} = $1 \times 10^{-18} > K_{sp}$; ZnS precipitates

22.28 $[OH^-]^2 = \frac{K_{sp}\ Mg(OH)_2}{[Mg^{2+}]} = \frac{1 \times 10^{-11}}{0.020} = 5 \times 10^{-10}$

$[OH^-] = 2 \times 10^{-5}$ M; $[H^+] = 5 \times 10^{-10}$ M; pH = 9.3

22.29 $K = K_{sp}\ Zn(OH)_2 \times K_f\ Zn(OH)_4^{2-} = (5 \times 10^{-17})(3 \times 10^{15}) = 0.2$

22.30 $K = \frac{K_{sp}\ Fe(OH)_2}{K_w^2} = \frac{1 \times 10^{-15}}{(1 \times 10^{-14})^2} = 1 \times 10^{13}$

less soluble, since K_{sp} for $Fe(OH)_2$ is smaller

CHAPTER 23

BASIC SKILLS

1. Given the formula of a species, determine the oxidation number of each atom.
2. Given a redox equation, select the oxidizing agent; the reducing agent.
3. Given the formulas of products and reactants, balance a redox equation by the half-equation method.
4. Given a balanced equation or half-equation in acidic solution, write the corresponding balanced equation in basic solution.
5. Contrast electrolytic with voltaic cells; identify anode and cathode in either type of cell and indicate the flow of current through all parts of the cell.
6. Predict the products when an ionic solute is electrolyzed in water solution.
7. Relate the number of electrons or coulombs passing through an electrolytic cell to the amounts of products formed at the electrodes.
8. Relate the amount of energy used in an electrolysis (joules or kilowatt hours) to the number of coulombs and volts.

LECTURE NOTES

This is the first of two chapters devoted to the principles of electrochemistry. It emphasizes oxidation number, balancing of redox equations, and the processes taking place in electrolytic and voltaic cells.

Students often find redox equations difficult to balance on first exposure. With practice, however, most of them become adept at it. The half-equation method is by no means the only one available. It does, however, have one important advantage. The process of breaking a redox equation into two half-equations is one which is used over and over again in electrochemistry.

The material in this chapter will certainly require two lectures, probably $2\frac{1}{2}$.

LECTURE 1

I Oxidation and Reduction

A. Oxidation = loss of electrons. Reduction = gain of electrons

$Zn(s) + Cu^{2+}(aq) \longrightarrow Zn^{2+}(aq) + Cu(s)$

$Zn(s) \longrightarrow Zn^{2+}(aq) + 2\ e^-$; oxidation

$Cu^{2+}(aq) + 2\ e^- \longrightarrow Cu(s)$; reduction

oxidation and reduction must occur simultaneously

B. Oxidation number - measure of "charge" carried by an atom in an ionic or molecular species. Assigned according to arbitrary rules.

1. Oxidation number of element in elementary substance (e.g., Cl_2, Fe) is zero.
2. Oxidation number of element in monatomic ion is the charge of that ion. Oxidation no. of Fe is +2 in Fe^{2+}, +3 in Fe^{3+}.
3. Group 1 elements: oxidation number = +1
   Group 2 elements: oxidation number = +2
   Oxidation number of F = -1
   Oxidation number of hydrogen almost always +1
   Oxidation number of O is usually -2
4. Sum of oxidation numbers of all atoms in a molecule = 0; in a polyatomic ion, the sum is the charge of the ion.

   H_2SO_4 +2 + oxid. no. S + 4(-2) = 0; oxid. no. S = +6

   $Cr_2O_7^{2-}$ 2(oxid. no. Cr) + 7(-2) = -2; oxid. no. Cr = +6

C. Oxidation = increase in oxidation number; reduction = decrease in oxidation number

$HCl(g) + HNO_3(l) \longrightarrow NO_2(g) + \frac{1}{2}\ Cl_2(g) + H_2O(l)$

HCl is oxidized; oxid. no. Cl increases from -1 to 0

HNO_3 is reduced; oxid. no N decreases from +5 to +4

Oxidizing agent brings about oxidation and is itself reduced. Reducing agent brings about reduction and is itself oxidized. In the above equation, HCl is the reducing agent, HNO_3 the reducing agent.

II <u>Balancing Redox Equations</u>

A. $Cr_2O_7^{2-}(aq) + I^-(aq) \longrightarrow Cr^{3+}(aq) + I_2(s)$; acidic solution

1. Split into two half-equations

   $I^-(aq) \longrightarrow I_2(s)$; oxidation

   $Cr_2O_7^{2-}(aq) \longrightarrow Cr^{3+}(aq)$; reduction

2. Balance half-equations separately

   a. $2\ I^-(aq) \longrightarrow I_2(s)$ balance element whose oxid.no.changes

   $2\ I^-(aq) \longrightarrow I_2(s) + 2e^-$ balance charge by adding e^-

   b. $Cr_2O_7^{2-}(aq) \longrightarrow 2Cr^{3+}(aq)$

   $Cr_2O_7^{2-}(aq) \longrightarrow 2Cr^{3+}(aq) + 7H_2O$; balance O by adding H_2O

   $Cr_2O_7^{2-}(aq) + 14H^+(aq) \longrightarrow 2Cr^{3+}(aq) + 7H_2O$

   balance H by adding H^+ ions

$Cr_2O_7^{2-}(aq) + 14H^+(aq) + 6e^- \longrightarrow 2Cr^{3+}(aq) + 7\ H_2O$

3. Combine half-equations so that electrons cancel.

$$3\left[2I^-(aq) \longrightarrow I_2(s) + 2e^-\right]$$
$$Cr_2O_7^{2-}(aq) + 14H^+(aq) + 6e^- \longrightarrow 2Cr^{3+}(aq) + 7H_2O$$
$$\overline{6I^-(aq) + Cr_2O_7^{2-}(aq) + 14H^+ \longrightarrow 3I_2(s) + 2Cr^{3+}(aq) + 7H_2O}$$

B. To balance in basic solution, "neutralize" H^+ ions by adding OH^- ions.

$$6I^-(aq) + Cr_2O_7^{2-}(aq) + 14H^+(aq) \longrightarrow 3I_2(s) + 2Cr^{3+}(aq) + 7H_2O$$
$$+\ 14\ OH^- \qquad\qquad + 14\ OH^-$$
$$\overline{6I^-(aq) + Cr_2O_7^{2-}(aq) + 7H_2O \longrightarrow 3I_2(s) + 2Cr^{3+}(aq) + 14\ OH^-(aq)}$$

LECTURE 2

I Balanced Redox Equations (cont.)

A. Use in stoichiometry

$$Cr_2O_7^{2-}(aq) + 6I^-(aq) + 14H^+(aq) \longrightarrow 2Cr^{3+}(aq) + 3I_2(s) + 7\ H_2O$$

Suppose 22.0 mL of 0.150 M $K_2Cr_2O_7$ is required to react with a sample weighing 5.00 g. Percent of I^- in sample?

$$\text{moles } Cr_2O_7^{2-} = 0.0220\text{ L} \times \frac{0.150\text{ mol } K_2Cr_2O_7}{1\text{ L}} \times \frac{1\text{ mol } Cr_2O_7^{2-}}{1\text{ mol } K_2Cr_2O_7}$$
$$= 0.00330\text{ mol } Cr_2O_7^{2-}$$

$$\text{mass } I^- = 0.00330\text{ mol } Cr_2O_7^{2-} \times \frac{6\text{ mol } I^-}{1\text{ mol } Cr_2O_7^{2-}} \times \frac{126.9\text{ g } I^-}{1\text{ mol } I^-}$$
$$= 2.51\text{ g } I^-$$

$$\%\ I^- = \frac{2.51\text{ g}}{5.00\text{ g}} \times 100 = 50.2$$

II Electrolytic Cells

Electrical energy used to bring about nonspontaneous redox reaction such as:

$$2Na^+ + 2\ Cl^- \longrightarrow 2Na(s) + Cl_2(g)$$

oxidation occurs at anode: $2Cl^- \longrightarrow Cl_2(g) + 2e^-$

reduction occurs at cathode: $2Na^+(aq) + 2e^- \longrightarrow 2Na(s)$

anions move to anode, cations move to cathode

A. Products of electrolysis in water solution
1. Cathode
   a. cation reduced to metal; usually occurs with transition metal cations.
   b. H^+ ions reduced to H_2; occurs with strong acids.
   c. H_2O molecules reduced to H_2; occurs with Group 1, Group 2 metals, Al.

$$2H_2O + 2e^- \longrightarrow H_2(g) + 2\ OH^-(aq)$$

2. Anode

a. anion oxidized to nonmetal; Cl^-, Br^-, I^-

b. OH^- ions oxidized to O_2: occurs with strong bases

$$2\ OH^-(aq) \longrightarrow \tfrac{1}{2}\ O_2(g) + H_2O + 2\ e^-$$

c. H_2O molecules oxidized to O_2; NO_3^-, SO_4^{2-}, F^-

$$H_2O \longrightarrow \tfrac{1}{2}\ O_2(g) + 2\ H^+(aq) + 2e^-$$

3. Overall result

$NiCl_2$: $Ni + Cl_2$

NaCl: H_2, OH^-, Cl_2

$$2Cl^-(aq) + 2H_2O \longrightarrow Cl_2(g) + 2\ OH^-(aq) + H_2(g)$$

$CuSO_4$: Cu, O_2, H^+

$$Cu^{2+}(aq) + H_2O \longrightarrow Cu(s) + \tfrac{1}{2}\ O_2(g) + 2\ H^+(aq)$$

H_2SO_4: H_2, O_2

B. Amounts of products formed in electrolysis

1. $Ag^+(aq) + e^- \rightarrow Ag(s)$; 1 mol e^- = 96485 C $\rightarrow$ 1 mol Ag

$Cu^{2+}(aq) + 2e^- \rightarrow Cu(s)$; 2 mol e^- = 2(96485 C) $\rightarrow$ 1 mol Cu

no. of coulombs = no. of amperes x no. of seconds

no. of joules = no. of coulombs x no. of volts

2. How much silver is plated from $AgNO_3$ solution by a current of 2.60 A in one hour?

$$\text{mol } e^- = (2.60)(3600)C \times \frac{1 \text{ mol } e^-}{96485\ C} = 0.0970 \text{ mol } e^-$$

$$\text{mass Ag} = 0.0970 \text{ mol } e^- \times \frac{1 \text{ mol Ag}}{1 \text{ mol } e^-} \times \frac{107.9 \text{ g Ag}}{1 \text{ mol Ag}} = 10.5 \text{ gAg}$$

LECTURE $2\frac{1}{2}$

I Electrolytic Cells (cont.)

How many joules are required to produce 1.00 g of copper from $CuSO_4$ solution, using 2.00 V?

$$\text{no. coulombs} = 1.00 \text{ g Cu} \times \frac{1 \text{ mol Cu}}{63.6 \text{ g Cu}} \times \frac{2 \text{ mol } e^-}{1 \text{ mol Cu}} \times \frac{96485\ C}{1 \text{ mol } e^-}$$

$$= 3.03 \times 10^3\ C$$

$$\text{no. joules} = 3.03 \times 10^3\ C \times 2.00\ V = 6.06 \times 10^3\ J = 6.06\ kJ$$

$$1\ kWh = 3.6 \times 10^3\ kJ$$

II Voltaic Cells

Spontaneous reaction used to produce electrical energy

A. Salt bridge cells

$$Zn(s) + Cu^{2+}(aq) \longrightarrow Zn^{2+}(aq) + Cu(s)$$

must design cell to make electron transfer occur indirectly

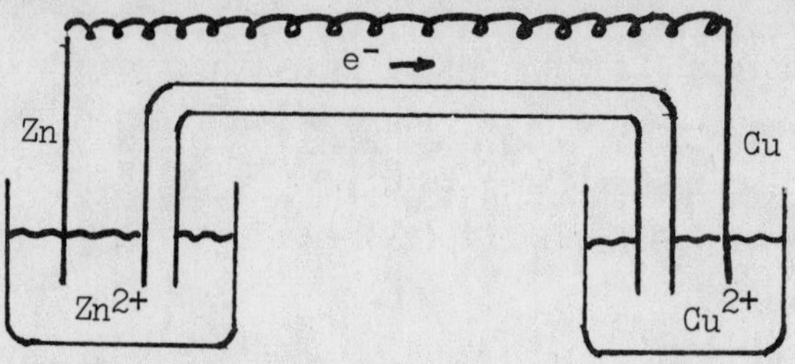

anode: $Zn(s) \longrightarrow Zn^{2+}(aq) + 2e^-$

cathode: $Cu^{2+}(aq) + 2\ e^- \longrightarrow Cu(s)$

salt bridge allows current to flow, but prevents contact between Zn and Cu^{2+}, which would shortcircuit cell.

$$Zn(s) + 2H^+(aq) \longrightarrow Zn^{2+}(aq) + H_2(g)$$

use inert, Pt electrode for $H^+ - H_2$ half-cell

B. Commercial voltaic cells

Lead storage battery:

anode: $Pb(s) + SO_4^{2-}(aq) \longrightarrow PbSO_4(s) + 2e^-$

cathode: $PbO_2(s) + 4H^+(aq) + SO_4^{2-}(aq) \longrightarrow 2PbSO_4(s) + 2H_2O$

$$Pb(s) + PbO_2(s) + 4H^+(aq) + 2SO_4^{2-}(aq) \longrightarrow 2PbSO_4(s) + 2H_2O$$

As cell discharges, conc. of H_2SO_4 drops, as does density. Battery can be recharged by reversing reaction, consuming electrical energy.

DEMONSTRATIONS

1. Electrolytic and voltaic cells: Test. Dem. 20, 81
2. Electroplating: Test. Dem. 15, 77
3. Electrolysis NaCl solution: Test. Dem. 78
4. Electrolysis of ZnI_2: Test. Dem. 129, 161

QUIZZES

Quiz 1

1. Balance the following redox equation:

   $Cu(s) + NO_3^-(aq) \longrightarrow Cu^{2+}(aq) + NO(g)$; acidic solution
2. Using the balanced equation obtained in (1), calculate the number of moles of NO produced from 1.00 g of copper.

Quiz 2

1. Balance the following redox equation:

   $Cl_2(g) + Ni(s) \longrightarrow Cl^-(aq) + NiO_2(s)$; basic solution

2. For the equation in (1), identify the species oxidized; the species reduced; the oxidizing agent; the reducing agent.

Quiz 3

1. Give the oxidation number of each atom in:

   a. HNO_3 b. MnO_4^- c. SO_2F_2

2. What mass of copper is plated from a $CuSO_4$ solution using a current of 3.45 A for two hours? (1 mol e^- = 96485 C)

Quiz 4

1. How many kilojoules are required to obtain 1.00 g of Al from Al_2O_3, using a voltage of 10.0 V? (1 mol e^- = 96485 C)

2. Draw a diagram of a voltaic cell in which the following reaction occurs:

   $Cu(s) + 2\ Ag^+(aq) \longrightarrow Cu^{2+}(aq) + 2\ Ag(s)$

   Label anode and cathode and indicate the direction of electron flow.

Quiz 5

1. Balance the following redox equation:

   $Cl_2(g) \longrightarrow Cl^-(aq) + ClO_3^-(aq)$; basic solution

2. Consider the electrolysis of $CaCl_2$ in water solution. Write a balanced equation for:

   a. the anode half-reaction
   b. the cathode half-reaction
   c. the overall reaction

PROBLEMS 1-30

23.1 a. N = +3, O = -2 b. H = +1, I = +5, O = -2
c. Te = +6, F = -1 d. N = +3, O = -2
e. Na = +1, Mo = +6, O = -2

23.2 a. N = +3, O = -2, F = -1 b. C = -2, H = +1, O = -2
c. Ru = +5, F = -1 d. N = -2, H = +1
e. H = +1, As = +5, O = -2

23.3 a. oxidation b. reduction c. reduction d. oxidation

23.4 a. oxidizing and reducing agent b. oxidizing agent
c. reducing agent d. oxidizing and reducing agent
e. oxidizing agent

23.5 a. Cr = +3 → 0, Al = 0 → +3, O = -2; N = +5 → +4, Sn = +2 → +4, H = +1, O = -2
b. Cr_2O_3, NO_3^- c. Al, Sn^{2+}

23.6 a. I^- is oxidized, Fe^{3+} is reduced; Fe^{3+} = oxidizing agent, I^- = reducing agent

b. H_2S is oxidized, ClO_3^- is reduced; ClO_3^- = oxidizing agent, H_2S = reducing agent

c. S^{2-} is oxidized, ClO_3^- is reduced; ClO_3^- = oxidizing agent, S^{2-} = reducing agent

23.7 $Cr_2O_3(s) + 2\ Al(s) \longrightarrow 2\ Cr(s) + Al_2O_3(s)$

$2[NO_3^-(aq) + 2\ H^+(aq) + e^- \longrightarrow NO_2(g) + H_2O]$

$Sn^{2+}(aq) \longrightarrow Sn^{4+}(aq) + 2\ e^-$

$2\ NO_3^-(aq) + 4\ H^+(aq) + Sn^{2+}(aq) \longrightarrow 2\ NO_2(g) + 2H_2O + Sn^{4+}$

23.8 a. $2\ [Fe^{3+}(aq) + e^- \longrightarrow Fe^{2+}(aq)]$

$2\ I^-(aq) \longrightarrow I_2(s) + 2\ e^-$

$2\ Fe^{3+}(aq) + 2\ I^-(aq) \longrightarrow 2\ Fe^{2+}(aq) + I_2(s)$

b. $4\ [ClO_3^-(aq) + 6\ H^+(aq) + 6\ e^- \longrightarrow Cl^-(aq) + 3\ H_2O]$

$3[H_2S(aq) + 4\ H_2O \longrightarrow SO_4^{2-}(aq) + 10\ H^+(aq) + 8\ e^-]$

$4\ ClO_3^-(aq) + 3\ H_2S(aq) \longrightarrow 4\ Cl^-(aq) + 3\ SO_4^{2-}(aq) + 6H^+$

c. $ClO_3^-(aq) + 6\ H^+(aq) + 6\ e^- \longrightarrow Cl^-(aq) + 3\ H_2O$

$3\ [S^{2-}(aq) \longrightarrow S(s) + 2\ e^-]$

$ClO_3^-(aq) + 6H^+(aq) + 3S^{2-}(aq) \longrightarrow Cl^-(aq) + 3H_2O + 3S(s)$

$ClO_3^-(aq) + 3H_2O + 3S^{2-}(aq) \longrightarrow Cl^-(aq) + 6\ OH^-(aq) + 3S(s)$

23.9 a. $3\ [Se(s) + 2\ H_2O \longrightarrow SeO_2(s) + 4\ H^+(aq) + 4\ e^-]$

$4\ [NO_3^-(aq) + 4\ H^+(aq) + 3\ e^- \longrightarrow NO(g) + 2\ H_2O]$

$3Se(s) + 4NO_3^-(aq) + 4H^+(aq) \longrightarrow 3SeO_2(s) + 4NO(g) + 2H_2O$

b. $2\ [Sn^{2+}(aq) \longrightarrow Sn^{4+}(aq) + 2\ e^-]$

$O_2(g) + 4\ H^+(aq) + 4\ e^- \longrightarrow 2\ H_2O$

$2Sn^{2+}(aq) + O_2(g) + 4H^+(aq) \longrightarrow 2Sn^{4+}(aq) + 2H_2O$

c. $3\ [H_2O_2(aq) \longrightarrow O_2(g) + 2\ H^+(aq) + 2\ e^-]$

$Cr_2O_7^{2-}(aq) + 14\ H^+(aq) + 6\ e^- \longrightarrow 2\ Cr^{3+}(aq) + 7\ H_2O$

$3H_2O_2(aq) + Cr_2O_7^{2-}(aq) + 8H^+(aq) \longrightarrow 3O_2(g) + 2Cr^{3+}(aq) + 7H_2O$

23.10 a. $NO_2^-(aq) + H_2O \longrightarrow NO_3^-(aq) + 2\ H^+(aq) + 2\ e^-$

$Br_2(l) + 2\ e^- \longrightarrow 2\ Br^-(aq)$

$NO_2^-(aq) + H_2O + Br_2(l) \longrightarrow NO_3^-(aq) + 2H^+(aq) + 2Br^-(aq)$

$NO_2^-(aq) + 2\ OH^-(aq) + Br_2(l) \longrightarrow NO_3^-(aq) + H_2O + 2Br^-(aq)$

23.10 b. $2\,[MnO_4^-(aq) + 8\,H^+(aq) + 5\,e^- \rightarrow Mn^{2+}(aq) + 4\,H_2O]$

$5\,[Cl^-(aq) + H_2O \rightarrow ClO^-(aq) + 2\,H^+(aq) + 2\,e^-]$

$2MnO_4^-(aq) + 5Cl^-(aq) + 6H^+(aq) \rightarrow 2Mn^{2+}(aq) + 5ClO^-(aq) + 3H_2O$

$2MnO_4^-(aq) + 5Cl^-(aq) + 3H_2O \rightarrow 2Mn^{2+}(aq) + 5ClO^-(aq) + 6\,OH^-(aq)$

c. $Cl_2(g) + 2\,e^- \rightarrow 2\,Cl^-(aq)$

$Cl_2(g) + 2\,H_2O \rightarrow 2\,ClO^-(aq) + 4\,H^+(aq) + 2\,e^-$

$2Cl_2(g) + 2H_2O \rightarrow 2Cl^-(aq) + 2ClO^-(aq) + 4H^+(aq)$

$Cl_2(g) + H_2O \rightarrow Cl^-(aq) + ClO^-(aq) + 2H^+(aq)$

$Cl_2(g) + 2\,OH^-(aq) \rightarrow Cl^-(aq) + ClO^-(aq) + H_2O$

23.11 $4\,[Au(CN)_2^-(aq) + e^- \rightarrow Au(s) + 2\,CN^-(aq)]$

$2\,H_2O \rightarrow O_2(g) + 4\,H^+(aq) + 4\,e^-$

$4Au(CN)_2^-(aq) + 2H_2O \rightarrow 4Au(s) + 8CN^-(aq) + O_2(g) + 4H^+(aq)$

$4Au(CN)_2^-(aq) + 4OH^-(aq) \rightarrow 4Au(s) + 8CN^-(aq) + O_2(g) + 2H_2O$

23.12 a. oxidation, oxidation, reduction, reduction

b. $Fe^{2+}(aq) \rightarrow Fe^{3+}(aq) + e^-$

$NO_2(g) + H_2O \rightarrow NO_3^-(aq) + 2\,H^+(aq) + e^-$

$Al^{3+}(aq) + 3\,e^- \rightarrow Al(s)$

$2\,ClO^-(aq) + 4\,H^+(aq) + 2\,e^- \rightarrow Cl_2(g) + 2\,H_2O$

c. $3Fe^{2+}(aq) + Al^{3+}(aq) \rightarrow 3Fe^{3+}(aq) + Al(s)$

$2Fe^{2+}(aq) + 2ClO^-(aq) + 4H^+(aq) \rightarrow 2Fe^{3+}(aq) + Cl_2(g) + 2H_2O$

$3NO_2(g) + 3H_2O + Al^{3+}(aq) \rightarrow 3NO_3^-(aq) + 6H^+(aq) + Al(s)$

$2NO_2(g) + 2ClO^-(aq) \rightarrow 2NO_3^-(aq) + Cl_2(g)$

23.13 a. oxidation, oxidation, oxidation, reduction

b. $S^{2-}(aq) \rightarrow S(s) + 2\,e^-$

$Cr(OH)_3(s) + 5\,OH^-(aq) \rightarrow CrO_4^{2-}(aq) + 3e^- + 4\,H_2O$

$Cl^-(aq) + 6\,OH^-(aq) \rightarrow ClO_3^-(aq) + 6e^- + 3\,H_2O$

$Fe(OH)_3(s) + 3\,e^- \rightarrow Fe(s) + 3\,OH^-(aq)$

c. $3S^{2-}(aq) + 2Fe(OH)_3(s) \rightarrow 3S(s) + 2Fe(s) + 6\,OH^-(aq)$

$Cr(OH)_3(s) + 2OH^-(aq) + Fe(OH)_3(s) \rightarrow CrO_4^{2-}(aq) + Fe(s) + 4H_2O$

$Cl^-(aq) + 2Fe(OH)_3(s) \rightarrow ClO_3^-(aq) + 3H_2O + 2\,Fe(s)$

23.14 a. $MnO_4^-(aq) + 8\,H^+(aq) + 5\,e^- \longrightarrow Mn^{2+}(aq) + 4\,H_2O$

$C_2O_4^{2-}(aq) \longrightarrow 2\,CO_2(g) + 2\,e^-$

$2MnO_4^-(aq) + 16H^+(aq) + 5C_2O_4^{2-}(aq) \longrightarrow 2Mn^{2+}(aq) + 8H_2O + 10\,CO_2(g)$

b. no. moles $C_2O_4^{2-} = 0.00216\text{ mol }MnO_4^- \times \dfrac{5\text{ mol }C_2O_4^{2-}}{2\text{ mol }MnO_4^-}$

$= 0.00540\text{ mol }C_2O_4^{2-}$

conc. $C_2O_4^{2-} = \dfrac{0.00540\text{ mol}}{0.0252\text{ L}} = 0.214\text{ M}$

23.15 a. no. moles $H^+ = 10.0\text{ g Cu} \times \dfrac{1\text{ mol Cu}}{63.55\text{ g Cu}} \times \dfrac{4\text{ mol }H^+}{1\text{ mol Cu}}$

$= 0.629\text{ mol }H^+ = 0.629\text{ mol }HNO_3$

volume $HNO_3 = \dfrac{0.629\text{ mol}}{16.0\text{ mol/L}} = 0.0393\text{ L} = 39.3\text{ mL}$

b. $10.0\text{ g Cu} \times \dfrac{1\text{ mol Cu}}{63.55\text{ g Cu}} \times \dfrac{2\text{ mol }NO_2}{1\text{ mol Cu}} \times \dfrac{46.01\text{ g }NO_2}{1\text{ mol }NO_2}$

$= 14.5\text{ g }NO_2$

23.16 a. moles $Fe^{2+} = 0.0250\text{ L} \times 0.0400\text{ mol/L} = 0.00100\text{ mol }Fe^{3+}$

moles $Cr_2O_7^{2-} = 0.00100\text{ mol }Fe^{3+} \times \dfrac{1\text{ mol }Cr_2O_7^{2-}}{6\text{ mol }Fe^{3+}}$

$= 0.000167\text{ mol }Cr_2O_7^{2-}$

$M = \dfrac{0.000167\text{ mol}}{0.0193\text{ L}} = 0.00865\text{ mol/L}$

b. no. moles e^- = no. moles $Fe^{2+} = 0.00100$ mol

23.17 a. $I_2(s)$, $H_2(g)$ b. $Cl_2(g)$, $Cu(s)$ c. $O_2(g)$, $H_2(g)$

d. $O_2(g)$, $Ni(s)$

23.18 a. $Zn^{2+}(aq) + 2\,e^- \longrightarrow Zn(s)$; cathode

$2\,Br^-(aq) \longrightarrow Br_2(l) + 2\,e^-$; anode

$Zn^{2+}(aq) + 2\,Br^-(aq) \longrightarrow Zn(s) + Br_2(l)$

b. $2\,H_2O + 2\,e^- \longrightarrow H_2(g) + 2\,OH^-(aq)$; cathode

$H_2O \longrightarrow \frac{1}{2}\,O_2(g) + 2\,H^+(aq) + 2\,e^-$; anode

$H_2O \longrightarrow H_2(g) + \frac{1}{2}\,O_2(g)$

c. $2\,H_2O + 2\,e^- \longrightarrow H_2(g) + 2\,OH^-(aq)$; cathode

$H_2O \longrightarrow \frac{1}{2}\,O_2(g) + 2\,H^+(aq) + 2\,e^-$; anode

$H_2O \longrightarrow H_2(g) + \frac{1}{2}\,O_2(g)$

23.19

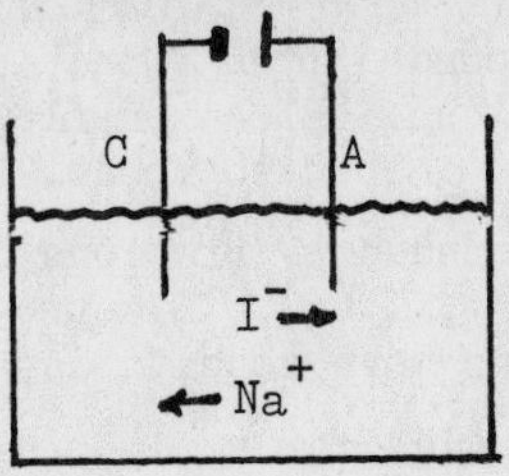

23.20 a. $8.80 \times 10^3\ C \times \frac{1\ mol\ e^-}{96485\ C} \times \frac{1\ mol\ Al}{3\ mol\ e^-} \times \frac{26.98\ g\ Al}{1\ mol\ Al} = 0.820\ gAl$

b. $0.500\ g\ Al \times \frac{1\ mol\ Al}{26.98\ g\ Al} \times \frac{3\ mol\ e^-}{1\ mol\ Al} \times \frac{96485\ C}{1\ mol\ e^-} = 5360\ C$

$t = \frac{5360\ C}{25.0\ C/s} = 214\ s$

23.21 a. no. coulombs = $2.50\ C/s \times 480\ s = 1.20 \times 10^3\ C$

mass Zn = $1.20 \times 10^3\ C \times \frac{1\ mol\ e^-}{96485\ C} \times \frac{1\ mol\ Zn}{2\ mol\ e^-} \times \frac{65.38\ g\ Zn}{1\ mol\ Zn}$
= 0.407 g

b. $18\ g\ Zn \times 0.20 \times \frac{1\ mol\ Zn}{65.38\ g\ Zn} \times \frac{2\ mol\ e^-}{1\ mol\ Zn} \times \frac{96485\ C}{1\ mol\ e^-}$

$= 1.06 \times 10^4\ C$

$t = \frac{1.06 \times 10^4\ C}{2.50\ C/s} = 4.2 \times 10^3\ s$

23.22 a. $1.00\ g\ Ni \times \frac{1\ mol\ Ni}{58.70\ g\ Ni} \times \frac{2\ mol\ e^-}{1\ mol\ Ni} \times \frac{96485\ C}{1\ mol\ e^-} = 3.29 \times 10^3\ C$

b. no. joules = $(3.29 \times 10^3\ C)(2.61\ V) = 8.59 \times 10^3\ J$

$= 8.59 \times 10^3\ J \times \frac{1\ kWh}{3.6 \times 10^6\ J} = 0.00239\ kWh$

23.23 a. no. coulombs = $2.00\ C/s \times 3600\ s = 7200\ C$

mass Pb = $7200\ C \times \frac{1\ mol\ e^-}{96485\ C} \times \frac{1\ mol\ Pb}{2\ mol\ e^-} \times \frac{207.2\ g\ Pb}{1\ mol\ Pb}$

= 7.73 g Pb

b. no. kWh = $\frac{7200 \times 12}{3.6 \times 10^6} = 0.024\ kWh$

23.24 a. $2\ Cr(s) + 3\ Co^{2+}(aq) \longrightarrow 2\ Cr^{3+}(aq) + 3\ Co(s)$

b. $Ni(s) + Cu^{2+}(aq) \longrightarrow Ni^{2+}(aq) + Cu(s)$

c. $I_2(s) + 2\ Br^-(aq) \longrightarrow 2\ I^-(aq) + Br_2(l)$

23.25 a. Pt cathode surrounded by Fe^{3+} and Fe^{2+} ions; Pb anode surrounded by Pb^{2+} ions. Electrons flow from Pb to Pt; cations move to cathode, anions to anode.

23.25 b. Pt cathode surrounded by H^+ ions and $H_2(g)$; Mg anode surrounded by Mg^{2+} ions. Electron flow from Mg to Pt; cations move to cathode, anions to anode.

c. Pt cathode surrounded by Br^- ions, $Br_2(l)$; Pt anode surrounded by I^- ions, $I_2(s)$. Electrons flow from anode to cathode. Cations move to cathode, anions to anode.

23.26 a. $Zn(s) \longrightarrow Zn^{2+}(aq) + 2\ e^-$

b. $Pb(s) + SO_4^{2-}(aq) \longrightarrow PbSO_4(s) + 2\ e^-$

c. $Zn(s) + 2\ MnO_2(s) \longrightarrow ZnO(s) + Mn_2O_3(s)$

23.27 anode: $Pb(s) + 2\ Cl^-(aq) \longrightarrow PbCl_2(s) + 2\ e^-$

cathode: $PbO_2(s) + 4H^+(aq) + 2Cl^-(aq) \longrightarrow PbCl_2(s) + 2H_2O$

$$Pb(s) + PbO_2(s) + 4H^+(aq) + 4Cl^-(aq) \longrightarrow 2PbCl_2(s) + 2H_2O$$

23.28 a. gain of electrons reduces oxidation number
b. H_2O molecules more readily reduced than Na^+ ions
c. picks up electrons from another species
d. must do so to maintain electrical neutrality

23.29 a. other, competing reactions occur, including the reduction of water
b. can be oxidized to Co^{3+} or reduced to Co
c. electrical energy is consumed in electrolysis, produced in a voltaic cell

23.30

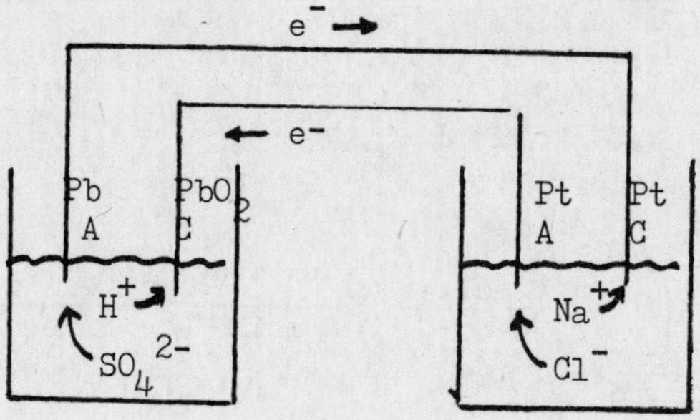

CHAPTER 24

BASIC SKILLS

1. Use standard electrode potentials (Table 24.1) to:
   a. compare the relative strengths of different oxidizing agents; different reducing agents.
   b. calculate a cell voltage at standard concentrations.
   c. decide whether or not a given redox reaction will occur spontaneously at standard concentrations.
2. Use the Nernst equation (Equation 24.3) to calculate:
   a. the voltage of a cell, given E^o_{tot} and the concentrations of all species.
   b. the concentration of one species, given those of all other species involved, E and E^o_{tot}.
3. Describe the electrochemical mechanism of corrosion of iron.

LECTURE NOTES

The most difficult portion of this chapter involves the Nernst equation. Students have a great deal of difficulty using this equation to calculate the concentration of one species (e.g., H^+), knowing E and E^o_{tot}. The principal problem here is their unfamiliarity with logarithms. Other items in the chapter, such as applications of E^o_{tot}, ΔG^o and K, seem to go over quite well.

This chapter will require two lectures at a minimum. If you spend any lecture time upon corrosion, you should plan on $2\frac{1}{2}$ lectures.

LECTURE 1

I Standard Voltages

A. E^o_{tot} = cell voltage when all species involved are at standard concentrations (1 atm for gases, 1 M for solutes in water)

$Zn(s) + 2H^+(aq, 1M) \longrightarrow Zn^{2+}(aq, 1M) + H_2(g, 1\ atm)$

$E^o_{tot} = +0.76\ V = E^o_{ox}\ Zn \rightarrow Zn^{2+} + E^o_{red}\ H^+ \rightarrow H_2$

Cannot measure either E^o_{ox} or E^o_{red} independently. Set:

$2H^+(aq, 1M) + 2e^- \longrightarrow H_2(g, 1\ atm);\quad E^o_{red} = 0.00\ V$

Thus: $Zn(s) \longrightarrow Zn^{2+}(aq) + 2e^-$; $E^o_{ox} = +0.76$ V

Can now calculate other E^o values:

$Zn(s) + Cu^{2+}(aq) \longrightarrow Zn^{2+}(aq) + Cu(s)$; $E^o_{tot} = +1.10$ V

$1.10 \text{ V} = E^o_{ox}\ Zn \rightarrow Zn^{2+} + E^o_{red}\ Cu^{2+} \rightarrow Cu$

$E^o_{red}\ Cu^{2+} \rightarrow Cu = 1.10 \text{ V} - E^o_{ox}\ Zn \rightarrow Zn^{2+}$

$= 1.10 \text{ V} - 0.76 \text{ V} = +0.34 \text{ V}$

Table 24.1 lists values of E^o_{red} for many different species. Can get E^o_{ox} values by changing signs. For example:

$Cu(s) \longrightarrow Cu^{2+}(aq) + 2e^-$; $E^o_{ox}\ Cu = -E^o_{red}\ Cu^{2+} = -0.34$ V

$Zn(s) \longrightarrow Zn^{2+}(aq) + 2e^-$; $E^o_{ox}\ Zn = -E^o_{red}\ Zn^{2+} = +0.76$ V

B. Relative strengths of oxidizing and reducing agents

1. Oxidizing agents (left column, Table 24.1). The larger (more positive) the value of E^o_{red}, the stronger the oxidizing agent.

$Li^+(aq) + e^- \longrightarrow Li(s)$ $E^o_{red} = -3.05$ V

$2H^+(aq) + 2e^- \longrightarrow H_2(g)$ $E^o_{red} = 0.00$ V

$Cl_2(g) + 2e^- \longrightarrow 2Cl^-(aq)$ $E^o_{red} = +1.36$ V

oxidizing agents become stronger moving down left column

2. Reducing agents (right column, Table 24.1). The larger the value of E^o_{ox}, the stronger the reducing agent.

$Li(s) \longrightarrow Li^+(aq) + e^-$ $E^o_{ox} = +3.05$ V

$H_2(g) \longrightarrow 2H^+(aq) + 2e^-$ $E^o_{ox} = 0.00$ V

$2Cl^-(aq) \longrightarrow Cl_2(g) + 2e^-$ $E^o_{ox} = -1.36$ V

reducing agents become weaker moving down right column

C. Calculation of standard cell voltage

$E^o_{tot} = E^o_{ox} + E^o_{red}$

1. $Ni(s) + Br_2(l) \longrightarrow Ni^{2+}(aq) + 2\ Br^-(aq)$

$E^o_{tot} = E^o_{red}\ Br_2 + E^o_{ox}\ Ni = +1.07 \text{ V} + 0.25 \text{ V} = +1.32 \text{ V}$

Since calculated voltage is positive, this reaction can occur in voltaic cell.

2. $2Br^-(aq) + 2H_2O \longrightarrow Br_2(l) + H_2(g) + 2\ OH^-(aq)$

$E^o_{tot} = E^o_{red}\ H_2O + E^o_{ox}\ Br^- = -0.83 \text{ V} - 1.07 \text{ V} = -1.90 \text{ V}$

Must apply at least 1.90 V in electrolytic cell to make reaction go.

D. Determination of whether redox reaction will occur in laboratory. Reaction goes if calculated E^o_{tot} is positive.

1. $Ni(s) + Br_2(l) \longrightarrow Ni^{2+}(aq) + 2\ Br^-(aq)$

$E^o_{tot} = +1.32$ V, so nickel metal reacts directly with bromine. Ni^{2+} ions do not react with Br^- ions.

2. Will a redox reaction occur if bromine is added to a solution of tin(II) chloride?

Possible oxidations

$Sn^{2+}(aq) \longrightarrow Sn^{4+}(aq) + 2\,e^-;\quad E^o_{ox} = -0.15\ V$

$2\,Cl^-(aq) \longrightarrow Cl_2(g) + 2\,e^-;\quad E^o_{ox} = -1.36\ V$

Possible reductions:

$Sn^{2+}(aq) + 2\,e^- \longrightarrow Sn(s);\quad E^o_{red} = -0.14\ V$

$Br_2(l) + 2\,e^- \longrightarrow 2\,Br^-(aq);\quad E^o_{red} = +1.07\ V$

Reaction that occurs:

$Sn^{2+}(aq) + Br_2(l) \longrightarrow Sn^{4+}(aq) + 2\,Br^-(aq);\quad E^o_{tot} = +0.92V$

LECTURE 2

I Effect of Concentration Upon Voltage

A. Qualitative. Voltage is measure of reaction spontaneity. Hence:

- voltage is increased by increasing concentrations of reactants or decreasing concentrations of products
- voltage is decreased by decreasing concentration of reactants or increasing concentration of products.

$Zn(s) + Cu^{2+}(aq) \longrightarrow Zn^{2+}(aq) + Cu(s)$

E decreases as $\frac{\text{conc. } Zn^{2+}}{\text{conc. } Cu^{2+}}$ increases

B. Quantitative. Nernst equation:

$aA + bB \longrightarrow cC + dD$

$$E = E^o_{tot} - \frac{0.0591}{n}\log_{10}\frac{(\text{conc. C})^c(\text{conc. D})^d}{(\text{conc. A})^a(\text{conc. B})^b}$$

n = number of electrons transferred in equation

use pressure for gases, molarities for species in water

1. $Zn(s) + Cu^{2+}(aq) \longrightarrow Zn^{2+}(aq) + Cu(s)$

$$E = +1.10\ V - \frac{0.0591}{2}\log_{10}\frac{(\text{conc. } Zn^{2+})}{(\text{conc. } Cu^{2+})}$$

Note that voltage decreases as conc. Zn^{2+}/ conc. Cu^{2+} increases

2. $Cl_2(g) + 2\,Br^-(aq) \longrightarrow 2\,Cl^-(aq) + Br_2(l)$

$$E = +0.29\ V - \frac{0.0591}{2}\log_{10}\frac{(\text{conc. } Cl^-)^2}{(P\ Cl_2)(\text{conc. } Br^-)^2}$$

Calculate voltage when conc. Br^- = 1 M, P Cl_2 = 1 atm, conc. Cl^- = 0.01 M.

$$E = +0.29\ V - \frac{0.0591}{2}\log_{10}\frac{(0.01)^2}{(1)(1)^2} = +0.41\ V$$

C. Use of Nernst equation to determine concentration of ion in solution

$Zn(s) + 2H^+(aq) \longrightarrow Zn^{2+}(aq) + H_2(g)$

$$E = +0.76\ V - \frac{0.0591}{2} \log_{10} \frac{(P\ H_2)(\text{conc. } Zn^{2+})}{(\text{conc. } H^+)^2}$$

Suppose conc. Zn^{2+} = 1 M, P H_2 = 1 atm:

$$E = +0.76\ V - \frac{0.0591}{2} \log_{10} \frac{1}{(\text{conc. } H^+)^2}$$

$$= +0.76\ V + 0.0591 \log_{10}(\text{conc. } H^+)$$

Measure voltage; calculate conc. H^+. Suppose E = 0.20 V

$$\log_{10}(\text{conc. } H^+) = \frac{-0.56}{0.0591} = -9.5;\ \text{conc. } H^+ = 3 \times 10^{-10}\ M$$

II Relation between E^o_{tot}, ΔG^o and K

A. ΔG^o (in kJ) = $-96.5\ n\ E^o_{tot}$

$$\log_{10} K = \frac{n\ E^o_{tot}}{0.0591}$$

Note that if E^o_{tot} is +, ΔG^o is −, $\log_{10}$ K is +, K > 1

LECTURE 2½

I Relation between E^o_{tot}, ΔG^o and K

$Cl_2(g) + 2Br^-(aq) \longrightarrow 2Cl^-(aq) + Br_2(l)$

E^o_{tot} = 1.36 V − 1.07 V = +0.29 V

$\Delta G^o = -2(96.5)(+0.29) = -56$ kJ

$$\log_{10} K = \frac{2(+0.29)}{0.0591} = +9.8;\ K = 6 \times 10^9$$

II Corrosion of Iron and Steel

A. Half-reactions

$Fe(s) \longrightarrow Fe^{2+}(aq) + 2\ e^-$ anode

$\frac{1}{2}O_2(g) + H_2O + 2e^- \longrightarrow 2\ OH^-(aq)$ cathode

$Fe(s) + \frac{1}{2}O_2(g) + H_2O \longrightarrow Fe(OH)_2(s)$

later converted to $Fe(OH)_3$ in air

B. Corrosion occurs more readily when

- iron is in contact with a less active metal (Sn, Cu), which acts as cathode.
- there is a difference in O_2 concentration. With bridge support, corrosion occurs just below water line, where conc. O_2 is relatively low.
- ions are present. Seawater enhances corrosion.

Corrosion occurs less readily when
- metal is painted; cuts off supply of O_2 and H_2O
- iron is in contact with a more active metal (Zn, Mg), which acts as a sacrificial anode

DEMONSTRATIONS

1. Measurement of electrode potentials: Test. Dem. 196
2. Zn - H^+ cell: Test. Dem. 141
3. Formation of Sn, Pb and Ag trees: Test. Dem. 53, 127
4. Removal of silver tarnish with Al: Test. Dem. 31
5. Concentration cells: Test. Dem. 144
6. Controlled corrosion of iron: Test. Dem. 171
7. Cathodic protection: Test. Dem. 104; J. Chem. Educ. 58 505, 802 (1981)

QUIZZES

Quiz 1

1. Using Table 24.1, arrange the reducing agents in the list below in order of increasing strength; do the same with the oxidizing agents.

   NO_3^-, Fe^{2+}, AgBr, SO_4^{2-}, Ag, Zn

2. For the reaction: $Al(s) + 3Ag^+(aq) \longrightarrow Al^{3+}(aq) + 3Ag(s)$ calculate:

   a. E^o_{tot} b. ΔG^o c. K

Quiz 2

1. Write a balanced equation for the reaction of nitric acid with silver to give NO(g) and Ag^+. Calculate E^o_{tot} for this reaction, using Table 24.1.
2. For the corrosion of iron in air, write a balanced equation for:

   a. the anode half-reaction b. the cathode half-reaction
   c. the overall reaction

Quiz 3

1. Consider the reaction:

   $Cu(s) + 4H^+(aq) + SO_4^{2-}(aq) \longrightarrow Cu^{2+}(aq) + SO_2(g) + 2H_2O$

   Referring to Table 24.1, calculate:

   a. E^o_{tot} b. ΔG^o c. K
   d. E when P SO_2 = 1 atm, conc. Cu^{2+} = conc. SO_4^{2-} = 1 M, pH = 3.0

Quiz 4

1. Using Table 24.1, decide whether or not a reaction will occur when a solution of tin(II) bromide is treated with nitric acid. Write a balanced equation for any reaction that occurs.

2. Explain in your own words why iron in contact with copper corrodes while iron in contact with zinc does not.

Quiz 5

1. Consider the reaction:

$$Co(s) + 2H^+(aq) \longrightarrow Co^{2+}(aq) + H_2(g)$$

Using Table 24.1, calculate

a. E^o_{tot} b. ΔG^o c. the pH at which E = 0 when all other species are at standard concentrations

2. Suppose that, in Question 1c, the concentration of Co^{2+} were 0.10 M instead of 1 M and $P\ H_2$ were 729 mmHg instead of 1 atm. Would the voltage be greater or less than zero? Explain your reasoning.

PROBLEMS 1-30

24.1 $F^- < Au < Fe^{2+} < SO_2 < K$

24.2 oxidizing agents: $Fe^{2+} < NO_3^- < MnO_4^-$

reducing agents: $Fe^{2+} < I^- < Ca$

24.3 a. species between Cd and Pb in right column of Table 24.1
b. species between I_2 and Br_2 in left column of Table 24.1
c. species between Au + $4Cl^-$ and Au in right column of Table 24.1

24.4 a. $E^o_{tot} = E^o_{ox}\ Al + E^o_{red}\ NO_3^- = +1.66\ V + 0.96\ V = +2.62\ V$
b. $E^o_{tot} = E^o_{ox}\ Cu + E^o_{red}\ NO_3^- = +0.22\ V + 0.01\ V = +0.23\ V$
c. $E^o_{tot} = E^o_{ox}\ Cu + E^o_{red}\ I_2 = -0.34\ V + 0.53\ V = +0.19\ V$

24.5 a. $E^o_{tot} = E^o_{ox}\ Cr^{2+} + E^o_{red}\ Sn^{4+} = +0.41\ V + 0.15\ V = +0.56\ V$
b. $E^o_{tot} = E^o_{ox}\ Fe + E^o_{red}\ H_2O = +0.88\ V - 0.83\ V = +0.05\ V$
c. $E^o_{tot} = E^o_{ox}\ Fe(OH)_2 + E^o_{red}\ O_2 = +0.56\ V + 0.40\ V = +0.96\ V$

24.6 a. $E^o_{tot} = E^o_{ox}\ H_2 + E^o_{red}\ Cl_2 = 0.00\ V + 1.36\ V = +1.36\ V$
b. $E^o_{tot} = E^o_{ox}\ Sn + E^o_{red}\ Pb^{2+} = +0.14\ V - 0.13\ V = +0.01\ V$
c. $E^o_{tot} = E^o_{ox}\ I^- + E^o_{red}\ Cl_2 = -0.53\ V + 1.36\ V = +0.83\ V$

24.7 a. +1.00 V b. -0.66 V c. +1.56 V

24.8 a. $E^o_{tot} = E^o_{red}\ Ni^{2+} + E^o_{ox}\ Cl^- = -0.25\ V - 1.36\ V = -1.61\ V$; apply 1.61 V

24.8 b. $E^o_{tot} = E^o_{ox}\ I^- + E^o_{red}\ H_2O = -0.53\ V - 0.83\ V = -1.36\ V$
apply 1.36 V

24.9 a. $E^o_{tot} = E^o_{red}\ AuCl_4^- + E^o_{ox}\ Fe^{2+} = +1.00\ V - 0.77\ V = +0.23\ V$
spontaneous

b. $E^o_{tot} = E^o_{red}\ NO_3^- + E^o_{ox}\ Cl^- = +0.96\ V - 1.36\ V = -0.40\ V$
nonspontaneous

c. $E^o_{tot} = E^o_{ox}\ Cu + E^o_{red}\ H^+ = -0.34\ V + 0.00\ V = -0.34\ V$
nonspontaneous

24.10 a. $E^o_{tot} = E^o_{ox}\ Al + E^o_{red}\ Cu^{2+} = +1.66\ V + 0.34\ V = +2.00\ V$
spontaneous

b. $E^o_{tot} = E^o_{ox}\ Cd + E^o_{red}\ NO_3^- = +0.40\ V + 0.96\ V = +1.36\ V$
spontaneous

c. $E^o_{tot} = E^o_{red}\ Sn^{2+} + E^o_{ox}\ Pb = -0.14\ V + 0.13\ V = -0.01\ V$
nonspontaneous

24.11 a. $E^o_{tot} = E^o_{ox}\ Co = +0.28\ V$; reacts

b. $E^o_{tot} = E^o_{ox}\ Au = -1.50\ V$; does not react

c. $E^o_{tot} = E^o_{ox}\ Mg = +2.37\ V$; reacts

d. $E^o_{tot} = E^o_{ox}\ Cu = -0.34\ V$; does not react

24.12 a. no reaction (neither Ca^{2+} nor NO_3^- can be oxidized)

b. two reactions: $Br_2(l) + 2I^-(aq) \longrightarrow 2Br^-(aq) + I_2(s)$

$E^o_{tot} = +0.54\ V$

$Br_2(l) + 2Fe^{2+}(aq) \longrightarrow 2Br^-(aq) + 2Fe^{3+}(aq)$

$E^o_{tot} = +0.30\ V$

c. no reaction (F^- cannot be oxidized by Br_2)

24.13 a. $$E = E^o_{tot} - \frac{0.0591}{2} \log_{10} \frac{(\text{conc. } Zn^{2+})(P\ H_2)}{(\text{conc. } H^+)^2}$$

b. $$E = +0.76\ V - \frac{0.0591}{2} \log_{10} \frac{(1.0)(1.0)}{(1.0 \times 10^{-3})^2}$$

$$= +0.76\ V - \frac{0.0591}{2}(6.00) = +0.58\ V$$

24.14 $MnO_2(s) + 4\ H^+(aq) + 2\ e^- \longrightarrow Mn^{2+}(aq) + 2\ H_2O$

$$E_{red} = +1.23\ V - \frac{0.0591}{2} \log_{10} \frac{1}{(1.0 \times 10^{-4})^4} = +0.76\ V$$

24.15 a. $$E = E^o_{tot} - \frac{0.0591}{2} \log_{10} \frac{(\text{conc. } Zn^{2+})}{(\text{conc. } Cd^{2+})}$$

$$= +0.36\ V - \frac{0.0591}{2} \log_{10} \frac{0.50}{0.020} = +0.32\ V;$$ spontaneous

24.15 b. $E = E^{o}_{tot} - \frac{0.0591}{2} \log_{10} \frac{(\text{conc. } Sn^{4+})}{(\text{conc. } Sn^{2+})(\text{conc. } Cu^{2+})}$

$= +0.19 \text{ V} - \frac{0.0591}{2} \log_{10} \frac{0.010}{(0.10)(1)} = +0.22 \text{ V}$; spontaneous

24.16 $E = E^{o}_{tot} - \frac{0.0591}{2} \log_{10} \frac{(\text{conc. } Ni^{2+})}{(\text{conc. } Cu^{2+})}$

$0.00 \text{ V} = 0.59 \text{ V} - \frac{0.0591}{2} \log_{10} \frac{1}{(\text{conc. } Cu^{2+})}$

$-0.59 = \frac{0.0591}{2} \log_{10}(\text{conc. } Cu^{2+})$; conc. $Cu^{2+} = 1 \times 10^{-20}$ M

24.17 a. $E^{o}_{tot} = E^{o}_{red}\ O_2 + E^{o}_{ox}\ Cl^- = +1.23 \text{ V} - 1.36 \text{ V} = -0.13 \text{ V}$

b. $E = -0.13 \text{ V} - \frac{0.0591}{4} \log_{10} \frac{(P\ Cl_2)^2}{(1)(1)}$

$0.13 = -\frac{0.0591}{2} \log_{10} P\ Cl_2$; $P\ Cl_2 = 4 \times 10^{-5}$ atm

24.18 a. $E^{o}_{tot} = E^{o}_{ox}\ Pb + E^{o}_{red}\ H^+ = +0.13 \text{ V} + 0.00 \text{ V} = +0.13 \text{ V}$

b. $E = +0.13 \text{ V} - \frac{0.0591}{2} \log_{10} \frac{(P\ H_2)(\text{conc. } Pb^{2+})}{(\text{conc. } H^+)^2}$

$0.21 = 0.13 - \frac{0.0591}{2} \log_{10}(\text{conc. } Pb^{2+})$

$\log_{10}(\text{conc. } Pb^{2+}) = \frac{-2(0.08)}{0.0591} = -2.7$; conc. $Pb^{2+} = 2 \times 10^{-3}$ M

c. $K_{sp} = [Pb^{2+}] \times [Cl^-]^2 = (2 \times 10^{-3})(0.10)^2 = 2 \times 10^{-5}$

24.19 $2Ag(s) + Cu^{2+}(aq) \longrightarrow 2Ag^+(aq) + Cu(s)$

$E = E^{o}_{tot} - \frac{0.0591}{2} \log_{10} \frac{(\text{conc. } Ag^+)^2}{(\text{conc. } Cu^{2+})}$

$+0.22 = -0.46 - \frac{0.0591}{2} \log_{10} \frac{(\text{conc. } Ag^+)^2}{0.10}$

$\log_{10} \frac{(\text{conc. } Ag^+)^2}{0.10} = -\frac{2(0.68)}{0.0591} = -23.0$;

$(\text{conc. } Ag^+)^2 = 0.10 \times 1 \times 10^{-23} = 1 \times 10^{-24}$

conc. $Ag^+ = 1 \times 10^{-12}$ M

K_{sp} AgBr $= [Ag^+] \times [Br^-] = (1 \times 10^{-12})(0.10) = 1 \times 10^{-13}$

24.20 a. $\Delta G^o = -96.5(2)(0.00) = 0.0$ kJ

$\log_{10} K = \frac{2(0.00)}{0.0591} = 0$; $K = 1$

b. $\Delta G^o = -96.5(2)(+0.35) = -68$ kJ

$\log_{10} K = \frac{2(+0.35)}{0.0591} = 12$; $K = 10^{12}$

24.20 c. $\Delta G^\circ = +68$ kJ; $K = 10^{-12}$

24.21 a. $E^\circ_{tot} = \dfrac{-\Delta G^\circ}{96.5n} = \dfrac{21.6}{96.5} = 0.223$ V

b. $E^\circ_{tot} = 21.6/193.0 = 0.112$ V

c. $E^\circ_{tot} = 21.6/289.5 = 0.0746$ V

24.22 $E^\circ_{tot} = \dfrac{0.0591}{n} \log_{10} K = \dfrac{0.0591}{2}(-6.00) = -0.177$ V

24.23 $E^\circ_{tot} = E^\circ_{red}\ Fe^{3+} + E^\circ_{ox}\ Sn^{2+} = +0.77\ V - 0.15\ V = +0.62$ V

$\Delta G^\circ = -96.5(2)(+0.62) = -1.2 \times 10^2$ kJ

$\log_{10} K = \dfrac{2(+0.62)}{0.0591} = 21$; $K = 10^{21}$

24.24 a. $\Delta G^\circ = -96.5(3)(+2.62) = -758$ kJ

b. $\Delta G^\circ = -96.5(2)(+0.23) = -44$ kJ

c. $\Delta G^\circ = -96.5(2)(+0.19) = -37$ kJ

(taking the reaction to be:

$Cu(s) + 2I^-(aq) \longrightarrow Cu^{2+}(aq) + I_2(s)$)

24.25 a. $\log_{10} K = \dfrac{2(+0.56)}{0.0591} = +19$; $K = 10^{19}$

b. $\log_{10} K = \dfrac{2(+0.05)}{0.0591} = +1.7$; $K = 50$

c. $\log_{10} K = \dfrac{4(0.96)}{0.0591} = +65$; $K = 10^{65}$ (assuming 1 mol O_2 reacts)

24.26 no; E°_{tot} is negative, ΔG° is positive

24.27 a. O_2 takes part in reaction; water solution necessary for current flow

b. prevents accumulation of dirt which promotes corrosion by setting up a concentration cell.

24.28 a. silver reacts with H_2S to form very insoluble (stable) Ag_2S

b. Al_2O_3 adheres to surface

24.29 a. Al ($E^\circ_{ox} = +1.66$ V) reacts with H^+

b. forms $Cu(OH)_2 \cdot CuCO_3$

24.30 cell reaction: $2Ag(s) + Cl_2(g) \longrightarrow 2Ag^+(aq) + 2Cl^-(aq)$

a. increases voltage; Cl_2 is a reactant

b. no effect

c. decreases voltage; Cl^- is a product

d. increases voltage by precipitating AgCl, lowering conc. of Ag^+

CHAPTER 25

BASIC SKILLS

1. Describe how the major transition metals (Figure 25.1) are obtained from their ores.
2. Discuss and write balanced equations for the redox reactions by which transition metals dissolve in hydrochloric acid; in nitric acid; in aqua regia.
3. Write formulas for the aquo complex ions of the major transition metals; write balanced equations to explain the acidity of these ions.
4. Write balanced equations for the precipitation of sulfides and hydroxides of transition metal ions by S^{2-} ions, OH^- ions, H_2S, or NH_3.
5. Write balanced equations to explain why transition metal hydroxides dissolve in acid; in NaOH; in NH_3.
6. Apply the principles introduced in Chapters 23 and 24 to predict the stability toward oxidation, reduction, or disproportionation of transition metal cations.
7. Discuss the redox chemistry of the oxyanions of chromium and manganese.

LECTURE NOTES

This is a descriptive chapter which illustrates the principles of oxidation-reduction reactions discussed in Chapters 23 and 24. It also reviews material from Chapter 4 (metallurgical processes), Chapter 21 (complex ions) and Chapter 22 (qualitative analysis). The treatment of uses of the transition metals (pp 750-752) is introduced primarily to hold the interest of students. Unless you spend a great deal of time on that material, this chapter can readily be covered in two lectures.

LECTURE 1

I Transition Metals

A. Metallurgy: most often occur as sulfides or oxides

1. Sulfides: roast in air to form either the metal or its oxide.

$$HgS(s) + O_2(g) \longrightarrow Hg(g) + SO_2(g)$$

$$2ZnS(s) + 3\ O_2(g) \longrightarrow 2ZnO(s) + 2\ SO_2(g)$$

2. Oxides: reduce with coke (CO) or Al

$Fe_2O_3(s) + 3CO(g) \longrightarrow 2Fe(l) + 3CO_2(g)$

$Fe_2O_3(s) + 2Al(s) \longrightarrow 2Fe(l) + Al_2O_3(s)$

B. Oxidation of metals to cations

1. Dilute HCl: effective with all metals having positive E^o_{ox}

$Mn(s) + 2H^+(aq) \longrightarrow Mn^{2+}(aq) + H_2(g)$

2. Concentrated HNO_3; NO_3^- ion is oxidizing agent

$Cu(s) + 4H^+(aq) + 2NO_3^-(aq) \longrightarrow Cu^{2+}(aq) + 2NO_2(g) + 2H_2O$

3. Aqua regia (3 volumes 12 M HCl, 1 volume 16 M HNO_3)
NO_3^- is oxidizing agent, Cl^- is complexing agent

$Hg(l) + 4H^+(aq) + 2NO_3^-(aq) + 4Cl^-(aq) \longrightarrow HgCl_4^{2-}(aq) + 2NO_2(g) + 2H_2O$

C. Uses

1. Coinage. US coins are mostly Cu-Ni alloys. Penny: 98% Zn covered with Cu.
2. Photography - film covered with light-sensitive silver halide

exposure: $AgBr(s) \longrightarrow AgBr^*(s)$; a few Ag^+ ions are reduced to silver

development: $AgBr^*(s) + e^- \longrightarrow Ag(s) + Br^-$; organic reducing agent used to reduce Ag^+ ions that were exposed, forming a black image

fixing: $AgBr(s) + 2S_2O_3^{2-}(aq) \longrightarrow Ag(S_2O_3)_2^{3-}(aq) + Br^-(aq)$
removes undeveloped AgBr

II <u>Transition Metal Ions</u>

A. Ordinarily exist as complex ions in water solution

$Cr(H_2O)_6^{3+}$	$Ni(H_2O)_6^{2+}$	$Cu(H_2O)_4^{2+}$
violet	green	blue

These ions act as weak acids in water:

$Cr(H_2O)_6^{3+}(aq) \rightleftharpoons H^+(aq) + Cr(H_2O)_5(OH)^{2+}(aq)$

$K_a = 1 \times 10^{-4}$

pH of 0.1 M $Cr(NO_3)_3$ solution?

$\frac{[H^+]^2}{0.1} \approx 1 \times 10^{-4}$; $[H^+]^2 = 1 \times 10^{-5}$

$[H^+] = 3 \times 10^{-3}$ M; pH = 2.5

B. Precipitation of sulfides, hydroxides

1. Sulfides: precipitate with sulfide ions or H_2S

$Cu^{2+}(aq) + H_2S(aq) \longrightarrow CuS(s) + 2H^+(aq)$

2. Hydroxides: precipitate with NaOH or NH_3

$Fe^{3+}(aq) + 3\ OH^-(aq) \longrightarrow Fe(OH)_3(s)$

$$Fe^{3+}(aq) + 3NH_3(aq) + 3H_2O \rightarrow Fe(OH)_3(s) + 3NH_4^+(aq)$$

All hydroxides dissolve in strong acid:

$$Fe(OH)_3(s) + 3H^+(aq) \rightarrow Fe^{3+}(aq) + 3H_2O$$

Many dissolve in NaOH or NH_3:

$$Zn(OH)_2(s) + 2\ OH^-(aq) \rightarrow Zn(OH)_4^{2-}(aq)$$

$$Zn(OH)_2(s) + 4\ NH_3(aq) \rightarrow Zn(NH_3)_4^{2+}(aq) + 2\ OH^-(aq)$$

LECTURE 2

I Transition Metal Cations (cont.)

A. Oxidation-reduction reactions

May be unstable in water solution because of:

1. Reaction with water

$$Co^{3+}(aq) + e^- \rightarrow Co^{2+}(aq) \quad ; E^o_{red} = +1.82\ V$$

$$H_2O \rightarrow \tfrac{1}{2}\ O_2(g) + 2H^+(aq) + 2e^- \quad E^o_{ox} = -1.23\ V$$

$$2Co^{3+}(aq) + H_2O \rightarrow 2Co^{2+}(aq) + \tfrac{1}{2}O_2(g) + 2H^+(aq);\ E^o_{tot} = +0.59\ V$$

However, complexes of Co^{3+} are stable

2. Disproportionation

$$Cu^+(aq) + e^- \rightarrow Cu(s) \quad ; E^o_{red} = +0.52\ V$$

$$Cu^+(aq) \rightarrow Cu^{2+}(aq) + e^- \quad ; E^o_{ox} = -0.15\ V$$

$$2Cu^+(aq) \rightarrow Cu^{2+}(aq) + Cu(s);\ E^o_{tot} = +0.37\ V,\ K = 2 \times 10^6$$

3. Oxidation by dissolved oxygen

$$Fe^{2+}(aq) \rightarrow Fe^{3+}(aq) + e^- \quad ; E^o_{ox} = -0.77\ V$$

$$\tfrac{1}{2}O_2(g) + 2H^+(aq) + 2e^- \rightarrow H_2O \quad E^o_{red} = +1.23\ V$$

$$2Fe^{2+}(aq) + \tfrac{1}{2}O_2(g) + 2H^+(aq) \rightarrow 2Fe^{3+}(aq) + H_2O$$

$$E^o_{tot} = +0.46\ V$$

II Oxyanions of Cr, Mn

A. CrO_4^{2-} and $Cr_2O_7^{2-}$ ions

$$2CrO_4^{2-}(aq) + 2H^+(aq) \rightleftharpoons Cr_2O_7^{2-}(aq) + H_2O$$

yellow   red

Dichromate ion is stable in acidic solution, chromate ion in basic solution

$Cr_2O_7^{2-}$ is a strong oxidizing agent in acidic solution:

$$Cr_2O_7^{2-}(aq) + 14H^+(aq) + 6e^- \rightarrow 2Cr^{3+}(aq) + 7H_2O$$

$$E^o_{red} = +1.33\ V$$

$$E_{red} = 1.33\ V - \frac{0.0591}{6} \log_{10} \frac{(\text{conc. } Cr^{3+})^2}{(\text{conc. } Cr_2O_7^{2-})(\text{conc. } H^+)^{14}}$$

Suppose conc. Cr^{3+} = conc. $Cr_2O_7^{2-}$ = 1 M

$$E_{red} = 1.33\ V - \frac{0.0591}{6} \log_{10} \frac{1}{(conc.\ H^+)^{14}}$$

$$= 1.33\ V + \frac{0.0591}{6} \times 14 \times \log_{10}(conc.\ H^+)$$

$$= 1.33\ V + 0.14 \log_{10}(conc.\ H^+) = 1.33\ V - 0.14\ pH$$

E_{red} decreases from 1.33 V in 1 M acid to +0.35 V in neutral soln.

B. MnO_4^- and MnO_4^{2-} ions

1. MnO_4^- is a very powerful (and dangerous) oxidizing agent

$$\underset{\text{purple}}{MnO_4^-(aq)} + 8H^+(aq) + 5e^- \rightarrow Mn^{2+}(aq) + 4H_2O$$

$$E^o_{red} = +1.52\ V$$

2. MnO_4^{2-} is unstable in acid; disproportionates

$$MnO_4^{2-}(aq) \rightarrow MnO_4^-(aq) + e^-$$

$$MnO_4^{2-}(aq) + 4H^+(aq) + 2e^- \rightarrow MnO_2(s) + 2H_2O$$

$$\underset{\text{green}}{3MnO_4^{2-}(aq)} + 4H^+(aq) \rightarrow \underset{\text{purple}}{2MnO_4^-(aq)} + \underset{\text{brown}}{MnO_2(s)} + 2H_2O$$

DEMONSTRATIONS

1. Roasting of sulfide ores: Test. Dem. 25
2. Photographic process: Test. Dem. 18, 82, 141
3. Formation of silver mirror: Test. Dem. 28, 94; J. Chem. Educ. 58 655 (1981)
4. Equilibrium between Cu^+ and Cu^{2+}: J. Chem. Educ. 50 A59 (1973)
5. CrO_4^{2-} - $Cr_2O_7^{2-}$ equilibrium: Test. Dem. 182
6. Decomposition of $(NH_4)_2Cr_2O_7$: Test. Dem. 5, 53; J. Chem. Educ. 61 908 (1984)
7. Oxidation states of Mn: J. Chem. Educ. 54 302 (1977)
8. Reaction of $KMnO_4$ with glycerine: Test. Dem. 168
9. MnO_4^- and MnO_4^{2-} ions: Test. Dem. 46, 134, 175

QUIZZES

Quiz 1

1. Write a balanced redox equation for the reaction of silver with nitric acid to form NO(g).
2. Write a balanced equation for the
   a. formation of a precipitate when NH_3 is added to a solution of $Fe(NO_3)_3$.
   b. dissolving of the precipitate in (a) when HCl is added.

Quiz 2

1. Given:

   $MnO_4^-(aq) + 8H^+(aq) + 5e^- \rightarrow Mn^{2+}(aq) + 4H_2O$; $E^o_{red} = +1.52$ V

   calculate E_{red} when conc. MnO_4^- = conc. Mn^{2+} = 1.0 M, pH = 2.0

2. Write a balanced equation for the
   a. formation of a precipitate when solutions of $NiCl_2$ and NaOH are mixed.
   b. dissolving of the precipitate in (a) when NH_3 is added.

Quiz 3

1. Given: E^o_{red} $Au^{3+} \rightarrow Au^+$ = +1.40 V; E^o_{red} $Au^+ \rightarrow Au$ = +1.69 V

   calculate K for the reaction: $3Au^+(aq) \rightleftharpoons Au^{3+}(aq) + 2Au(s)$

2. Write a balanced equation to explain why:
   a. a solution containing the $Cr(H_2O)_6^{3+}$ ion is acidic.
   b. a solution containing the CrO_4^{2-} ion turns red when acid is added.

Quiz 4

1. The acid dissociation constant of the $Fe(H_2O)_6^{2+}$ ion is 2 x 10^{-7}. Calculate the pH of a 0.2 M $FeCl_2$ solution.

2. Write a balanced equation for the reaction that takes place when
   a. mercury(II) sulfide is heated in air.
   b. manganese dissolves in 6 M HCl.

Quiz 5

1. Consider the half-reaction:

   $Cr_2O_7^{2-}(aq) + 14H^+(aq) + 6e^- \rightarrow 2Cr^{3+}(aq) + 7H_2O$; $E^o_{red} = +1.33$ V

   At what pH does E_{red} become 1.00 V when all species other than H^+ are at unit concentration?

2. When sulfide ions are added to a solution of $Cr(NO_3)_3$, a green precipitate of $Cr(OH)_3$ forms. Write a balanced equation for the reaction involved.

PROBLEMS 1-30

25.1 a. $Cu_2S(s) + O_2(g) \longrightarrow 2Cu(s) + SO_2(g)$

b. $NiO(s) + CO(g) \longrightarrow Ni(s) + CO_2(g)$

c. $Fe_2O_3(s) + 2Al(s) \longrightarrow 2Fe(s) + Al_2O_3(s)$

25.2 a. $AgBr(s) + 2S_2O_3^{2-}(aq) \longrightarrow Ag(S_2O_3)_2^{3-}(aq) + Br^-(aq)$

b. $Fe(H_2O)_6^{3+}(aq) \longrightarrow Fe(H_2O)_5(OH)^{2+}(aq) + H^+(aq)$

c. $Cu(OH)_2(s) + 4NH_3(aq) \longrightarrow Cu(NH_3)_4^{2+}(aq) + 2\ OH^-(aq)$

25.3 a. $Ni^{2+}(aq) + H_2S(aq) \longrightarrow NiS(s) + 2H^+(aq)$

b. $Fe^{3+}(aq) + 3NH_3(aq) + 3H_2O \longrightarrow Fe(OH)_3(s) + 3NH_4^+(aq)$

c. $Co^{2+}(aq) + 2\ OH^-(aq) \longrightarrow Co(OH)_2(s)$

25.4 a. $Co(H_2O)_6^{2+}(aq) \longrightarrow H^+(aq) + Co(H_2O)_5(OH)^+(aq)$

b. $Zn(OH)_2(s) + 2\ H^+(aq) \longrightarrow Zn^{2+}(aq) + 2\ H_2O$

c. $Mn(s) + 2H_2O(l) \longrightarrow Mn(OH)_2(s) + H_2(g)$

25.5 a. $Cu^{2+}(aq) + 2NH_3(aq) + 2H_2O \longrightarrow Cu(OH)_2(s) + 2NH_4^+(aq)$

b. $Cu(OH)_2(s) + 4NH_3(aq) \longrightarrow Cu(NH_3)_4^{2+}(aq) + 2\ OH^-(aq)$

25.6 a. $Cu^{2+}(aq) + 4Cl^-(aq) \longrightarrow CuCl_4^{2-}(aq)$

b. $Fe^{3+}(aq) + SCN^-(aq) \longrightarrow Fe(SCN)^{2+}(aq)$

c. $Co(s) + 2H^+(aq) \longrightarrow Co^{2+}(aq) + H_2(g)$

25.7 a. $Ni(s) + 2H^+(aq) \longrightarrow Ni^{2+}(aq) + H_2(g)$

b. $3Cu(s) + 8H^+(aq) + 2NO_3^-(aq) \longrightarrow 3Cu^{2+}(aq) + 2NO(g) + 4H_2O$

25.8 $Hg(l) + 4Cl^-(aq) \longrightarrow HgCl_4^{2-}(aq) + 2e^-$

$2[NO_3^-(aq) + 2H^+(aq) + e^- \longrightarrow NO_2(g) + H_2O]$

$Hg(l) + 2NO_3^-(aq) + 4Cl^-(aq) + 4H^+(aq) \longrightarrow HgCl_4^{2-}(aq) + 2H_2O + 2NO_2(g)$

25.9 a. $Cu(s) \longrightarrow Cu^{2+}(aq) + 2e^-$

$2[NO_3^-(aq) + 2H^+(aq) + e^- \longrightarrow NO_2(g) + H_2O]$

$Cu(s) + 2NO_3^-(aq) + 4H^+(aq) \longrightarrow Cu^{2+}(aq) + 2NO_2(g) + 2H_2O$

b. $4[Cr(OH)_3(s) + H_2O \longrightarrow CrO_4^{2-}(aq) + 5H^+(aq) + 3e^-]$

$3[O_2(g) + 4H^+(aq) + 4e^- \longrightarrow 2H_2O]$

$4Cr(OH)_3(s) + 3\ O_2(g) \longrightarrow 4CrO_4^{2-}(aq) + 8H^+(aq) + 2H_2O$

$4Cr(OH)_3(s) + 3\ O_2(g) + 8\ OH^-(aq) \longrightarrow 4CrO_4^{2-}(aq) + 10H_2O$

25.10 a. $3[Cd(s) \longrightarrow Cd^{2+}(aq) + 2e^-]$

$2[NO_3^-(aq) + 4H^+(aq) + 3e^- \longrightarrow NO(g) + 2H_2O]$

$3Cd(s) + 2NO_3^-(aq) + 8H^+(aq) \longrightarrow 3Cd^{2+}(aq) + 2NO(g) + 4H_2O$

b. $2[Mn^{2+}(aq) + 4H_2O \longrightarrow MnO_4^-(aq) + 8H^+(aq) + 5e^-]$

$5[BiO_3^-(aq) + 6H^+(aq) + 2e^- \longrightarrow Bi^{3+}(aq) + 3H_2O]$

$2Mn^{2+}(aq) + 5BiO_3^-(aq) + 14H^+(aq) \longrightarrow 2MnO_4^-(aq) + 5Bi^{3+}(aq) + 7H_2O$

25.11 Mn, Cr, Zn, Cd

25.12 $2Cu^+(aq) \longrightarrow Cu^{2+}(aq) + Cu(s)$

$E^o_{tot} = E^o_{red}\ Cu^+ + E^o_{ox}\ Cu^+ = +0.52\ V - 0.15\ V = +0.37\ V$

$3Au^+(aq) \longrightarrow Au^{3+}(aq) + 2Au(s)$

$E^o_{tot} = E^o_{red}\ Au^+ + E^o_{ox}\ Au^+ = +1.69\ V - 1.40\ V = +0.29\ V$

25.13 a. $Ni(s) + Zn^{2+}(aq) \longrightarrow Ni^{2+}(aq) + Zn(s)$

$E^o_{tot} = E^o_{ox}\ Ni + E^o_{red}\ Zn^{2+} = +0.25\ V - 0.76\ V = -0.51\ V$

nonspontaneous

b. $Cr^{2+}(aq) + Mn^{3+}(aq) \longrightarrow Cr^{3+}(aq) + Mn^{2+}(aq)$

$E^o_{tot} = E^o_{ox}\ Cr^{2+} + E^o_{red}\ Mn^{3+} = +0.41\ V + 1.51\ V = +1.92\ V$

spontaneous

c. $Hg^{2+}(aq) + Hg(l) \longrightarrow Hg_2^{\ 2+}(aq)$

$E^o_{tot} = E^o_{red}\ Hg^{2+} + E^o_{ox}\ Hg = +0.92\ V - 0.79\ V = +0.13\ V$

spontaneous

25.14 a. $E^o_{tot} = E^o_{red}\ Co^{3+} + E^o_{ox}\ H_2O = +1.82\ V - 1.23\ V = +0.59\ V$

spontaneous

b. $E^o_{tot} = E^o_{ox}\ Cr^{2+} + E^o_{red}\ I_2 = +0.41\ V + 0.53\ V = +0.94\ V$

spontaneous

25.15 a. Cr^{2+} b. Au^+

25.16 Co^{3+}, Au^{3+}, Au^+ and Mn^{3+}

25.17 a. $E = +0.37\ V - \frac{0.0591}{1} \log_{10}\frac{(\text{conc. } Cu^{2+})}{(\text{conc. } Cu^{+})^2}$

$= +0.37\ V - \frac{0.0591}{1} \log_{10} \frac{(1 \times 10^{-4})}{(1 \times 10^{-4})^2}$

$= +0.37\ V - 0.24\ V = +0.13\ V$

b. $0.00 = 0.37 - \frac{0.0591}{1} \log_{10} \frac{1}{(\text{conc. } Cu^{+})^2}$

$= 0.37 + 2(0.0591)\log_{10}(\text{conc. } Cu^{+})$

$\log_{10}(\text{conc. } Cu^{+}) = \frac{-0.37}{2(0.0591)}$; conc. $Cu^+ = 7 \times 10^{-4}$ M

25.18 $Cr_2O_7^{\ 2-}(aq) + 14H^+(aq) + 6e^- \longrightarrow 2Cr^{3+}(aq) + 7H_2O$

$E_{red} = E^o_{red} - \frac{0.0591}{6} \log_{10} \frac{1}{(\text{conc. } H^{+})^{14}}$

$= +1.33\ V - \frac{0.0591}{6} \log_{10} \frac{1}{(6.0)^{14}} = +1.44\ V$

25.19 $E^o_{tot} = E^o_{red}\ O_2(g) + E^o_{ox}\ Co(NH_3)_6^{\ 2+}(aq) = +0.4\ V - 0.1\ V$

$= +0.3\ V$

$E = +0.3\ V - \frac{0.0591}{2} \log_{10}(\text{conc. } OH^{-})^2$

$= +0.3\ V - 0.0591 \log_{10}(\text{conc. } OH^{-})$

a. $E = +0.3\ V - 0.0591 \log_{10}(1.0 \times 10^{-10}) = +0.9\ V$

b. $E = +0.3\ V - 0.0591 \log_{10}(1.0 \times 10^{-7}) = +0.7\ V$

25.20 a. $\Delta G^o = -96.5(2)(-0.51) = 98$ kJ

b. $\Delta G^o = -96.5(1)(+1.92) = -185$ kJ

c. $\Delta G^o = -96.5(1)(+0.13) = -13$ kJ

25.21 a. $\log_{10}K = \frac{2(+0.59)}{0.0591} = 20$; $K = 10^{20}$

b. $\log_{10}K = \frac{2(+0.94)}{0.0591} = 32$; $K = 10^{32}$

25.22 a. $3Fe^{2+}(aq) \longrightarrow 2Fe^{3+}(aq) + Fe(s)$

$E^o_{tot} = -0.77\text{ V} - 0.44\text{ V} = -1.21\text{ V}$

$\log_{10}K = \frac{2(-1.21)}{0.0591} = -40.9$; $K = 1 \times 10^{-41}$

b. $[Fe^{3+}]^2 = K \times [Fe^{2+}]^3 = 1 \times 10^{-41}(0.10)^3 = 1 \times 10^{-44}$

$[Fe^{3+}] = 1 \times 10^{-22}$ M

25.23 a. $E^o_{tot} = E^o_{ox}\ Fe + E^o_{red}\ Fe^{3+} = +0.44\text{ V} + 0.77\text{ V} = +1.21\text{ V}$

$\Delta G^o = -96.5(2)(1.21) = -234$ kJ

b. $E^o_{tot} = E^o_{ox}\ Hg + E^o_{red}\ Hg^{2+} = -0.79\text{ V} + 0.92\text{ V} = +0.13\text{ V}$

$\Delta G^o = -96.5(1)(0.13) = -13$ kJ

25.24 $E^o_{tot} = \frac{-\Delta G^o}{(96.5)(2)} = \frac{124.7}{193.0} = +0.646\text{ V}$

$\log_{10}K = \frac{2(0.646)}{0.0591} = 21.9$; $K = 8 \times 10^{21}$

25.25 See discussion Section 25.1

25.26 See discussion Section 25.1

25.27 a. zinc, cadmium, mercury

b. Au (also, to a lesser extent, Ag and Cu)

c. manganese

d. MnO_4^-

25.28 a. NiS roasted to NiO, which is reduced by heating with C

b. Reduce MnO_2 with C or Al

c. ZnS roasted to ZnO, which is reduced by heating with C

25.29 a. $Cr(H_2O)_6^{3+}$ and $Cr(H_2O)_5Cl^{2+}$ have different colors

b. Cl^- ions form $AuCl_4^-$ complex, NO_3^- acts as oxidizing agent

c. $Zn(H_2O)_4^{2+}(aq) \rightleftharpoons H^+(aq) + Zn(H_2O)_3(OH)^+(aq)$

25.30 $\frac{[H^+]^2}{0.10} \approx 2 \times 10^{-7}$; $[H^+]^2 = 2 \times 10^{-8}$

$[H^+] = (2 \times 10^{-8})^{\frac{1}{2}}$; pH = 3.8

CHAPTER 26

BASIC SKILLS

1. Assign names to the oxyacids and oxyanions of chlorine, sulfur, nitrogen, and other nonmetals.
2. Compare, as to relative acid strength:
   a. oxyacids of similar molecular formula containing different central nonmetal atoms.
   b. oxyacids of the same nonmetal differing in the number of oxygen atoms per molecule.
3. Decide whether a given species containing Cl, N or S can act as an oxidizing agent; a reducing agent.
4. Using the Nernst equation, relate the oxidizing and reducing strengths of oxyanions to pH.
5. Use standard potential diagrams (Figures 26.3, 26.4, 26.6) to decide whether or not a given redox reaction will occur at standard concentrations.
6. State the principal oxidation states of the elements Cl, N, S and P and give an example of a species in each state.
7. Calculate the ratio $[HB]/[B^-]$ in a solution of a weak acid (e.g. HClO, HNO_2) at a particular pH.

LECTURE NOTES

This chapter, like the preceding one, is descriptive; no new principles are introduced. Most of the reactions covered are of the redox type discussed in Chapters 23 and 24. In addition, some acid-base chemistry is considered in connection with such weak acids as HClO, HNO_2, H_2SO_3, and H_3PO_4. The chapter can be covered in two lectures. However, this can easily expand to three if you spend very much time on the descriptive chemistry of the oxidation states of chlorine, nitrogen, and sulfur. In the outlines that follow, we have compromised on 2½ lectures.

LECTURE 1

I Nonmetals

A. Oxygen and Fluorine

1. Elements are strong oxidizing agents

$$F_2(g) + 2e^- \longrightarrow 2F^-(aq);\; E^o_{red} = +2.87\ V$$

$O_3(g) + 2H^+(aq) + 2e^- \longrightarrow O_2(g) + H_2O;\ E^o_{red} = +2.07\ V$

$\frac{1}{2}O_2(g) + 2H^+(aq) + 2e^- \longrightarrow H_2O;\ E^o_{red} = +1.23\ V$

2. Hydrogen peroxide, H_2O_2; oxid. no. O = -1

   strong oxid. agent:

   $H_2O_2(aq) + 2H^+(aq) + 2e^- \longrightarrow 2H_2O;\ E^o_{red} = +1.77\ V$

   weak reducing agent:

   $H_2O_2(aq) \longrightarrow O_2(g) + 2H^+(aq) + 2e^-;\ E^o_{ox} = -0.68\ V$

   disproportionation:

   $2H_2O_2(aq) \longrightarrow O_2(g) + 2H_2O;\ E^o_{tot} = +1.09\ V$

B. Oxyacids and Oxyanions

1. Nomenclature

H_2SO_4	sulfur<u>ic</u> acid	SO_4^{2-}	sulf<u>ate</u> ion
H_2SO_3	sulfur<u>ous</u> acid	SO_3^{2-}	sulf<u>ite</u> ion
$HClO_4$	<u>per</u>chlor<u>ic</u> acid	ClO_4^-	<u>per</u>chlor<u>ate</u> ion
$HClO_3$	chlor<u>ic</u> acid	ClO_3^-	chlor<u>ate</u> ion
$HClO_2$	chlor<u>ous</u> acid	ClO_2^-	chlor<u>ite</u> ion
HClO	<u>hypo</u>chlor<u>ous</u> acid	ClO^-	<u>hypo</u>chlor<u>ite</u> ion

2. Acid strength
   a. increases with increasing electronegativity of central atom:

      HClO ($K_a = 3 \times 10^{-8}$) vs HIO ($K_a = 2 \times 10^{-11}$)

   b. increases with increasing oxidation number of central atom:

      $HClO_4 \gg HClO$

3. Oxidizing and reducing properties
   a. Species in highest oxidation state (ClO_4^-, NO_3^-, SO_4^{2-}) can act only as oxidizing agents.
   b. species in intermediate oxidation states (ClO_3^-, NO_2^-, SO_3^{2-}) can act as either oxidizing or reducing agents.
   c. species in lowest oxidation state (Cl^-, N^{3-}, S^{2-}) can act only as reducing agents.

4. Effect of pH on oxidizing, reducing strength.
   a. Oxidizing agents stronger in acidic solution

      $NO_3^-(aq) + 2H^+(aq) + e^- \longrightarrow NO_2(g) + H_2O;\ E^o_{red} = +0.78\ V$

      $$E_{red} = +0.78\ V - \frac{0.0591}{1} \log_{10} \frac{(P\ NO_2)}{(conc.\ NO_3^-)(conc.\ H^+)^2}$$

      If P NO_2 = 1 atm, conc. NO_3^- = 1 M:

      $$E_{red} = +0.78\ V - 0.0591 \log_{10} \frac{1}{(conc.\ H^+)^2}$$

      $$= +0.78\ V - 0.1182\ pH$$

b. Reducing agents stronger in basic solution

$$NO_2^-(aq) + H_2O \longrightarrow NO_3^-(aq) + 2H^+(aq) + e^-$$

$$E_{ox} = E^o_{ox} + 0.0591\ pH$$

LECTURE 2

I Chlorine

A. Oxidation states

+7 $HClO_4$, ClO_4^- 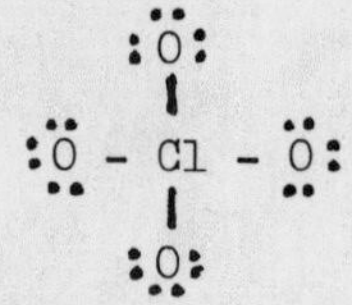

+5 $HClO_3$, ClO_3^- 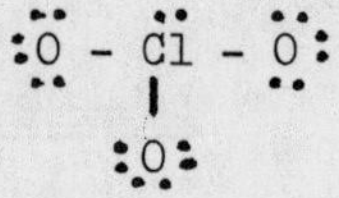

+3 $HClO_2$, ClO_2^- 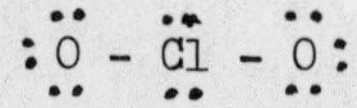

+1 $HClO$, ClO^- 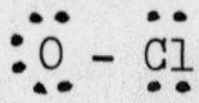

B. Disproportionation of intermediate states

$$ClO_4^- \xrightarrow{+1.19V} ClO_3^- \xrightarrow{+1.21V} HClO_2;\quad E^o_{tot} = +0.02\ V$$

$$ClO_4^- \xrightarrow{+0.36V} ClO_3^- \xrightarrow{+0.33V} ClO_2^-;\quad E^o_{tot} = -0.03\ V$$

ClO_3^- spontaneously disproportionates in acidic solution (1 M H^+), but not in basic solution, at standard concentrations.

C. Hypochlorous acid, hypochlorites

$$Cl_2(g) + H_2O \longrightarrow HClO(aq) + H^+(aq) + Cl^-(aq)$$

$$E^o_{tot} = +1.36\ V - 1.63\ V = -0.27\ V$$

$$\log_{10}K = \frac{1(-0.27)}{0.0591} = -4.6;\ K = 3 \times 10^{-5}$$

$$Cl_2(g) + 2\ OH^-(aq) \longrightarrow ClO^-(aq) + Cl^-(aq) + H_2O$$

$$E^o_{tot} = +0.95\ V;\ K = 1 \times 10^{16}$$

II Nitrogen

A. Oxidation states

	Acidic Solution	Basic Solution
+5	NO_3^-	NO_3^-
+4	$NO_2(g)$	$NO_2(g)$
+3	HNO_2	NO_2^-
+2	$NO(g)$	$NO(g)$
+1	$N_2O(g)$	$N_2O(g)$
0	$N_2(g)$	$N_2(g)$
-3	NH_4^+	NH_3

HNO_2: $K_a = 4.5 \times 10^{-4}$; $\frac{[HNO_2]}{[NO_2^-]} = \frac{H^+}{(4.5 \times 10^{-4})}$

In 1 M H^+: $\frac{[HNO_2]}{[NO_2^-]} = 2.2 \times 10^3$; HNO_2 is principal species

In 1 M OH^-, $[H^+] = 1.0 \times 10^{-14}$ M,

$\frac{[HNO_2]}{[NO_2^-]} = \frac{1.0 \times 10^{-14}}{4.5 \times 10^{-4}} = 2.2 \times 10^{-11}$; NO_2^- is main species

B. Nitric Acid

Strong acid; strong oxidizing agent in acidic solution. Oxidizes inactive metals such as Cu and Ag:

$$Ag(s) + 2H^+(aq) + NO_3^-(aq) \longrightarrow Ag^+(aq) + NO_2(g) + H_2O$$

Metal sulfides:

$$NO_3^-(aq) + 2H^+(aq) + e^- \longrightarrow NO_2(g) + H_2O$$

$$Bi_2S_3(s) \longrightarrow 2Bi^{3+}(aq) + 3S(s) + 6e^-$$

$$Bi_2S_3(s) + 6NO_3^-(aq) + 12H^+(aq) \longrightarrow 2Bi^{3+}(aq) + 3S(s) + 6H_2O + 6NO_2(g)$$

Can also be reduced, in dilute solution, to NO, N_2, NH_4^+

LECTURE 2½

I Sulfur

A. Oxidation states

	Acidic Solution	Basic Solution
+6	HSO_4^- ($K_a = 1.2 \times 10^{-2}$)	SO_4^{2-}
+4	H_2SO_3 ($K_a = 1.7 \times 10^{-2}$)	SO_3^{2-}
	HSO_3^- ($K_a = 5.6 \times 10^{-8}$)	
+2	$S_2O_3^{2-}$	$S_2O_3^{2-}$
0	S(s)	S(s)
-2	H_2S ($K_a = 1 \times 10^{-7}$)	HS^-, S^{2-}

Consider +4 S:

$$\frac{[H_2SO_3]}{[HSO_3^-]} = \frac{[H^+]}{1.7 \times 10^{-2}} \quad ; \quad \frac{[HSO_3^-]}{[SO_3^{2-}]} = \frac{[H^+]}{5.6 \times 10^{-8}}$$

in neutral solution: $\frac{[H_2SO_3]}{[HSO_3^-]} = 6 \times 10^{-6}$; $\frac{[HSO_3^-]}{[SO_3^{2-}]} = 2$

$[HSO_3^-]$ is principal species

B. H_2SO_4

1. Strong acid $H_2SO_4(aq) \longrightarrow H^+(aq) + HSO_4^-(aq)$

2. Relatively weak oxidizing agent

$$SO_4^{2-}(aq) + 4H^+(aq) + 2e^- \longrightarrow SO_2(g) + 2H_2O; \; E^\circ_{red} = +0.20 \text{ V}$$

3. Powerful dehydrating agent; reaction with water is exothermic.

II Phosphorus

+5 H_3PO_4, $H_2PO_4^-$, HPO_4^{2-}, PO_4^{3-}

+3 H_3PO_3, $H_2PO_3^-$, HPO_3^{2-}

+1 H_3PO_2, $H_2PO_2^-$

Lewis structures of acids:

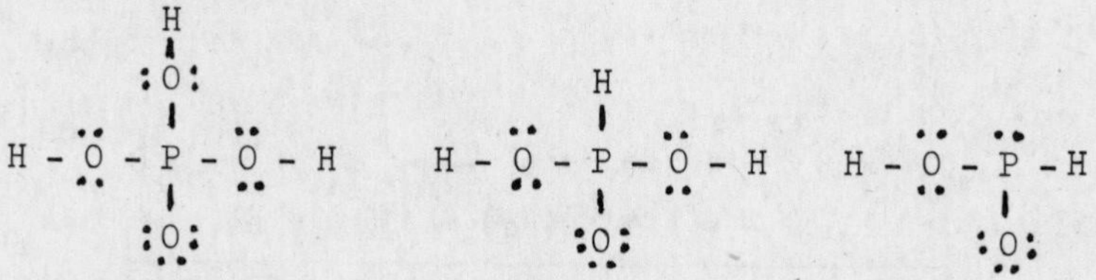

phosphoric acid   phosphorous acid   hypophosphorous acid

DEMONSTRATIONS

1. Ozone as an oxidizing agent: Test. Dem. 58
2. Oxidizing strength of H_2O_2: Test. Dem. 213
3. Reaction of Cl_2 with Br^-, I^-: Test. Dem. 5, 213
4. Reaction of ClO_3^- with SO_3^{2-}: J. Chem. Educ. <u>60</u> 994 (1983)
5. Oxidizing properties of HNO_3: Test. Dem. 99
6. Reaction of H_2SO_4 with sugar: Test. Dem. 35
7. Reducing action of SO_2: Test. Dem. 42

QUIZZES

<u>Quiz 1</u>

1. Write Lewis structures for each of the following:
   a. ClO_4^- b. HNO_3 c. $S_2O_3^{2-}$

2. Write a balanced equation for:
   a. the disproportionation of H_2O_2
   b. the reaction of chlorine gas with water
   c. the dissociation of sulfuric acid in water

Quiz 2

1. Give the formula of a species in which
   a. the oxidation number of chlorine is +3
   b. the oxidation number of nitrogen is +4
   c. the oxidation number of phosphorus is +5

2. K_a of HClO is 3×10^{-8}. Calculate the ratio $[HClO]/[ClO^-]$ at pH

   a. 3.00   b. 7.00

Quiz 3

1. Give the formulas of the following species:
   a. phosphorous acid   b. hypochlorite ion
   c. hydrogen sulfite ion

2. Write a balanced redox equation for the reaction of NiS with nitric acid; the products include SO_2 and NO_2.

Quiz 4

1. Explain in terms of structure why:
   a. HClO is a stronger acid than HBrO
   b. $HClO_4$ is a stronger acid than HClO

2. Given:

   $NO_3^-(aq) + H_2O + 2e^- \longrightarrow NO_2^-(aq) + 2\ OH^-(aq)$; $E^o_{red} = -0.87$ V

   Calculate E_{red} when conc. NO_3^- = conc. NO_2^- = 1.0 M and the pH is 7.0.

Quiz 5

1. Referring to Figure 26.6:
   a. calculate E^o_{tot} for $3S(s) + 2H_2O \longrightarrow 2H_2S(aq) + SO_2(g)$
   b. show, by calculation, which species should disproportionate in basic solution
   c. write a balanced redox equation for the disproportionation of the $S_2O_3^{2-}$ ion in acidic solution.

PROBLEMS 1-30

26.1 a. bromous acid   b. bromate   c. perbromate
d. hypobromous acid

26.2 a. nitrous acid, HNO_2 b. chlorous acid, $HClO_2$
c. perbromic acid, $HBrO_4$ d. phosphoric acid, H_3PO_4

26.3 a. selenic acid   b. selenite ion

26.4 a. $HBrO_3$ b. HClO   c. H_2SO_3 d. HIO_2

26.5 a. either   b. reducing agent   c. either   d. either

26.6 oxid. agents: NO_3^-, NO_2, HNO_2, NO, N_2O, N_2, NH_3OH^+, $N_2H_5^+$
red. agents: NO_2, HNO_2, NO, N_2O, N_2, NH_3OH^+, $N_2H_5^+$, NH_4^+

26.7 $\frac{[NO_2^-]}{[HNO_2]} = \frac{K_a}{[H^+]} = 1$; $[H^+] = K_a = 4.5 \times 10^{-4}$ M; pH = 3.35

26.8 a. $\frac{[H_2SO_3]}{[HSO_3^-]} = \frac{[H^+]}{K_a\ H_2SO_3} = 1$; $[H^+] = 1.7 \times 10^{-2}$ M; pH = 1.77

b. $\frac{[HSO_3^-]}{[SO_3^{2-}]} = \frac{[H^+]}{K_a\ HSO_3^-} = 2.0$; $[H^+] = 1.1 \times 10^{-7}$ M; pH = 6.96

c. $\frac{[H_2SO_3]}{[HSO_3^-]} = \frac{[H^+]}{K_a\ H_2SO_3} = 0.50$; $[H^+] = 8.5 \times 10^{-3}$ M; pH = 2.07

26.9 a. $5[Cl_2(g) + 2e^- \longrightarrow 2Cl^-(aq)]$

$Cl_2(g) + 6H_2O \longrightarrow 2ClO_3^-(aq) + 12H^+(aq) + 10e^-$

$6Cl_2(g) + 6H_2O \longrightarrow 10Cl^-(aq) + 2ClO_3^-(aq) + 12H^+(aq)$

$3Cl_2(g) + 3H_2O \longrightarrow 5Cl^-(aq) + ClO_3^-(aq) + 6H^+(aq)$

$3Cl_2(g) + 6\ OH^-(aq) \longrightarrow 5Cl^-(aq) + ClO_3^-(aq) + 3H_2O$

b. $CuS(s) \longrightarrow Cu^{2+}(aq) + S(s) + 2e^-$

$2[NO_3^-(aq) + 2H^+(aq) + e^- \longrightarrow NO_2(g) + H_2O]$

$CuS(s) + 2NO_3^-(aq) + 4H^+(aq) \longrightarrow Cu^{2+}(aq) + S(s) + 2NO_2(g) + 2H_2O$

c. $4[Zn(s) \longrightarrow Zn^{2+}(aq) + 2e^-]$

$NO_3^-(aq) + 10H^+(aq) + 8e^- \longrightarrow NH_4^+(aq) + 3H_2O$

$4Zn(s) + NO_3^-(aq) + 10H^+(aq) \longrightarrow 4Zn^{2+}(aq) + NH_4^+(aq) + 3H_2O$

26.10 a. $Br_2(l) + 2I^-(aq) \longrightarrow 2Br^-(aq) + I_2(s)$

b. $SCN^-(aq) + 4H_2O \longrightarrow SO_4^{2-}(aq) + HCN(aq) + 7H^+(aq) + 6e^-$

$6[NO_3^-(aq) + 2H^+(aq) + e^- \longrightarrow NO_2(g) + H_2O]$

$SCN^-(aq) + 6NO_3^-(aq) + 5H^+(aq) \longrightarrow SO_4^{2-}(aq) + HCN(aq) + 6NO_2(g) + 2H_2O$

26.11 a. $HClO(aq) + H_2O \longrightarrow HClO_2(aq) + 2H^+(aq) + 2e^-$

$2HClO(aq) + 2H^+(aq) + 2e^- \longrightarrow Cl_2(g) + 2H_2O$

$3HClO(aq) \longrightarrow HClO_2(aq) + Cl_2(g) + H_2O$

b. $2ClO^-(aq) + 4H^+(aq) + 2e^- \longrightarrow Cl_2(g) + 2H_2O$

$ClO^-(aq) + H_2O \longrightarrow ClO_2^-(aq) + 2H^+(aq) + 2e^-$

$3ClO^-(aq) + 2H^+(aq) \longrightarrow Cl_2(g) + ClO_2^-(aq) + H_2O$

$3ClO^-(aq) + H_2O \longrightarrow Cl_2(g) + ClO_2^-(aq) + 2\ OH^-(aq)$

26.11 c. $3[N_2(g) + 2H_2O \rightarrow 2NO(g) + 4H^+(aq) + 4e^-]$

$2[N_2(g) + 8H^+(aq) + 6e^- \rightarrow 2NH_4^+(aq)]$

$5N_2(g) + 6H_2O + 4H^+(aq) \rightarrow 6NO(g) + 4NH_4^+(aq)$

26.12 a. $Cr_2O_7^{2-}(aq) + 14H^+(aq) + 6e^- \rightarrow 2Cr^{3+}(aq) + 7H_2O$

$3[2I^-(aq) \rightarrow I_2(s) + 2e^-]$

$Cr_2O_7^{2-}(aq) + 14H^+(aq) + 6I^-(aq) \rightarrow 2Cr^{3+}(aq) + 3I_2(s) + 7H_2O$

b. $Cr_2O_7^{2-}(aq) + 14H^+(aq) + 6e^- \rightarrow 2Cr^{3+}(aq) + 7H_2O$

$6[Fe^{2+}(aq) \rightarrow Fe^{3+}(aq) + e^-]$

$Cr_2O_7^{2-}(aq) + 14H^+(aq) + 6Fe^{2+}(aq) \rightarrow 2\ Cr^{3+}(aq) + 7H_2O + 6Fe^{3+}(aq)$

c. $Cr_2O_7^{2-}(aq) + 14H^+(aq) + 6e^- \rightarrow 2Cr^{3+}(aq) + 7H_2O$

$3[HNO_2(aq) + H_2O \rightarrow NO_3^-(aq) + 3H^+(aq) + 2e^-]$

$Cr_2O_7^{2-}(aq) + 5H^+(aq) + 3HNO_2(aq) \rightarrow 2Cr^{3+}(aq) + 4H_2O + 3NO_3^-(aq)$

d. $Cr_2O_7^{2-}(aq) + 14H^+(aq) + 6e^- \rightarrow 2Cr^{3+}(aq) + 7H_2O$

$3[H_2S(aq) \rightarrow S(s) + 2H^+(aq) + 2e^-]$

$Cr_2O_7^{2-}(aq) + 8H^+(aq) + 3H_2S(aq) \rightarrow 2Cr^{3+}(aq) + 7H_2O + 3S(s)$

26.13 a. BrO_4^- b. Br_2 c. Br^- d. Br_2

26.14 a. $E^o_{tot} = +1.77\ V - 1.82\ V = -0.05\ V$; no

b. $E^o_{tot} = +1.77\ V - 1.36\ V = +0.41\ V$; yes

c. $E^o_{tot} = +1.77\ V - 0.77\ V = +1.00\ V$; yes

d. $E^o_{tot} = +1.77\ V - 0.15\ V = +1.62\ V$; yes

26.15 a. $E^o_{tot} = +1.21\ V - 1.19\ V = +0.02\ V$; yes

b. $E^o_{tot} = +1.63\ V - 1.43\ V = +0.20\ V$; yes

c. $E^o_{tot} = +1.47\ V - 1.63\ V = -0.16\ V$; no

d. cannot disproportionate

26.16 a. $E^o_{tot} = +0.78\ V - 1.10\ V = -0.32\ V$; no

b. $E^o_{tot} = -0.87\ V - 0.89\ V = -1.76\ V$; no

c. impossible; two reductions

d. $E^o_{tot} = -1.00\ V + 1.59\ V = +0.59\ V$; yes

26.17 a. $E^o_{tot} = E^o_{red}\ ClO_3^- + E^o_{ox}\ NO_2^- = +0.33\ V - 0.01\ V = +0.32\ V$

b. $E^o_{tot} = E^o_{red}\ SO_4^{2-} + E^o_{ox}\ Cl^- = +0.20\ V - 1.36\ V = -1.16\ V$

26.17 c. $E^{o}_{tot} = E^{o}_{red}\ H_3PO_4 + E^{o}_{ox}\ Cl^- = -0.28\ V - 1.49\ V = -1.77\ V$

26.18 a. $E_{red} = +0.78\ V - \frac{0.0591}{1} \log_{10} \frac{1}{(\text{conc. } H^+)^2}$

$= +0.78\ V + 2(0.0591)\log_{10}(\text{conc. } H^+) = +0.78\ V$

b. $E_{red} = +0.78\ V - 0.83\ V = -0.05\ V$

c. $E_{red} = +0.78\ V - 1.65\ V = -0.87\ V$

26.19 $E = E^{o}_{tot} - \frac{0.0591}{2} \log_{10} \frac{(\text{conc. } ClO_3^-)(\text{conc. } ClO^-)}{(\text{conc. } ClO_2^-)^2}$

$= +0.33\ V - \frac{0.0591}{2} \log_{10} \frac{(0.10)(0.10)}{(0.10)^2} = +0.33\ V$

26.20 $0.00 = +0.33 - \frac{0.0591}{2} \log_{10} \frac{1}{(\text{conc. } ClO_2^-)^2}$

$= +0.33 + 0.0591 \log_{10}(\text{conc. } ClO_2^-)$

$\log_{10}(\text{conc. } ClO_2^-) = \frac{-0.33}{0.0591} = -5.6$; conc. $ClO_2^- = 3 \times 10^{-6} M$

26.21 a. $\Delta G^{o} = -96.5(2)(+1.64) = -317$ kJ

$\log_{10}K = \frac{2(1.64)}{0.0591}$; $K = 3 \times 10^{55}$

b. $\Delta G^{o} = -96.5(2)(+0.29) = -56$ kJ

$\log_{10}K = \frac{2(0.29)}{0.0591}$; $K = 6 \times 10^{9}$

c. $\Delta G^{o} = =-96.5(2)(+1.29) = -249$ kJ

$\log_{10}K = \frac{2(1.29)}{0.0591}$; $K = 5 \times 10^{43}$

26.22 a. $\Delta G^{o} = -96.5(2)(+0.32) = -62$ kJ

b. $\Delta G^{o} = -96.5(2)(-1.16) = +224$ kJ

c. $\Delta G^{o} = -96.5(2)(-1.77) = +342$ kJ

26.23 $E^{o}_{tot} = -98.0/4(96.5) = -0.254\ V$

$E^{o}_{tot} = -1.74\ V + 1.49\ V = -0.25\ V$

26.24 a. +5 b. +1 c. +3 d. +7

26.25 a. H_3PO_3 b. H_3PO_4 c. PH_3

26.26 a. react with $Cl_2(g)$ b. heat at 350°C
c. bubble O_2 through solution

26.27 a. HNO_2 b. HS^-, H_2S c. S, SO_2 d. HClO
e. H_2SO_3, SO_2

26.28 a. $S_2O_3^{2-}$ b. SO_3^{2-}, HSO_3^- c. H_2SO_4, H_2SO_3, H_2S

26.29 a. $\left(H - \underset{\cdot\cdot}{\overset{\cdot\cdot}{O}} - \overset{:\overset{\cdot\cdot}{O}:}{\underset{:\underset{\cdot\cdot}{O}:}{P}} - \underset{\cdot\cdot}{\overset{\cdot\cdot}{O}} - H \right)^{-}$ b. $\left(H - \underset{\cdot\cdot}{\overset{\cdot\cdot}{O}} - \overset{:\overset{\cdot\cdot}{O}:}{\underset{:\underset{\cdot\cdot}{O}:}{S}} - \overset{\cdot\cdot}{\underset{\cdot\cdot}{O}}: \right)^{-}$

c. $\left(:\overset{\cdot\cdot}{\underset{\cdot\cdot}{O}} - \overset{:\overset{\cdot\cdot}{O}:}{\underset{:\underset{\cdot\cdot}{O}:}{I}} - \overset{\cdot\cdot}{\underset{\cdot\cdot}{O}}: \right)^{-}$

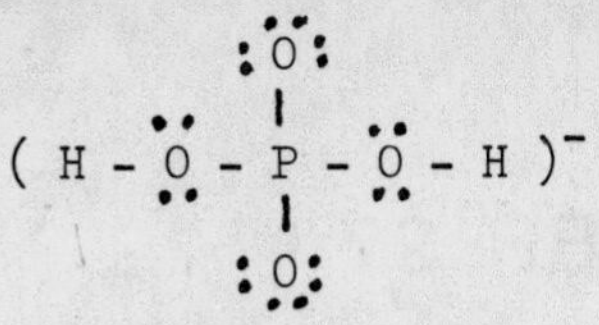

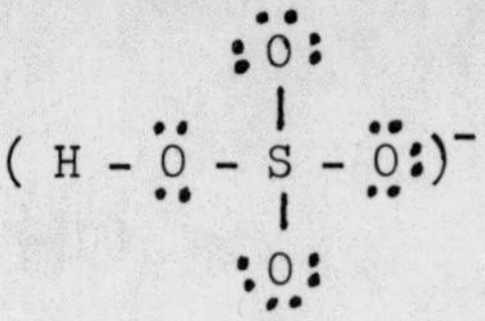

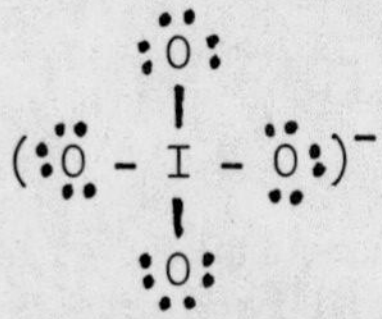

26.30 a. F^- too difficult to oxidize
b. does not form hazardous product
c. removes H_2O

CHAPTER 27

BASIC SKILLS

1. Write a balanced equation for a nuclear reaction, given the identities of all but one of the reactants and products.
2. Use the first order rate law and the expression for the half-life to relate the amount of a radioactive species to elapsed time.
3. Given a table of nuclear masses (Table 27.5), calculate Δm for a nuclear reaction and relate it to the energy change, ΔE.
4. Write typical nuclear equations to represent fission and fusion processes and discuss the characteristics of these reactions.

LECTURE NOTES

This material typically is covered at the end of the school year. It serves to hold the interest of restless students. More than any other topic covered in general chemistry, nuclear reactions relate directly to issues that are meaningful to students. Your challenge is to present the material in an impartial, informative, but stimulating way.

The principles presented in Chapter 27 can readily be covered in two lectures. If more time is available, it can profitably be spent in a more detailed discussion of such topics as the effect of radiation on human beings, the pros and cons of fission reactors, and the prospects for fusion reactors. The outline below, which assumes a modest amount of time spent on these topics, is designed for $2\frac{1}{2}$ lectures. That's how much time we typically spend on this chapter.

<u>LECTURE 1</u>

I <u>Radioactivity</u>

A. Natural: three different types of radiation

1. alpha: ${}^{4}_{2}He$ nuclei released. Atomic number decreases by 2 units, mass number by 4.

$$ {}^{238}_{92}U \longrightarrow {}^{4}_{2}He + {}^{234}_{90}Th $$

2. beta: ${}^{0}_{-1}e$ emitted. Atomic number increases by 1 unit, mass number unchanged.

$$ {}^{234}_{90}Th \longrightarrow {}^{0}_{-1}e + {}^{234}_{91}Pa $$

3. gamma: high energy radiation. No change in atomic or mass number.

Radioactive series: U-238 decays by a series of steps to form Pb-206

$$^{238}_{92}U \longrightarrow ^{206}_{82}Pb + ?\ ^{4}_{2}He + ?\ ^{0}_{-1}e$$

must be 8 alphas and 6 betas to balance equation

B. Induced radioactivity. Stable nucleus is bombarded by high energy particle (neutron, alpha particle, etc.). Unstable nucleus formed undergoes radioactive decay.

$$^{27}_{13}Al + ^{1}_{0}n \longrightarrow ^{28}_{13}Al \longrightarrow ^{28}_{14}Si + ^{0}_{-1}e$$

If isotope formed has too few neutrons, will get positron decay:

$$^{11}_{6}C \longrightarrow ^{11}_{5}B + ^{0}_{1}e$$

Bombardment reactions can be used to make very heavy elements

$$^{238}_{92}U + ^{1}_{0}n \longrightarrow ^{239}_{92}U \longrightarrow ^{239}_{94}Pu + 2\ ^{0}_{-1}e$$

$$^{239}_{94}Pu + ^{4}_{2}He \longrightarrow ^{242}_{96}Cm + ^{1}_{0}n$$

C. Effects of radiation
rad = absorption of 0.01 J per kilogram tissue
no. of rems = n(no. of rads); n = 1 for beta, 10 for alpha radiation

background ≈	0.13 rem/yr
chest X-ray	0.05-0.2 rem
fallout	0.004 rem
nuclear plant	0.0002 rem

II <u>Rate of Radioactive Decay</u>

1st order rate law:

$$\log_{10} \frac{X_o}{X} = \frac{kt}{2.30}$$; X_o = original amount, X = amount at time t

$$k = \frac{0.693}{t_{\frac{1}{2}}}$$

half-life of U-238 is 4.5×10^9 yr. How long does it take for 60% of a U-238 sample to decay?

$k = 0.693/(4.5 \times 10^9\ \text{yr}) = 1.5 \times 10^{-10}/\text{yr}$

$X = 0.40\ X_o$; $X_o/X = 1/0.40$

$$\log_{10} \frac{1}{0.40} = \frac{(1.5 \times 10^{-10})t}{2.30}$$; $t = 6.1 \times 10^9$ yr

LECTURE 2

I Rate of Radioactive Decay (cont.)

A. Age of rocks: measure relative amounts of Pb-206, U-238

0.50 mol Pb-206, 0.50 mol U-238; 4.5×10^9 yr
0.60 mol Pb-206, 0.40 mol U-238; 6.1×10^9 yr

B. Organic material: measure C-14 content

$${}^{14}_{7}N + {}^{1}_{0}n \longrightarrow {}^{14}_{6}C + {}^{1}_{1}H$$

$${}^{14}_{6}C \longrightarrow {}^{14}_{7}N + {}^{0}_{-1}e; \; t_{\frac{1}{2}} = 5720 \text{ yr}$$

In a living plant or animal, these two processes are in equilibrium, C-14 content is constant. When plant or animal dies, first process stops, C-14 content declines.

Suppose fragment of Shroud of Turin shows C-14 content 0.780 times that of living plant. Age of shroud?

$k = 0.693/5720 \text{ yr} = 1.21 \times 10^{-4} \text{ yr}$

$$\log_{10} \frac{1}{0.780} = \frac{1.21 \times 10^{-4} t}{2.30}; \quad t = 2050 \text{ yr}$$

II Mass-Energy Relations

A. Relation between ΔE and Δm ; $\Delta E = c^2 \Delta m$

ΔE (in J) = $(3.0 \times 10^8)^2$ Δm (in kg)

ΔE (in kJ) = 9.0×10^{10} Δm (in g)

B. Calculations

$${}^{239}_{94}Pu \longrightarrow {}^{4}_{2}He + {}^{235}_{92}U$$

Δm per mole: 234.9934 g + 4.0015 g - 239.0006 g = -0.0057 g

ΔE per mole: $-0.0057 \text{ g} \times 9.0 \times 10^{10} \frac{\text{kJ}}{\text{g}} = -5.1 \times 10^8 \text{ kJ}$

ΔE per gram = $\frac{-5.1 \times 10^8 \text{ kJ}}{239 \text{ g}} = -2.1 \times 10^6 \text{ kJ/g}$

compare to about 0-50 kJ/g for ordinary chemical reactions

C. Average mass per nuclear particle

${}^{2}_{1}H$: 2.01355/2 = 1.00678

${}^{239}_{94}Pu$: 239.0006/239 = 1.00000

${}^{56}_{26}Fe$: 55.92066/56 = 0.99858

Fusion of very light nuclei or fission of very heavy nuclei leads to decrease in mass and hence to evolution of energy (Figure 27.3)

LECTURE $2\frac{1}{2}$

I <u>Fission</u>

$$^{235}_{92}U + ^{1}_{0}n \longrightarrow ^{90}_{37}Rb + ^{144}_{55}Cs + 2\,^{1}_{0}n$$

Note that: 1. Many different isotopes are formed
2. More neutrons are produced than consumed. This leads to chain reaction. In nuclear reactor, excess neutrons are absorbed by cadmium rods.
3. Nuclei produced have too many neutrons and hence are intensely radioactive:

$$^{90}_{37}Rb \longrightarrow ^{0}_{-1}e + ^{90}_{38}Sr$$

This is the principal danger associated with nuclear reactors

II <u>Fusion</u>

$$2\,^{2}_{1}H \longrightarrow ^{4}_{2}He;\ \Delta m = -0.02560\ g;\ \Delta E = -2.3 \times 10^{9}\ kJ$$

per gram: $\Delta E = \dfrac{-2.3 \times 10^{9}\ kJ}{4.0\ g} = -5.7 \times 10^{8}\ kJ/g$

ΔE per gram is about 10 times that for fission, 200 times that for radioactivity. Unfortunately, activation energy is very high, since two deuterons strongly repel each other. T required is of the order of 10^{9}°C.

DEMONSTRATIONS

1. Detection of radiation by cloud chamber: Test. Dem. 21, 90
2. Beta ray properties: Test. Dem. 21

QUIZZES

<u>Quiz 1</u>

1. Write a balanced nuclear equation for:
   a. alpha emission by Ra-226
   b. beta emission by Th-234
2. The half-life of C-14 is 5720 yr. What fraction of a C-14 sample is left after 2020 yr?

<u>Quiz 2</u>

1. A certain radioactive series starts with Np-237 and ends with Bi-209. Both alpha and beta particles are evolved in the several steps. Write a balanced nuclear equation for the overall reaction.
2. Calculate, using Table 27.5, ΔE per gram of reactant for:

$$^{19}_{9}F \longrightarrow ^{18}_{8}O + ^{1}_{1}H$$

Quiz 3

1. Complete the following nuclear equations:

   a. ${}^{14}_{7}N$ + ____ $\longrightarrow$ ${}^{17}_{8}O + {}^{1}_{1}H$

   b. ${}^{10}_{5}B + {}^{1}_{1}H$ $\longrightarrow$ ____

2. The half-life of U-238 is 4.5×10^{9} yr. Estimate the age of a rock in which there is 0.521 mol U-238 and 0.479 mol Pb-206.

Quiz 4

1. Complete the following statements:

   a. Emission of an electron converts a ________ in the nucleus to a ________.

   b. Capture of an electron by a nucleus converts a ________ to a ________.

2. Calculate ΔE per gram of reactant for the following reaction, using Table 27.5.

   $3\ {}^{2}_{1}H \longrightarrow {}^{6}_{3}Li$

Quiz 5

1. Consider the reaction: ${}^{226}_{88}Ra \longrightarrow {}^{222}_{86}Rn$ + ________

   a. Complete the equation.

   b. Calculate ΔE, using Table 27.5.

   c. If the half-life of Ra-226 is 1590 yr, calculate the rate constant for the reaction.

PROBLEMS 1-30

27.1 ${}^{51}_{24}Cr \longrightarrow {}^{0}_{1}e + {}^{51}_{23}V$

27.2 a. ${}^{255}_{103}Lr \longrightarrow {}^{4}_{2}He + {}^{251}_{101}Md$

b. ${}^{77}_{36}Kr \longrightarrow {}^{0}_{1}e + {}^{77}_{35}Br$

c. ${}^{65}_{28}Ni \longrightarrow {}^{0}_{-1}e + {}^{65}_{29}Cu$

27.3 a. ${}^{235}_{92}U$ b. ${}^{231}_{91}Pa$

27.4 a. ${}^{121}_{51}Sb + {}^{4}_{2}He \longrightarrow {}^{124}_{52}Te + {}^{1}_{1}H$

b. ${}^{238}_{92}U + {}^{1}_{1}H \longrightarrow {}^{238}_{93}Np + {}^{1}_{0}n$ c. ${}^{27}_{13}Al + {}^{4}_{2}He \longrightarrow {}^{30}_{15}P + {}^{1}_{0}n$

27.5 a. ${}^{32}_{16}S + {}^{1}_{0}n \longrightarrow {}^{32}_{15}P + {}^{1}_{1}H$

b. ${}^{14}_{7}N + {}^{4}_{2}He \longrightarrow {}^{17}_{8}O + {}^{1}_{1}H$

c. ${}^{9}_{4}Be + {}^{4}_{2}He \longrightarrow {}^{1}_{0}n + {}^{12}_{6}C$

27.6 $^{238}_{92}U \rightarrow ^{4}_{2}He + ^{234}_{90}Th$; $^{234}_{90}Th \rightarrow ^{0}_{-1}e + ^{234}_{91}Pa$

$^{234}_{91}Pa \rightarrow ^{0}_{-1}e + ^{234}_{92}U$; $^{234}_{92}U \rightarrow ^{230}_{90}Th + ^{4}_{2}He$

$^{230}_{90}Th \rightarrow ^{4}_{2}He + ^{226}_{88}Ra$; $^{226}_{88}Ra \rightarrow ^{4}_{2}He + ^{222}_{86}Rn$

$^{222}_{86}Rn \rightarrow ^{4}_{2}He + ^{218}_{84}Po$

27.7 $^{238}_{92}U \longrightarrow ^{206}_{82}Pb + 8\ ^{4}_{2}He + 6\ ^{0}_{-1}e$

27.8 $10.0 \text{ mg} \times \left(\frac{1}{2}\right)^5 = 0.312 \text{ mg}$

27.9 a. $^{206}_{82}Pb$ b. $k = 0.693/138 \text{ d} = 0.00502/\text{d}$

c. $\log_{10} \frac{100.0}{10.0} = \frac{0.00502\ t}{2.30} = 1.00$; $t = 458 \text{ d}$

27.10 a. $k = 0.693/14.3 \text{ d} = 0.0485/\text{d}$

b. $\log \frac{100.0}{5.0} = \frac{0.0485\ t}{2.30} = 1.30$; $t = 61.6 \text{ d}$

27.11 $k = 0.693/433 \text{ yr} = 1.60 \times 10^{-3}/\text{yr}$

$\log_{10} \frac{100}{90} = \frac{(1.60 \times 10^{-3})t}{2.30} = 0.046$; $t = 66 \text{ yr}$

27.12 a. $\log_{10} \frac{600}{320} = \frac{k(24 \text{ h})}{2.30}$; $k = 0.026/\text{h}$

b. $t_{\frac{1}{2}} = 0.693/0.026 = 27 \text{ h}$

27.13 $k = 0.693/12.3 \text{ yr} = 0.0563/\text{yr}$

$\log_{10} \frac{1.00}{0.59} = \frac{0.0563\ t}{2.30} = 0.23$; $t = 9.4 \text{ yr}$

27.14 K-40 → Ar-40: $k = 0.693/1.26 \times 10^9 \text{ yr} = 5.50 \times 10^{-10}/\text{yr}$

U-238 → Pb-206: $k = 0.693/4.5 \times 10^9 \text{ yr} = 1.5 \times 10^{-10}/\text{yr}$

Rb-87 → Sr-87 : $k = 0.693/4.8 \times 10^{10} \text{ yr} = 1.4 \times 10^{-11}/\text{yr}$

$\log_{10} 5.13 = \frac{(5.50 \times 10^{-10})t}{2.30} = 0.710$; $t = 2.97 \times 10^9 \text{ yr}$

$\log_{10} 1.66 = \frac{(1.5 \times 10^{-10})t}{2.30} = 0.220$; $t = 3.4 \times 10^9 \text{ yr}$

$\log_{10} 1.049 = \frac{(1.4 \times 10^{-11})t}{2.30} = 0.0208$; $t = 3.4 \times 10^9 \text{ yr}$

some Ar may have escaped

27.15 No detectable amount of Pb formed ($t_{\frac{1}{2}} = 4.5 \times 10^9$ yr)

27.16 a. $225.9771\ g + 4.0015\ g - 229.9837\ g = -0.0051\ g$

b. $\Delta E = -0.0051\ g \times 9.00 \times 10^{10}\ \frac{kJ}{g} \times \frac{10^3\ J}{1\ kJ} = -4.6 \times 10^{11}\ J$

$$\Delta E = -\frac{4.6 \times 10^{11}\ J}{230} = -2.0 \times 10^{9}\ J$$

27.17 a. per mole:

$$\Delta m = 3(1.00867\ g) + 3(0.00055\ g) + 143.8817\ g + 88.8913\ g - 1.00867\ g - 234.9934\ g = -0.2014\ g$$

$$\Delta E = 9.00 \times 10^{10}\ \frac{kJ}{g} \times -0.2014\ g = -1.81 \times 10^{10}\ kJ$$

per gram: $\Delta E = -\frac{1.81 \times 10^{10}\ kJ}{235} = -7.70 \times 10^{7}\ kJ$

b. $7.70 \times 10^{7}\ kJ \times \frac{1\ g}{2.76\ kJ} \times \frac{1\ kg}{10^3\ g} = 2.79 \times 10^{4}\ kg$

27.18 a. per mole: $\Delta m = 3.01493\ g - 2.01355\ g - 1.00728\ g$
$= -0.00590\ g$

$$\Delta E = 9.00 \times 10^{10}\ \frac{kJ}{g} \times -0.00590\ g = -5.31 \times 10^{8}\ kJ$$

per gram: $\Delta E = -\frac{5.31 \times 10^{8}\ kJ}{3.02} = -1.76 \times 10^{8}\ kJ$

b. per mole: $\Delta m = 4.00150\ g + 2(1.00728\ g) - 2(3.01493\ g)$
$= -0.01380\ g$

$$\Delta E = 9.00 \times 10^{10}\ \frac{kJ}{g} \times -0.01380\ g = -1.24 \times 10^{9}\ kJ$$

per gram: $\Delta E = -\frac{1.24 \times 10^{9}\ kJ}{6.03} = -2.06 \times 10^{8}\ kJ$

27.19 $\Delta m = 4.00150\ g + 47.93588\ g - 51.92734\ g = +0.01004\ g$

ΔE is positive; reaction cannot occur

27.20 $^{238}_{92}U + ^{1}_{0}n \longrightarrow ^{239}_{93}Np + ^{0}_{-1}e$

per mole: $\Delta m = 239.0019\ g + 0.00055\ g - 238.0003\ g - 1.00867\ g = -0.0065\ g$

$$\Delta E = 9.00 \times 10^{10}\ \frac{kJ}{g} \times -0.0065\ g = -5.8 \times 10^{8}\ kJ$$

for sample: $\Delta E = -5.8 \times 10^{8}\ kJ \times 1.00 \times 10^{-3}$
$= -5.8 \times 10^{5}\ kJ$

27.21 mass per nuclear particle:

$^{2}_{1}H$ 2.01355/2 = 1.00678

$^{4}_{2}He$ 4.00150/4 = 1.00038 c < d < b < a

$^{59}_{27}Co$ 58.91837/59 = 0.9986164

$^{238}_{92}U$ 238.0003/238 = 1.000001

27.22 $^{14}_{6}C \longrightarrow 6\ ^{1}_{1}H + 8\ ^{1}_{0}n$

$\Delta m = 6(1.00728\ g) + 8(1.00867\ g) - 13.99995\ g = 0.11309\ g$

$\Delta E = 9.00 \times 10^{10}\ \frac{kJ}{g} \times 0.11309\ g = 1.02 \times 10^{10}\ kJ$

27.23 a. positron (too few neutrons) b. positron
b. electron (too many neutrons) d. electron

27.24 a. Pb b. metal c. larger

27.25 a. 3(15) + 10 = 55 rems
b. decrease in white blood cells

27.26 $\frac{30 - 4}{193} \times 100 = 13\%$

27.27 a. same nuclear structure b. same electronic structure
c. alpha particle is $^{4}_{2}He$ nucleus d. Δm too small

27.28 a. false; stays same b. false; decays rapidly c. true

27.29 Soil clothes with material containing radioactive isotope. Measure radioactivity after treatment with detergent.

27.30 See discussion Section 27.4

CHAPTER 28

BASIC SKILLS

1. Identify an organic compound containing a functional group as an alcohol, carboxylic acid, ester, or amine.
2. Draw all the structural isomers of a simple organic compound, given the molecular formula.
3. Draw all the structural isomers of a simple alkene or its halogen derivative.
4. Locate chiral centers in an organic molecule.
5. Given the structure of a monomer, write the structure of a portion of the addition polymer derived from it; carry out the reverse operation.
6. Given the structures of the two monomers involved, write the structure of a portion of the corresponding condensation polymer; carry out the reverse operation.
7. Describe the general structure of a carbohydrate; a protein.

LECTURE NOTES

This chapter is intended to give the student some idea of what organic chemistry is all about. We have tried to relate the material to topics with which students are familiar: the properties of alcohols, the making of soap, the preparation of plastics, - - . We have deliberately avoided the encyclopedic approach, mentioning only a few functional groups with selected examples of compounds within each group. No attempt is made to cover organic nomenclature in this chapter; the IUPAC nomenclature of hydrocarbons is, however, described in Appendix 3.

The material in this chapter will require between two and three lectures, depending upon how much time you choose to spend on natural polymers (carbohydrates and proteins). You will recall that hydrocarbons were covered in Chapter 13. Students may want to review that material before starting this chapter.

LECTURE 1

I Functional Groups in Organic Molecules

A. Alcohols: -OH group

1. CH_3OH (methyl alcohol). Made from destructive distillation

of wood or by synthesis: $CO(g) + 2\,H_2(g) \longrightarrow CH_3OH(l)$

2. C_2H_5OH (ethyl alcohol). Made by fermentation of sugars or hydration of ethylene:

   $C_2H_4(g) + H_2O(g) \longrightarrow C_2H_5OH(l)$

   Use in alcoholic beverages: % of alcohol varies from 4% in beer to 40% in brandy. Proof = 2 x volume % alcohol.

3. Other alcohols: $CH_2OH - CH_2OH$ ethylene glycol

   $CH_2OH - CHOH - CH_2OH$ glycerol

B. Acids: $-\overset{\displaystyle O}{\overset{\|}{C}}... $ i.e. $-C(=O)-OH$

Acetic acid (found in vinegar) $CH_3 - C(=O) - OH$

Treatment with base gives salt:

$CH_3COOH(aq) + OH^-(aq) \longrightarrow CH_3COO^-(aq) + H_2O$

Soaps are sodium salts of long-chain fatty acids

$CH_3(CH_2)_{16}COO^-, Na^+$ is sodium stearate, a soap component

C. Esters - formed by reaction of alcohol with carboxylic acid:

$R - C(=O) - OH + R' - OH \longrightarrow R - C(=O) - O - R' + H_2O$

$CH_3 - C(=O) - O - CH_2 - CH_3$ ethyl acetate (made from ethyl alcohol and acetic acid)

Fats are esters of long-chain carboxylic acids with glycerol

$R - C(=O) - O - CH_2$
$R' - C(=O) - O - CH$
$R'' - C(=O) - O - CH_2$

If R group contains double bond, fat is said to be "unsaturated". If there are no multiple carbon-carbon bonds, fat is saturated.

D. Amines: derived from ammonia by replacing H atoms with hydrocarbon groups.

CH_3NH_2	$(CH_3)_2NH$	$(CH_3)_3N$
methylamine	dimethylamine	trimethylamine

II Isomerism - different compounds have the same molecular formula

A. Structural isomerism: different bonding patterns and hence different structural formulas.

C_4H_{10}: $CH_3CH_2CH_2CH_3$ $\quad CH_3 - CH(CH_3) - CH_3$

C_4H_9Cl:

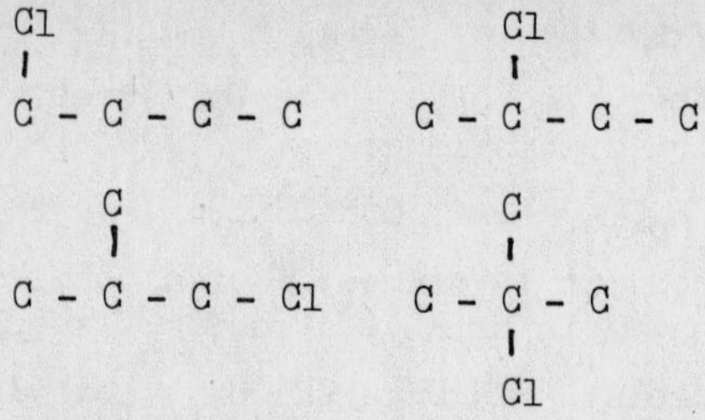

4 isomers in all

B. Geometric isomerism (cis and trans) Occurs with alkenes and their derivatives.

$C_2H_2Cl_2$:

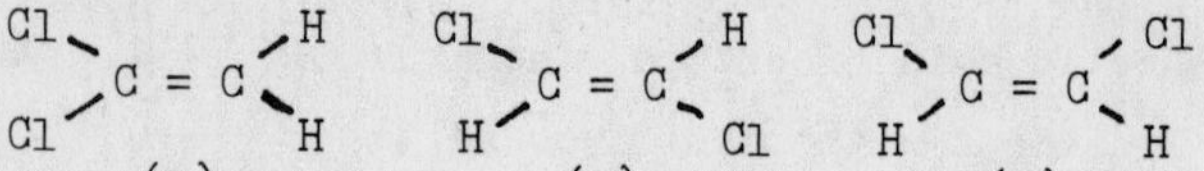

(1) (2) (3)

1 and 2 are structural isomers; 2 and 3 are geometric isomers

LECTURE 2

I Isomerism (cont.)

A. Optical isomers: differ in way they rotate plane polarized light and often in physiological properties. Compounds containing a chiral carbon atom (attached to four different groups) show optical isomerism.

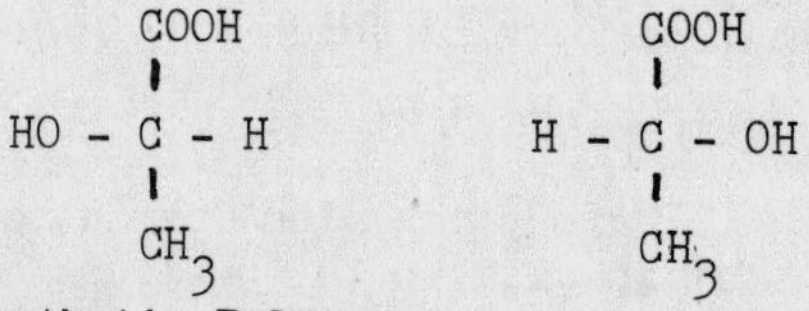

enantiomers of lactic acid

II Synthetic Polymers

A. Addition polymers. Alkene or derivative of alkene adds to itself to form a long-chain polymer

Polyethylene (formed from C_2H_4):

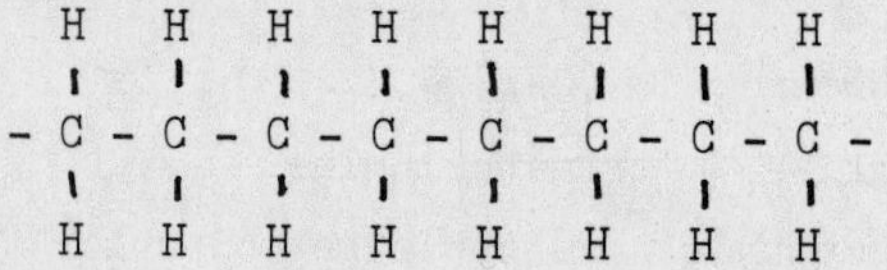

may contain 2000 or more C_2H_4 units

Polyvinyl chloride: made from C_2H_3Cl

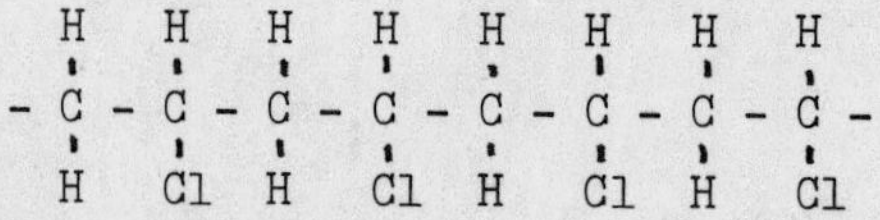

"head-to-tail" polymer

B. Condensation polymers

1. Polyesters: made from dicarboxylic acid and dialcohol

$$HO - CH_2 - CH_2 - OH \quad + \quad HOOC - \bigcirc - COOH$$

$$\downarrow$$

$$- O - CH_2 - CH_2 - O - \underset{\underset{O}{\|}}{C} - \bigcirc - \underset{\underset{O}{\|}}{C} - O - \quad + H_2O$$

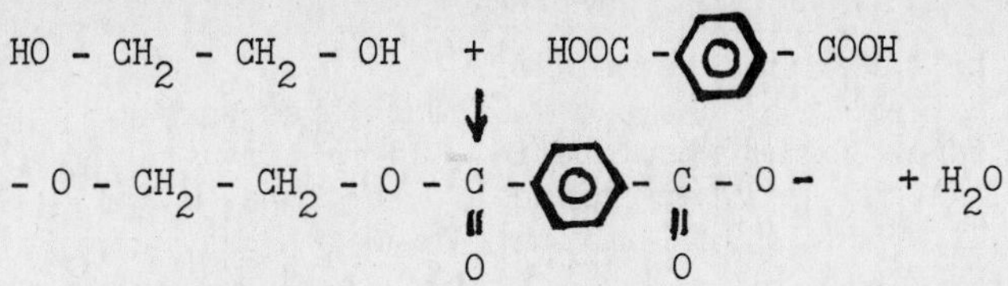

dacron

2. Polyamides: made from dicarboxylic acid and diamine

$$H_2N - (CH_2)_6 - NH_2 \quad + \quad HOOC - (CH_2)_4 - COOH$$

$$\downarrow$$

$$- \underset{\underset{H}{|}}{N} - (CH_2)_6 - \underset{\underset{H}{|}}{N} - \underset{\underset{O}{\|}}{C} - (CH_2)_4 - \underset{\underset{O}{\|}}{C} - \quad + \quad H_2O$$

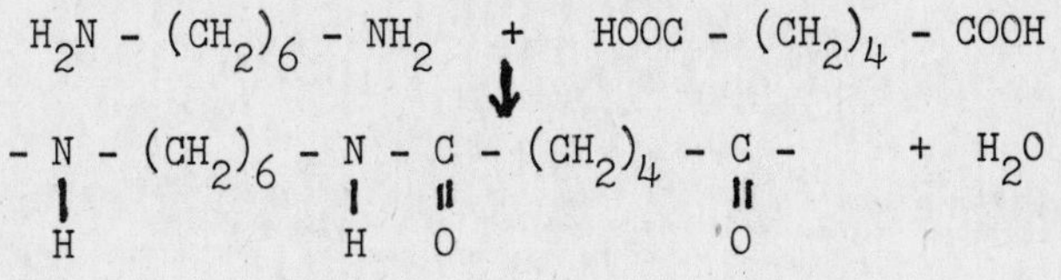

nylon

III Natural Polymers

A. Carbohydrates: general formula $C_n(H_2O)_m$

glucose $C_6H_{12}O_6$ See Figure 28.7

sucrose $C_{12}H_{22}O_{11}$

starch and cellulose are condensation polymers of glucose
(see Figure 28.9)

B. Proteins: polymers of alpha amino acids

$$H - \underset{\underset{NH_2}{|}}{\overset{\overset{H}{|}}{C}} - COOH \quad + \quad CH_3 - \underset{\underset{NH_2}{|}}{\overset{\overset{H}{|}}{C}} - COOH \qquad H - \underset{\underset{NH_2}{|}}{\overset{\overset{H}{|}}{C}} - \underset{O}{\overset{\|}{C}} - \underset{H}{\overset{|}{N}} - \underset{\underset{CH_3}{|}}{\overset{\overset{H}{|}}{C}} - COOH$$

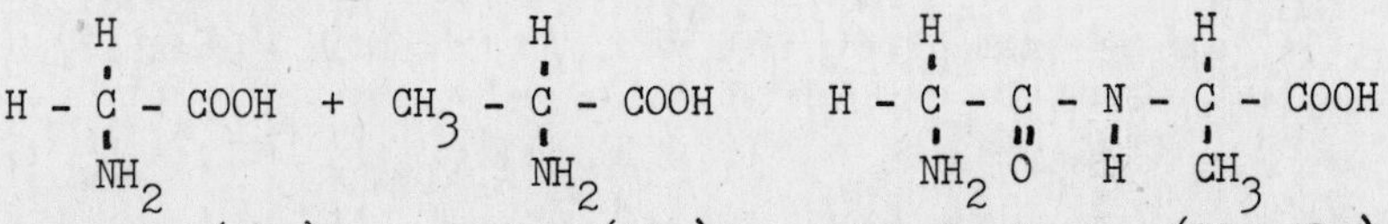

glycine (Gly)   alanine (Ala)   glycyl alanine (Gly-Ala)

Polymerization can continue to form a high molar mass protein

DEMONSTRATIONS

1. Generation of acetylene from CaC_2: Test. Dem. 47

2. Fermentation to produce alcohol: Test. Dem. 47

3. Formation of esters: Test. Dem. 48

4. Optical activity: J. Chem. Educ. 53 508 (1976) 55 319 (1978)

5. Preparation of nylon polymer: Test. Dem. 136, 164; J. Chem. Educ. 56 409 (1979)

QUIZZES

1. Write the structural formula of
    a. two different alcohols containing three carbon atoms
    b. the ester formed when acetic acid reacts with methyl alcohol
    c. the addition polymer formed by propylene, C_3H_6
    d. the dipeptide alanyl serine (Ala-Ser); use Table 28.4

1. Write the structural formulas of
    a. the acid and alcohol from which the following ester is formed

$$\begin{array}{ccccccc} & H & & & & H & \\ & | & & & & | & \\ CH_3 - & C & - & C - O - & C & - CH_3 \\ & | & & \| & & | & \\ & OH & & O & & CH_3 & \end{array}$$

    b. all the isomers of C_5H_9Cl

    c. the monomers from which the following polymer is formed

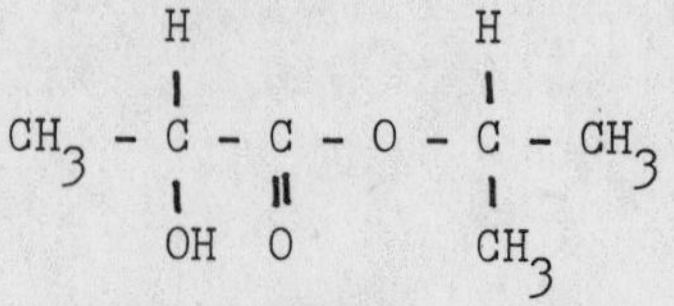

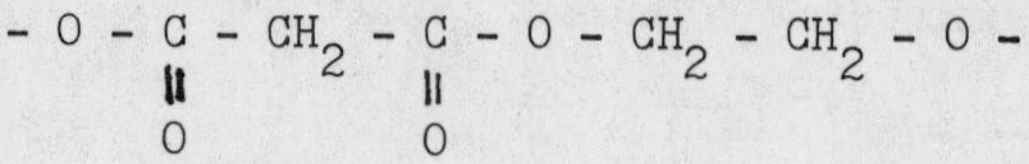

$$\begin{array}{ccccccc} - O - & C & - CH_2 - & C & - O - CH_2 - CH_2 - O - \\ & \| & & \| & \\ & O & & O & \end{array}$$

1. Draw all the isomers of the alkene C_4H_8

2. Draw the structure of an addition polymer formed from one of the isomers in (1).

1. Classify each of the following as an alcohol, carboxylic acid, and/or ester.

$$\begin{array}{ccc} & H & \\ & | & \\ a.\ H - & C - C - OH \\ & | \quad \| & \\ & OH \ O & \end{array} \qquad \begin{array}{ccc} & H & \\ & | & \\ b.\ H - & C - C - O - CH_3 \\ & | \quad \| & \\ & Cl \ O & \end{array}$$

$$c. \quad \bigcirc \quad \begin{array}{l} \overset{O}{\overset{\|}{- O - C - O - CH_3}} \\ - C - OH \\ \quad \| \\ \quad O \end{array}$$

2. Draw the structure of a condensation polymer made from the two monomers:

$$\begin{array}{ccc} & H & \\ & | & \\ CH_3 - & C - COOH \\ & | & \\ & NH_2 & \end{array} \qquad and \qquad \begin{array}{ccc} & CH_3 & \\ & | & \\ CH_2Cl - & C - COOH \\ & | & \\ & NH_2 & \end{array}$$

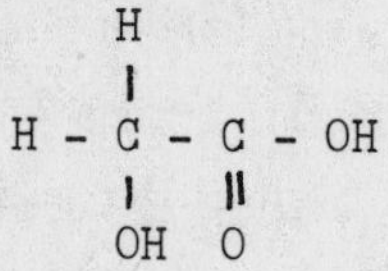

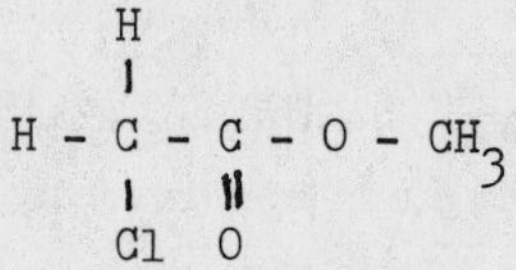

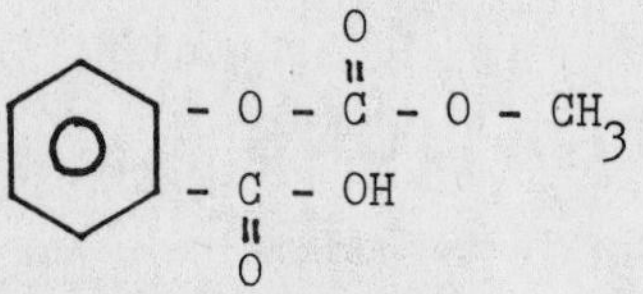

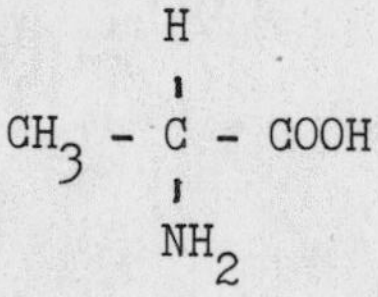

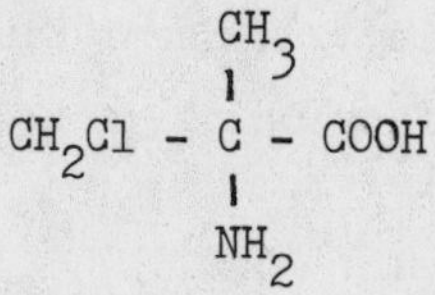

1. Referring to Table 28.4

    a. give the structure of Asp-Gly
    b. state the number of chiral carbon atoms in valine
    c. draw the structure of an isomer of glycine
    d. calculate the mass percent of carbon in tyrosine

PROBLEMS 1-30

28.1 $C_{14}H_9Cl_5$

28.2 a. alcohol   b. carboxylic acid, ester   c. alcohol, acid

28.3 a.
$$CH_3 - \overset{\overset{\displaystyle H}{|}}{\underset{\underset{\displaystyle OH}{|}}{C}} - CH_2 - CH_3 \quad (\text{or} \quad CH_3 - \overset{\overset{\displaystyle CH_3}{|}}{\underset{\underset{\displaystyle OH}{|}}{C}} - CH_3$$

    b.
$$CH_3 - CH_2 - CH_2 - CH_2 - \overset{\overset{\displaystyle \ }{\ }}{\underset{\underset{\displaystyle O}{\|}}{C}} - OH \quad (\text{and other isomers})$$

    c.
$$CH_3 - \overset{\overset{\displaystyle H}{|}}{\underset{\underset{\displaystyle CH_2CH_3}{|}}{C}} - O - \overset{\overset{\displaystyle \ }{\ }}{\underset{\underset{\displaystyle O}{\|}}{C}} - CH_2 - CH_2 - CH_2 - CH_3$$

28.4
$$CH_3 - CH_2 - \overset{\overset{\displaystyle \ }{\ }}{\underset{\underset{\displaystyle O}{\|}}{C}} - OH \quad , \quad Ca(OH)_2$$

28.5 a.
$$H - \overset{\overset{\displaystyle \ }{\ }}{\underset{\underset{\displaystyle O}{\|}}{C}} - O - CH_3 \quad \text{b.} \quad CH_3 - \overset{\overset{\displaystyle \ }{\ }}{\underset{\underset{\displaystyle O}{\|}}{C}} - O - CH_3$$

    c.
$$CH_3 - (CH_2)_6 - \overset{\overset{\displaystyle \ }{\ }}{\underset{\underset{\displaystyle O}{\|}}{C}} - O - CH_3$$

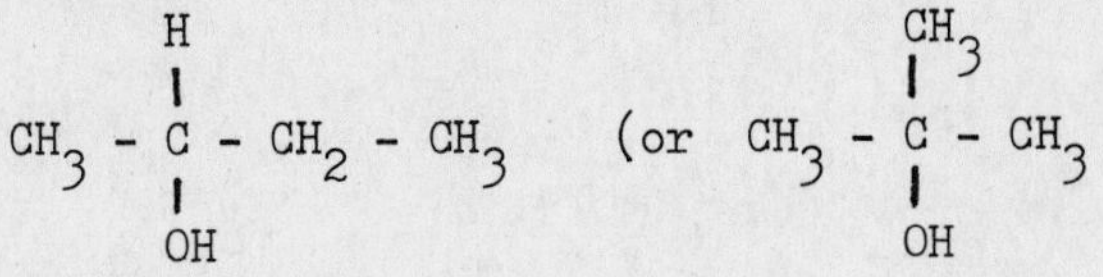

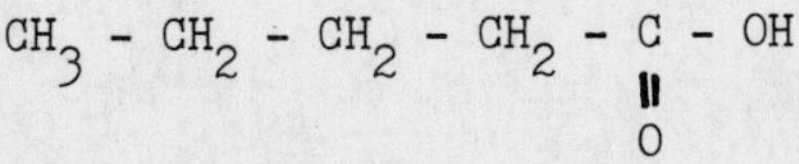

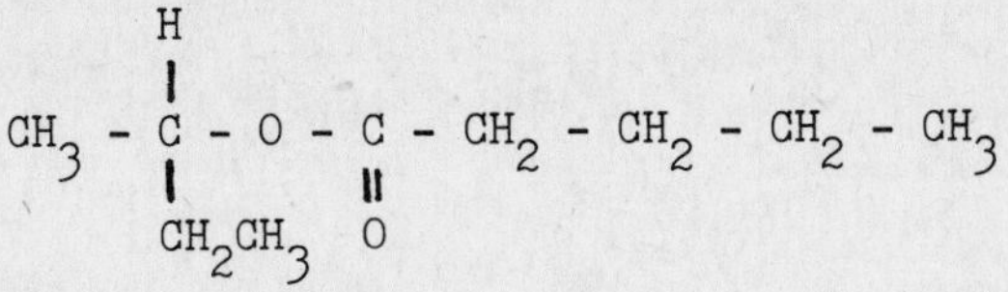

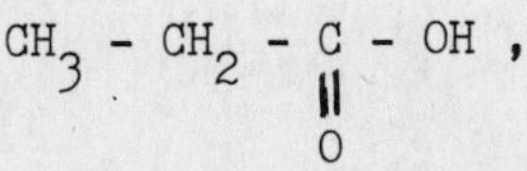

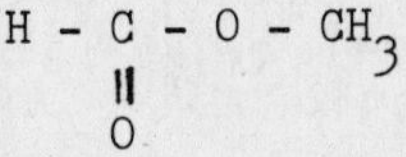

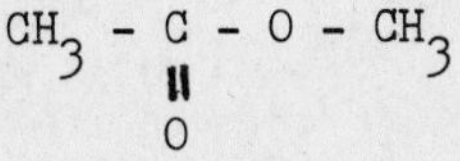

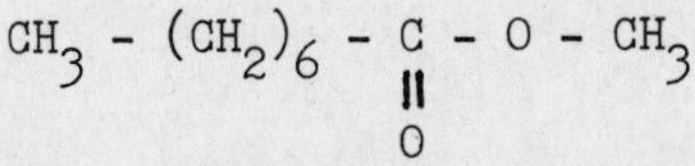

28.6 a.
$$CH_3 - CH_2 - CH_2 - NH_2 \quad \text{or} \quad CH_3 - \overset{\overset{\displaystyle H}{|}}{\underset{\underset{\displaystyle NH_2}{|}}{C}} - CH_3$$

    b.
$$CH_3 - CH_2 - \overset{\overset{\displaystyle \ }{\ }}{\underset{\underset{\displaystyle H}{|}}{N}} - CH_3$$

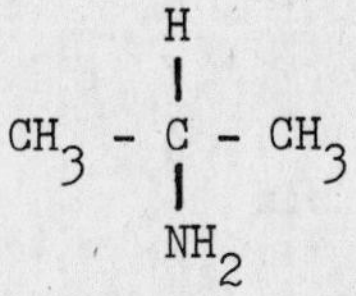

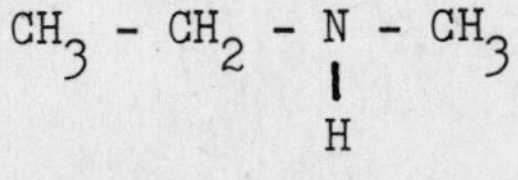

28.7 $C_8H_{10}O_2N_4$; molar mass = 194.20 g/mol

    $\% N = \dfrac{4(14.01)}{194.20} \times 100 = 28.86$

28.8

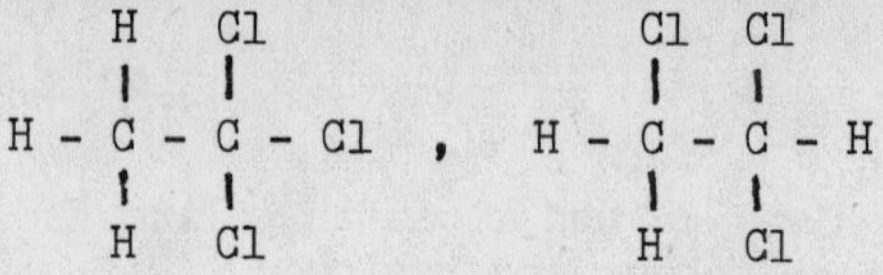

```
    H   Cl              Cl  Cl
    |   |               |   |
H - C - C - Cl  ,   H - C - C - H
    |   |               |   |
    H   Cl              H   Cl
```

28.9

28.10   C - C - C - C = C ,   C - C - C = C - C,   C = C - C - C ,
                                                              |
                                                              C

```
C - C = C - C  ,   C - C - C = C
    |                  |
    C                  C
```

28.11   C - C - C - C - C - OH ,   C - C - C - C - C ,
                                               |
                                               OH

```
C - C - C - C - C ,   C - C - C - C - OH
        |                     |
        OH                    C
```

```
        OH                OH
        |                 |
C - C - C - C ,   C - C - C - C  ,   HO - C - C - C - C ,
        |                 |                          |
        C                 C                          C
```

```
        C
        |
C - C - C -OH
        |
        C
```

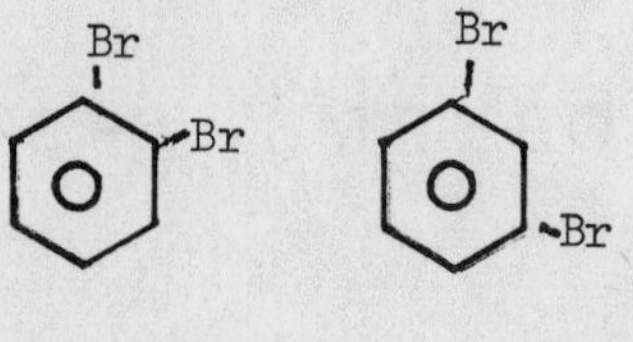

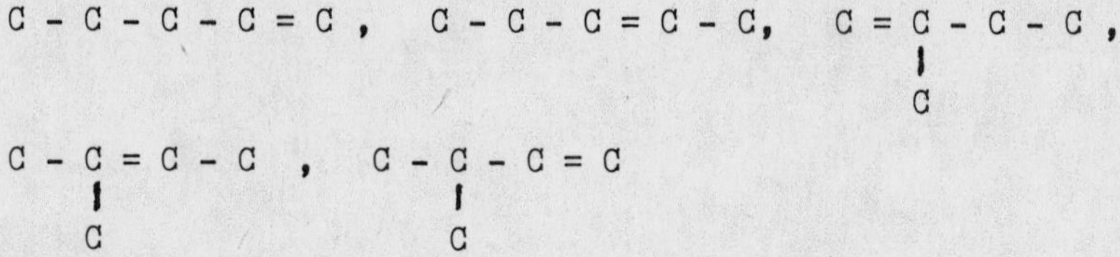

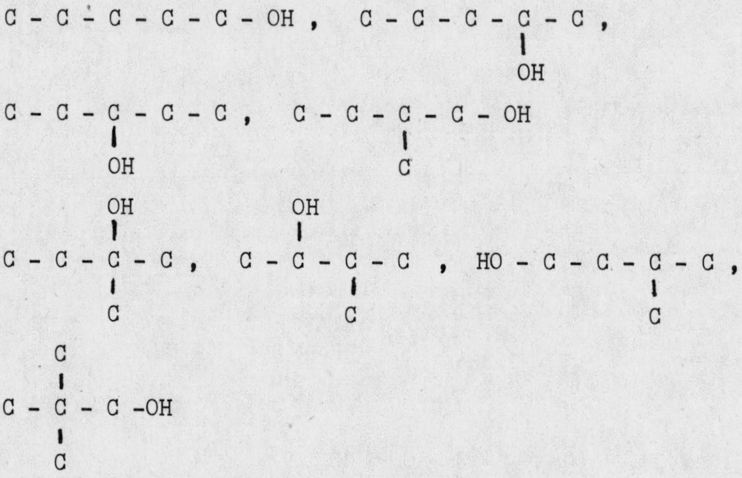

28.12

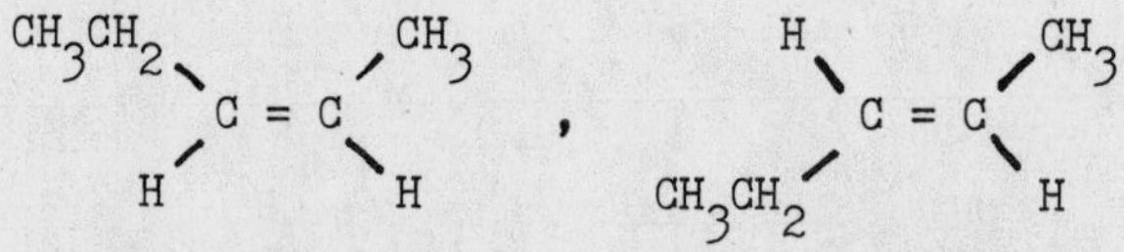

cis   trans

28.13

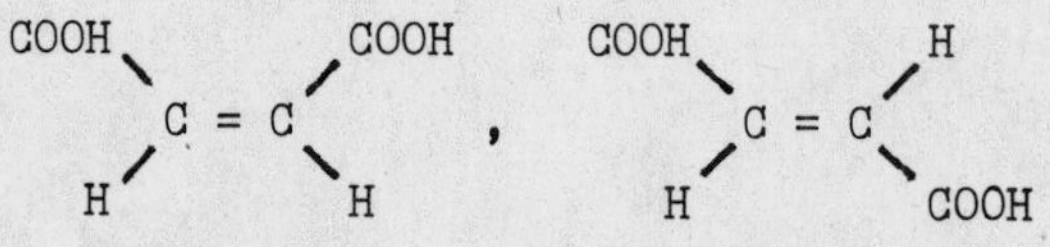

maleic acid   fumaric acid

28.14   5 in glucose, 4 in fructose

28.15 a. 2nd and 3rd C atoms from left   b. none
c. 3rd C atom from left

$$\begin{array}{cccc} Cl & Cl & Cl & Cl \\ | & | & | & | \end{array}$$

28.16 a. $- C - C - C - C -$ b. $3.2 \times 10^3 (165.82 \text{ g/mol})$

$$\begin{array}{cccc} | & | & | & | \\ Cl & Cl & Cl & Cl \end{array}$$

$$= 5.3 \times 10^5 \text{ g/mol}$$

c. simplest formula = CCl_2

% C = $\dfrac{12.01}{82.91} \times 100 = 14.49$; % Cl = 85.51

28.17 a.
$$\begin{array}{cccccccc} H & H & H & H & H & H & H & H \\ | & | & | & | & | & | & | & | \\ - C - & C - & C - & C - & C - & C - & C - & C - \\ | & | & | & | & | & | & | & | \\ H & CN & H & CN & H & CN & H & CN \end{array}$$

b.
$$\begin{array}{cccccccc} H & H & H & H & H & H & H & H \\ | & | & | & | & | & | & | & | \\ - C - & C - & C - & C - & C - & C - & C - & C - \\ | & | & | & | & | & | & | & | \\ H & CN & CN & H & H & CN & CN & H \end{array}$$

28.18 $\begin{array}{c} H \\ \diagdown \\ \diagup \\ H \end{array} C = C \begin{array}{c} H \\ \diagup \\ \diagdown \\ Cl \end{array}$ and $\begin{array}{c} H \\ \diagdown \\ \diagup \\ H \end{array} C = C \begin{array}{c} H \\ \diagup \\ \diagdown \\ \bigcirc \end{array}$

28.19 $- O - CH_2 - CH_2 - O - \underset{\underset{O}{\|}}{C} - \underset{\underset{O}{\|}}{C} - O - CH_2 - CH_2 - O - \underset{\underset{O}{\|}}{C} - \underset{\underset{O}{\|}}{C} -$

28.20 $- \underset{\underset{H}{|}}{N} - \bigcirc - \underset{\underset{O}{\|}}{C} - \underset{\underset{H}{|}}{N} - \bigcirc - \underset{\underset{O}{\|}}{C} -$

28.21 $H - \underset{\underset{H}{|}}{N} - CH_2 - \underset{\underset{O}{\|}}{C} - OH$

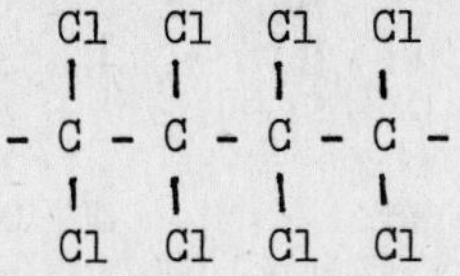

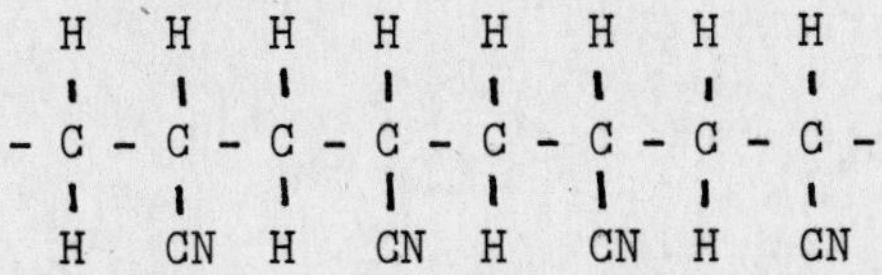

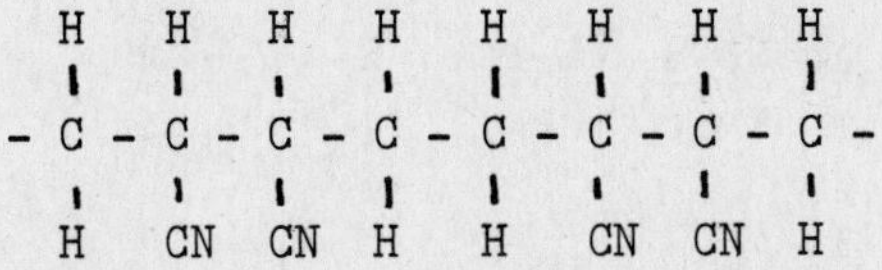

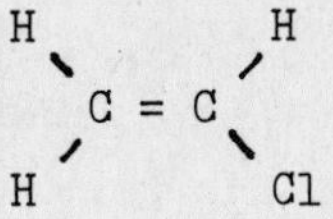

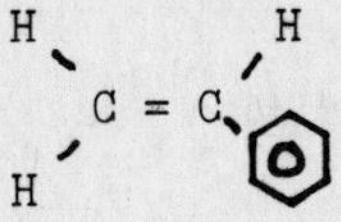

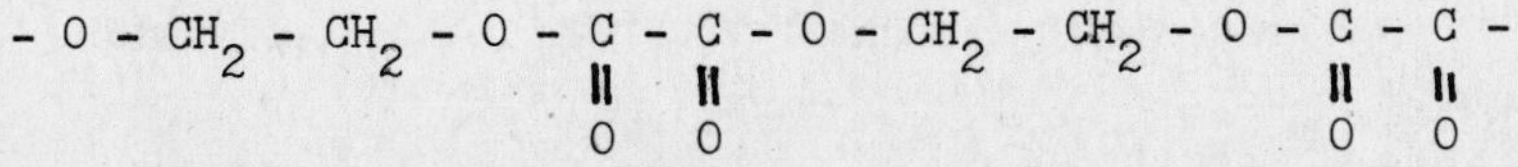

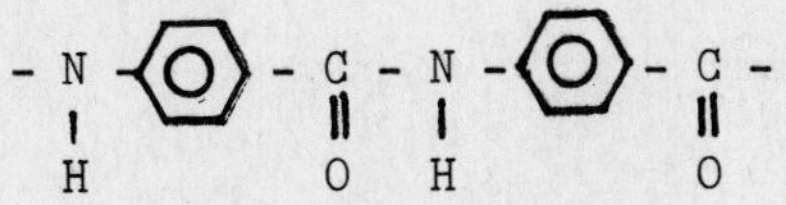

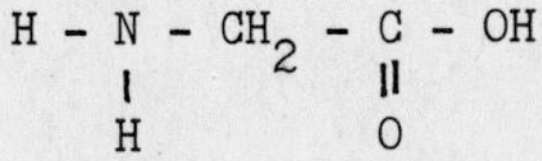

28.22 $C_{12}H_{22}O_{11}(aq) + H_2O \longrightarrow C_6H_{12}O_6(aq) + C_6H_{12}O_6(aq)$

   sucrose   glucose   fructose

28.23 a. molar mass $C_6H_{10}O_5$ = 162.14 g/mol

mass % C = $\dfrac{72.06}{162.14} \times 100 = 44.44$

mass % H = $\dfrac{10.08}{162.14} \times 100 = 6.217$

mass % O = 100.00 - 44.44 - 6.22 = 49.34

28.23 b. 1.62×10^6 g/mol

28.24
$$\underset{\underset{\displaystyle NH}{\overset{\displaystyle \|}{}}\underset{\displaystyle H}{\overset{\displaystyle \|}{}}}{H_2N - C - N} - CH_2 - CH_2 - CH_2 - \underset{\underset{\displaystyle NH_2}{\overset{\displaystyle H}{\overset{\displaystyle |}{}}\overset{\displaystyle O}{\overset{\displaystyle \|}{}}}}{C - C} - \underset{\displaystyle H}{\overset{\displaystyle H}{\overset{\displaystyle |}{}}}{N} - \underset{\underset{\displaystyle CH_2OH}{\overset{\displaystyle H}{\overset{\displaystyle |}{}}\overset{\displaystyle O}{\overset{\displaystyle \|}{}}}}{C - C} - OH$$

$$HO - CH_2 - \underset{NH_2}{\overset{H\;\;O}{C - C}} - \underset{H}{N} - \underset{COOH}{\overset{H}{C}} - CH_2 - CH_2 - CH_2 - \underset{H}{N} - \underset{NH}{C} - NH_2$$

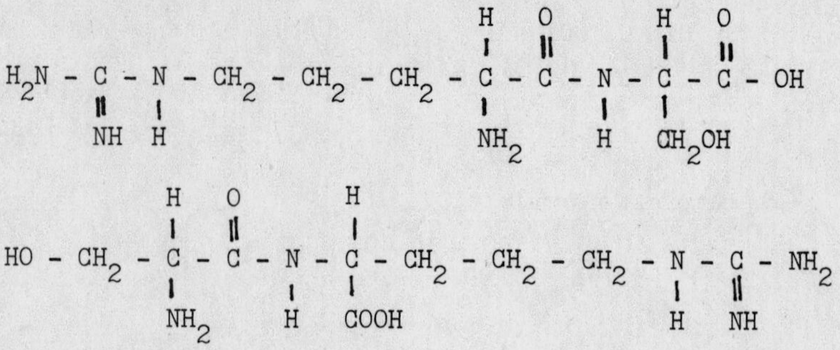

28.25 Val-Ala-Ala-Phe-Leu-Met

28.26 R groups usually too bulky

28.27 a. $CH_3 - NH_2$ b. $CH_3 - OH$ c. $H - \underset{\displaystyle O}{\overset{\displaystyle \|}{C}} - OH$

d. $H - \underset{\displaystyle O}{\overset{\displaystyle \|}{C}} - O - CH_3$

28.28 a. neither   b. addition   c. condensation   d. neither

28.29 a. synthetic polymer made in laboratory

b. polyester contains $- \underset{\displaystyle O}{\overset{\displaystyle \|}{C}} - O -$ linkage; polyamide $- \underset{\displaystyle O}{\overset{\displaystyle \|}{C}} - \underset{\displaystyle H}{\overset{\displaystyle |}{N}} -$

c. differ in orientation of H and OH groups on carbon 1 atom

28.30 a. many are toxic or carcinogenic to humans

b. made from wood

CHAPTER 29

Answers to Problems

29.1 a. $HgS(s) + O_2(g) \longrightarrow Hg(g) + SO_2(g)$

b. $Pb^{2+}(aq) + ClO^-(aq) + 2\ OH^-(aq) \longrightarrow PbO_2(s) + Cl^-(aq) + H_2O$

c. $2\ Ag(s) + S(s) \longrightarrow Ag_2S(s)$

29.2 a. $Ag^+(aq) + 2\ NH_3(aq) \longrightarrow Ag(NH_3)_2{}^+(aq)$

b. $Pb^{2+}(aq) + 2NH_3(aq) + 2H_2O \longrightarrow Pb(OH)_2(s) + 2NH_4{}^+(aq)$

c. $Hg_2Cl_2(s) + 2NH_3(aq) \longrightarrow Hg(l) + HgNH_2Cl(s) + NH_4{}^+(aq) + Cl^-(aq)$

29.3 $Ag^+(aq) + Cl^-(aq) \longrightarrow AgCl(s)$

$AgCl(s) + Cl^-(aq) \longrightarrow AgCl_2{}^-(aq)$

29.4 a. $Ag_2CO_3(s) + 2\ H^+(aq) \longrightarrow 2\ Ag^+(aq) + H_2CO_3(aq)$

b. $PbSO_4(s) + 3\ OH^-(aq) \longrightarrow Pb(OH)_3{}^-(aq) + SO_4{}^{2-}(aq)$

c. $AgI(s) + 2\ CN^-(aq) \longrightarrow Ag(CN)_2{}^-(aq) + I^-(aq)$

29.5 a. $AgBr(s) + 2\ NH_3(aq) \longrightarrow Ag(NH_3)_2{}^+(aq) + Br^-(aq)$

b. $Hg_2{}^{2+}(aq) \longrightarrow Hg(l) + Hg^{2+}(aq)$

c. $Ag_2SO_4(s) \longrightarrow 2\ Ag^+(aq) + SO_4{}^{2-}(aq)$

29.6 $PbCO_3(s) + 2\ H^+(aq) \longrightarrow Pb^{2+}(aq) + H_2CO_3(aq)$

$PbCO_3(s) + 3\ OH^-(aq) \longrightarrow Pb(OH)_3{}^-(aq) + CO_3{}^{2-}(aq)$

29.7 a. $AgCl(s) + 2\ NH_3(aq) \longrightarrow Ag(NH_3)_2{}^+(aq) + Cl^-(aq)$

b. $PbCl_2(s) \longrightarrow Pb^{2+}(aq) + 2\ Cl^-(aq)$

c. $Ag(NH_3)_2{}^+(aq) + 2H^+(aq) + Cl^-(aq) \longrightarrow AgCl(s) + 2NH_4{}^+(aq)$

29.8 $PbO(s) + H_2O \longrightarrow PbO_2(s) + 2\ H^+(aq) + 2\ e^-$

$ClO^-(aq) + 2\ H^+(aq) + 2\ e^- \longrightarrow Cl^-(aq) + H_2O$

$PbO(s) + ClO^-(aq) \longrightarrow PbO_2(s) + Cl^-(aq)$

29.9 $3\left[Ag(s) + 2\,Cl^-(aq) \longrightarrow AgCl_2^-(aq) + e^-\right]$

$$\frac{NO_3^-(aq) + 4\,H^+(aq) + 3\,e^- \longrightarrow NO(g) + 2\,H_2O}{3Ag(s) + 6Cl^-(aq) + NO_3^-(aq) + 4H^+(aq) \longrightarrow 3AgCl_2^-(aq) + NO(g) + 2H_2O}$$

29.10 a. $\dfrac{1.6 \times 10^{-10}}{1.2 \times 10^{-3}} = 1.3 \times 10^{-7}$ M

b. $[Ag^+]^2 = \dfrac{1 \times 10^{-11}}{1 \times 10^{-4}} = 1 \times 10^{-7}$; $[Ag^+] = 3 \times 10^{-4}$ M

c. $[Ag^+]^2 = \dfrac{1 \times 10^{-49}}{1 \times 10^{-20}} = 1 \times 10^{-29}$; $[Ag^+] = 3 \times 10^{-15}$ M

29.11 a. $[CrO_4^{2-}] = \dfrac{2 \times 10^{-12}}{(1.0 \times 10^{-2})^2} = 2 \times 10^{-8}$ M

b. $[Br^-] = \dfrac{5 \times 10^{-13}}{1.0 \times 10^{-2}} = 5 \times 10^{-11}$ M

c. $[S^{2-}] = \dfrac{1 \times 10^{-49}}{(1.0 \times 10^{-2})^2} = 1 \times 10^{-45}$ M

29.12 a. conc. Pb^{2+} x conc. $SO_4^{2-} = (1.0 \times 10^{-4})(1.0 \times 10^{-3})$

$= 1.0 \times 10^{-7} > 1 \times 10^{-8}$; yes

b. conc. $Pb^{2+} = \dfrac{0.50 \times 0.10 \text{ M}}{1.00} = 0.050$ M

conc. $Cl^- = \dfrac{0.50 \times 0.010 \text{ M}}{1.00} = 0.0050$ M

conc. Pb^{2+} x (conc. $Cl^-)^2 = (5.0 \times 10^{-2})(5.0 \times 10^{-3})^2$

$= 1.2 \times 10^{-6} < 1.7 \times 10^{-5}$; no

29.13 $\dfrac{[Pb(OH)_3^-]}{[Pb^{2+}] \times [OH^-]^3} = 1 \times 10^{14}$; $\dfrac{[Pb^{2+}]}{[Pb(OH)_3^-]} = \dfrac{1 \times 10^{-14}}{[OH^-]^3}$

a. ratio $= 1 \times 10^{-14}$

b. ratio $= \dfrac{1 \times 10^{-14}}{(1 \times 10^{-1})^3} = 1 \times 10^{-11}$

c. ratio $= \dfrac{1 \times 10^{-14}}{(1 \times 10^{-2})^3} = 1 \times 10^{-8}$

29.14 a. $K = K_{sp}\ AgI \times K_f\ Ag(CN)_2^- = (1 \times 10^{-16})(1 \times 10^{21})$
$= 1 \times 10^5$

b. $K = K_{sp}\ AgCN \times K_f\ Ag(CN)_2^- = (1 \times 10^{-14})(1 \times 10^{21})$
$= 1 \times 10^7$

c. $K = \dfrac{K_f\ Ag(CN)_2^-}{K_f\ Ag(NH_3)_2^+} = \dfrac{1 \times 10^{21}}{2 \times 10^7} = 5 \times 10^{13}$

29.15 a. $K = \dfrac{K_{sp}\ PbS}{K_{sp}\ PbCO_3} = \dfrac{1 \times 10^{-27}}{1 \times 10^{-13}} = 1 \times 10^{-14}$

b. $K = \dfrac{1 \times 10^{-16}}{1 \times 10^{-13}} = 1 \times 10^{-3}$

c. $K = \dfrac{1 \times 10^{-8}}{1 \times 10^{-13}} = 1 \times 10^5$

29.16 a. $K = K_{sp}\ Pb(OH)_2 \times K_f\ Pb(OH)_3^- = (4 \times 10^{-15})(1 \times 10^{14})$
$= 0.4$

b. $K = \dfrac{[Pb(OH)_3^-]}{[OH^-]} = \dfrac{s}{[OH^-]}$; $s = K \times [OH^-] = 0.04$ mol/L

29.17 a. $s = (K_{sp})^{\frac{1}{2}} = (1 \times 10^{-16})^{\frac{1}{2}} = 1 \times 10^{-8}$ M

b. $AgI(s) + 2\ S_2O_3^{2-}(aq) \rightleftharpoons Ag(S_2O_3)_2^{3-}(aq) + I^-(aq)$

$K = K_{sp}\ AgI \times K_f\ Ag(S_2O_3)_2^{3-} = (1 \times 10^{-16})(1 \times 10^{13})$
$= 1 \times 10^{-3}$

$\dfrac{s^2}{[S_2O_3^{2-}]^2} = 1 \times 10^{-3}$; $s = (1 \times 10^{-3})^{\frac{1}{2}} = 0.03$ M

29.18 $[OH^-]^2 = \dfrac{4 \times 10^{-15}}{1 \times 10^{-1}} = 4 \times 10^{-14}$; $[OH^-] = 2 \times 10^{-7}$ M

$[H^+] = \dfrac{1.0 \times 10^{-14}}{2 \times 10^{-7}} = 5 \times 10^{-8}$ M; pH = 7.3

29.19 $Ag^+(aq) + Cl^-(aq) \longrightarrow AgCl(s)$

$Pb^{2+}(aq) + 2\ Cl^-(aq) \longrightarrow PbCl_2(s)$

$Hg_2^{2+}(aq) + 2\ Cl^-(aq) \longrightarrow Hg_2Cl_2(s)$

$Pb^{2+}(aq) + CrO_4^{2-}(aq) \longrightarrow PbCrO_4(s)$

29.20 White ppt. in Step 1, completely dissolves in Step 2. Yellow ppt. in Step 3.

29.21 a. Will reprecipitate $PbCl_2$. Will not get a test for Pb^{2+} in Step 3; test for Hg_2^{2+} in Step 4 will be inconclusive.

b. $PbCl_2$ may obscure test for Hg_2^{2+}.

c. Will miss confirmatory test for Ag^+

29.22 1. Add NaOH to give precipitate consisting of Ag_2O, HgO and Hg and a solution containing $Pb(OH)_3^-$.

2. Test solution for lead by acidifying with HCl; white ppt. of $PbCl_2$ shows presence of Pb^{2+}.

3. Stir precipitate with dilute HNO_3; black ppt. of Hg(l) shows Hg_2^{2+} was present originally.

4. Test HNO_3 solution for Ag^+ by adding HCl to give white ppt. of AgCl.

29.23 only Hg_2^{2+}

29.24 Ag^+ present; Pb^{2+} and Hg_2^{2+} are in doubt (both $PbCl_2$ and Hg_2Cl_2 are insoluble in NH_3).

29.25 $Hg_2(NO_3)_2$ present; this is the only solid that is water-soluble.

AgCl absent; would not dissolve in 6 M NaOH.

$PbSO_4$, $PbCO_3$ in doubt; both would dissolve in 6 M NaOH.

29.26 a. $E^o_{tot} = E^o_{red}\ Pb^{2+} + E^o_{ox}\ Ag = -0.13\ V - 0.80\ V = -0.93\ V$; no

b. $E^o_{tot} = E^o_{red}\ Hg^{2+} + E^o_{ox}\ Hg = +0.92\ V - 0.79\ V = +0.13\ V$; yes

29.27 a (yellow); e (yellow-orange)

29.28 a. Heat with water; $PbCl_2$ should dissolve.

b. Observe color; Ag_2O is brown, AgCl white.

c. Add NH_3; Hg_2Cl_2 turns black.

d. Precipitate chlorides, add NH_3; AgCl dissolves.

29.29 Lead is used to make storage batteries; silver is used in jewelry; mercury is used in thermometers and barometers.

29.30 $HgS(s) + O_2(g) \longrightarrow Hg(g) + SO_2(g)$

29.31 a. $2\ PbS(s) + 3\ O_2(g) \longrightarrow 2\ PbO(s) + 2\ SO_2(g)$

b. $Pb(s) + 2\ H^+(aq) \longrightarrow Pb^{2+}(aq) + H_2(g)$

c. $Pb(s) + PbO_2(s) + 4H^+(aq) + 2SO_4{}^{2-}(aq) \longrightarrow 2PbSO_4(s) + 2H_2O$

29.32 a. $NH_3(aq) + H^+(aq) \longrightarrow NH_4{}^+(aq)$

b. $AgCl(s) + 2\ NH_3(aq) \longrightarrow Ag(NH_3)_2{}^+(aq) + Cl^-(aq)$

c. $Pb^{2+}(aq) + 2NH_3(aq) + 2H_2O \longrightarrow Pb(OH)_2(s) + 2NH_4{}^+(aq)$

29.33 $Pb^{2+}(aq) + 2\ OH^-(aq) \longrightarrow Pb(OH)_2(s)$

$Pb(OH)_2(s) + OH^-(aq) \longrightarrow Pb(OH)_3{}^-(aq)$

29.34 a. $2\ Ag^+(aq) + 2\ OH^-(aq) \longrightarrow Ag_2O(s) + H_2O$

b. $Hg_2{}^{2+}(aq) + 2\ OH^-(aq) \longrightarrow HgO(s) + Hg(l) + H_2O$

c. $Pb^{2+}(aq) + 3\ OH^-(aq) \longrightarrow Pb(OH)_3{}^-(aq)$

29.35 a. $Hg_2{}^{2+}(aq) + 2\ Cl^-(aq) \longrightarrow Hg_2Cl_2(s)$

b. $Pb(NO_3)_2(s) \longrightarrow Pb^{2+}(aq) + 2\ NO_3{}^-(aq)$

c. $2\ Hg(l) + 2\ Ag^+(aq) \longrightarrow Hg_2{}^{2+}(aq) + 2\ Ag(s)$

29.36 $AgCl(s) + 2\ NH_3(aq) \longrightarrow Ag(NH_3)_2{}^+(aq) + Cl^-(aq)$

$AgCl(s) + 2\ S_2O_3{}^{2-}(aq) \longrightarrow Ag(S_2O_3)_2{}^{3-}(aq) + Cl^-(aq)$

$AgCl(s) + 2\ CN^-(aq) \longrightarrow Ag(CN)_2{}^-(aq) + Cl^-(aq)$

29.37 a. $Pb^{2+}(aq) + 2\ Cl^-(aq) \longrightarrow PbCl_2(s)$

b. $Hg_2Cl_2(s) + 2NH_3(aq) \longrightarrow HgNH_2Cl(s) + Hg(l) + NH_4{}^+(aq) + Cl^-(aq)$

c. $PbCl_2(s) + CrO_4{}^{2-}(aq) \longrightarrow PbCrO_4(s) + 2\ Cl^-(aq)$

29.38 $Ag_2S(s) \longrightarrow 2\ Ag^+(aq) + S(s) + 2\ e^-$

$2\ [NO_3{}^-(aq) + 2\ H^+(aq) + e^- \longrightarrow NO_2(g) + H_2O]$

$Ag_2S(s) + 2NO_3{}^-(aq) + 4H^+(aq) \longrightarrow 2Ag^+(aq) + S(s) + 2NO_2(g) + 2H_2O$

29.39 $Hg(l) + 4\ Cl^-(aq) \longrightarrow HgCl_4{}^{2-}(aq) + 2\ e^-$

$$\underline{2\left[NO_3{}^-(aq) + 2\ H^+(aq) + e^- \longrightarrow NO_2(g) + H_2O\right]}$$

$$Hg(l) + 4Cl^-(aq) + 2NO_3{}^-(aq) + 4H^+(aq) \longrightarrow HgCl_4{}^{2-}(aq) + 2NO_2(g) + 2H_2O$$

29.40 a. $\dfrac{1.7 \times 10^{-5}}{(1.5 \times 10^{-2})^2} = 7.6 \times 10^{-2}$ M

b. $\dfrac{1 \times 10^{-8}}{3 \times 10^{-5}} = 3 \times 10^{-4}$ M

c. $\dfrac{4 \times 10^{-15}}{(1 \times 10^{-4})^2} = 4 \times 10^{-7}$ M

29.41 a. $\dfrac{1.7 \times 10^{-5}}{(1.0 \times 10^{-2})^2} = 1.7 \times 10^{-1}$ M

b. $\dfrac{1 \times 10^{-13}}{1.0 \times 10^{-2}} = 1 \times 10^{-11}$ M

c. $\dfrac{1 \times 10^{-16}}{1.0 \times 10^{-2}} = 1 \times 10^{-14}$ M

29.42 a. (conc. Ag^+)x(conc. Cl^-) = $(1.0 \times 10^{-5})(1.0 \times 10^{-3})$

$= 1.0 \times 10^{-8} > 1.6 \times 10^{-10}$; yes

b. conc. $Ag^+ = \dfrac{0.200 \times 0.0010\ M}{1.00} = 2.0 \times 10^{-4}$ M

conc. $CrO_4{}^{2-} = \dfrac{0.800 \times 0.0010\ M}{1.00} = 8.0 \times 10^{-4}$ M

(conc. $Ag^+)^2$(conc. $CrO_4{}^{2-}$) = $(2.0 \times 10^{-4})^2(8.0 \times 10^{-4})$

$= 3.2 \times 10^{-11} > 2 \times 10^{-12}$; yes

29.43 a. $\dfrac{[Ag^+]}{[Ag(NH_3)_2{}^+]} = \dfrac{1}{K_f \times [NH_3]^2} = \dfrac{1}{(2 \times 10^7)(1 \times 10^{-2})} = 5 \times 10^{-6}$

b. $\dfrac{[Ag^+]}{[Ag(S_2O_3)_2{}^{3-}]} = \dfrac{1}{K_f \times (0.1)^2} = \dfrac{1}{1 \times 10^{11}} = 1 \times 10^{-11}$

c. $\dfrac{[Ag^+]}{[Ag(CN)_2{}^-]} = \dfrac{1}{(1 \times 10^{21})(0.1)^2} = 1 \times 10^{-19}$

29.44 a. $K = K_{sp}\ Pb(OH)_2 \times K_f\ Pb(OH)_3^- = (4 \times 10^{-15})(1 \times 10^{14})$
$= 0.4$

b. $K = \dfrac{K_{sp}\ PbCO_3}{K_{sp}\ PbCrO_4} = \dfrac{1 \times 10^{-13}}{1 \times 10^{-16}} = 1 \times 10^3$

c. $K = \dfrac{K_{sp}\ PbCl_2}{K_{sp}\ PbS} = \dfrac{1.7 \times 10^{-5}}{1 \times 10^{-27}} = 2 \times 10^{22}$

29.45 a. $K = 1/K_{sp}\ AgCN = 1 \times 10^{14}$

b. $K = \dfrac{K_{sp}\ AgBr}{K_{sp}\ AgCN} = \dfrac{5 \times 10^{-13}}{1 \times 10^{-14}} = 50$

c. $K = \dfrac{K_{sp}\ AgI}{K_{sp}\ AgCN} = \dfrac{1 \times 10^{-16}}{1 \times 10^{-14}} = 1 \times 10^{-2}$

29.46 a. $K = K_{sp}\ AgCN \times K_f\ Ag(CN)_2^- = (1 \times 10^{-14})(1 \times 10^{21})$
$= 1 \times 10^7$

b. $AgCN(s) + CN^-(aq) \rightleftharpoons Ag(CN)_2^-(aq)$

$K = 1 \times 10^7 = \dfrac{s}{1 \times 10^{-6}}$; $s = 10$ M

29.47 a. $s = (1 \times 10^{-27})^{\frac{1}{2}} = 3 \times 10^{-14}$ M

b. $PbS(s) + 3\ OH^-(aq) \rightleftharpoons Pb(OH)_3^-(aq) + S^{2-}(aq)$

$K = K_{sp}\ PbS \times K_f\ Pb(OH)_3^- = (1 \times 10^{-27})(1 \times 10^{14})$
$= 1 \times 10^{-13}$

$\dfrac{s^2}{[OH^-]^3} = 1 \times 10^{-13}$; $s = (1 \times 10^{-13})^{\frac{1}{2}} = 3 \times 10^{-7}$ M

29.48 $Pb(OH)_2(s) + OH^-(aq) \rightleftharpoons Pb(OH)_3^-(aq)$

$K = K_{sp}\ Pb(OH)_2 \times K_f\ Pb(OH)_3^- = 0.4$

$0.4 = \dfrac{[Pb(OH)_3^-]}{[OH^-]} = \dfrac{0.010}{[OH^-]}$

$[OH^-] = \dfrac{0.010}{0.4} = 0.025$ M; $[H^+] = \dfrac{1.0 \times 10^{-14}}{2.5 \times 10^{-2}} = 4 \times 10^{-13}$ M

pH = 12.4

29.49 $AgCl(s) + 2\ NH_3(aq) \longrightarrow Ag(NH_3)_2^+(aq) + Cl^-(aq)$

$Hg_2Cl_2(s) + 2NH_3(aq) \longrightarrow Hg(l) + HgNH_2Cl(s) + NH_4^+(aq) + Cl^-(aq)$

$Ag(NH_3)_2^+(aq) + 2H^+(aq) + Cl^-(aq) \longrightarrow AgCl(s) + 2NH_4^+(aq)$

29.50 White ppt. in Step 1. Does not dissolve in Step 2; no precipitate in Step 3. Precipitate completely dissolves in Step 4, forms again in Step 5.

29.51 a. Will not precipitate Group I cations b. No effect

c. AgCl will not dissolve; Hg_2Cl_2 may not turn black

29.52 Add HCl to give chloride precipitate. Then add NH_3. AgCl will dissolve, $PbCl_2$ will remain unchanged, Hg_2Cl_2 will turn black.

29.53 Only Pb^{2+}; Ag^+ or Hg_2^{2+} would give precipitate (Ag_2O, HgO, Hg)

29.54 Pb^{2+} and Hg_2^{2+} present; Ag^+ in doubt

29.55 $AgNO_3$ absent; would dissolve in cold water

$PbCl_2$ present; dissolves in hot water

Ag_2S present; only black solid

Hg_2Cl_2 in doubt

29.56 a. $E^o_{tot} = E^o_{red}\ H^+ + E^o_{ox}\ Hg = 0.00\ V - 0.79\ V = -0.79\ V$; no

b. $E^o_{tot} = E^o_{red}\ Hg^{2+} + E^o_{ox}\ H_2 = 0.92\ V - 0.00\ V = 0.92\ V$; yes

29.57 Ag_2CrO_4, Pb_3O_4

29.58 a. Test solubility in water; Ag_2S insoluble.

b. Add NH_3; AgCl dissolves

c. Precipitate chlorides, then add NH_3; Hg_2Cl_2 turns black.

d. Add excess OH^-; $Pb(OH)_2$ dissolves

29.59 a. photography b. pigment c. medicine

29.60 $2\ PbS(s) + 3\ O_2(g) \longrightarrow 2\ PbO(s) + 2\ SO_2(g)$

*29.61 Produces 0.10 mol of Hg(l) per liter. Since Hg(l) and Ag^+ react in a 1:1 mole ratio, all of Ag^+ will be reduced; yes

*29.62

$$Ag_2CO_3(s) \rightleftharpoons 2\ Ag^+(aq) + CO_3^{2-}(aq) \qquad K_{sp}\ Ag_2CO_3$$

$$CO_3^{2-}(aq) + H^+(aq) \rightleftharpoons HCO_3^-(aq) \qquad 1/K_a\ HCO_3^-$$

$$HCO_3^-(aq) + H^+(aq) \rightleftharpoons H_2CO_3(aq) \qquad 1/K_a\ H_2CO_3$$

$$Ag_2CO_3(s) + 2H^+(aq) \rightleftharpoons 2Ag^+(aq) + H_2CO_3(aq) \qquad K$$

$$K = \frac{K_{sp}\ Ag_2CO_3}{(K_a\ HCO_3^-)(K_a\ H_2CO_3)} = \frac{1 \times 10^{-11}}{(4.8 \times 10^{-11})(4.2 \times 10^{-7})} = 5 \times 10^5$$

*29.63 conc. Ag^+ required to ppt. AgCl $= \frac{1.6 \times 10^{-10}}{1.0 \times 10^{-2}} = 1.6 \times 10^{-8} M$

conc. Ag^+ to ppt. $Ag_2CrO_4 = \left(\frac{2 \times 10^{-12}}{1 \times 10^{-3}}\right)^{\frac{1}{2}} = 4 \times 10^{-5}$ M

Hence white AgCl forms first

*29.64

$$Ag_2S(s) + 4NH_3(aq) \rightleftharpoons 2Ag(NH_3)_2^+(aq) + S^{2-}(aq)$$

$$K = K_{sp}\ Ag_2S \times (K_f\ Ag(NH_3)_2^+)^2$$

$$= (1 \times 10^{-49})(2 \times 10^7)^2 = 4 \times 10^{-35}$$

$$\frac{(2s)^2(s)}{[NH_3]^4} = 4 \times 10^{-35};\ s^3 = 1.3 \times 10^{-32};\quad s = 2 \times 10^{-11}\ M$$

*29.65 a. $E = -0.13\ V - 0.0591 \log_{10} \frac{(\text{conc. } Hg^{2+})}{(\text{conc. } Hg_2^{2+})}$

b. $0 = -0.13 - 0.0591 \log_{10}(\text{conc. } Hg^{2+})$

$\log_{10}(\text{conc. } Hg^{2+}) = \frac{-0.13}{0.0591} = -2.2$; conc. $Hg^{2+} = 0.006$ M

c. $E = -0.13\ V - 0.0591 \log_{10}(1 \times 10^{-26}) = +1.41$ V

d. yes

Test Questions, Multiple Choice

1. Of the Group I cations, _____ forms a stable complex with OH^-; _____ forms a stable complex with NH_3. To complete this sentence properly, the ions in the two successive blanks should be:

a. Ag^+ and Pb^{2+} b. Pb^{2+} and Ag^+ c. Ag^+ and Hg_2^{2+}

d. Hg_2^{2+} and Ag^+ e. Pb^{2+} and Hg_2^{2+}

2. How many of the following compounds dissolve in 1 M H^+?

$AgCl$, Ag_2CO_3, $Pb(OH)_2$, Ag_2S, Hg_2Cl_2

a. 1 b. 2 c. 3 d. 4 e. 5

3. Which one of the following has a bright yellow color?

a. $PbCl_2$ b. $AgCl$ c. $HgNH_2Cl$ d. $PbCrO_4$ e. $Ag(NH_3)_2^+$

4. The common oxidation states of lead in its compounds are _____ and _____; those of mercury are _____ and _____. To complete this sentence properly, the numbers that appear in the four successive blanks should be:

a. +1, +2, +3, +4 b. +1, +2, +1, +2 c. +2, +4, +1, +2

d. +1, +3, +1, +2 e. +3, +4, +1, +3

5. Which one of the following silver compounds has the highest water solubility?

a. Ag_2O b. $AgCl$ c. Ag_2CrO_4 d. $AgNO_3$ e. Ag_2CO_3

6. Treatment of Hg_2Cl_2 with NH_3 gives two insoluble products whose formulas are:

a. $HgCl_2$ and Hg c. $Hg(NH_3)_2Cl$ and Hg c. HgO and $HgCl_2$

d. Hg and $HgNH_2Cl$ e. $HgNH_3Cl$ and Hg

7. In Group I analysis, ammonia is used to

a. precipitate Pb^{2+}, Ag^+ and Hg_2^{2+}

b. dissolve $AgCl$, $PbCl_2$, and Hg_2Cl_2

c. dissolve $AgCl$ and give a precipitate with Hg_2Cl_2

d. dissolve Hg_2Cl_2 and give a precipitate with $AgCl$

e. none of the above

8. In Group I analysis, nitric acid is used to:

a. precipitate Hg_2^{2+} b. dissolve $AgCl$, $PbCl_2$ and Hg_2Cl_2

c. oxidize Pb^{2+} to Pb^{4+} d. destroy the $Ag(NH_3)_2^+$ complex ion

e. bring PbS into solution

9. A Group I unknown gives a chloride precipitate which is partially soluble in hot water. This means that:
   a. Pb^{2+} is absent; Ag^{+} and Hg_2^{2+} are in doubt
   b. Pb^{2+} is present; Ag^{+} and Hg_2^{2+} are in doubt
   c. only Pb^{2+} is present; the other ions are absent
   d. Pb^{2+} is absent; Ag^{+} and Hg_2^{2+} are present
   e. Pb^{2+} and Ag^{+} are present; Hg_2^{2+} is in doubt

10. A solid mixture contains AgCl, $Pb(NO_3)_2$, $PbSO_4$ and Hg_2Cl_2. Which one of the following solvents will completely dissolve the mixture?
   a. water b. 6 M NH_3 c. 6 M NaOH d. 6 M HCl
   e. none of the above

<u>Test Problems</u>

1. Write a balanced redox equation for the reaction of lead with aqua regia, a mixture of nitric and hydrochloric acids. Assume the products include $PbCl_2$ and NO.

2. Write balanced net ionic equations to explain the following observations:
   a. Silver chloride dissolves in 6 M NH_3.
   b. Lead sulfate dissolves in 6 M NaOH.
   c. Addition of chromate ions to a saturated solution of lead chloride gives a yellow precipitate.
   d. When a solution prepared by dissolving AgCl in ammonia is acidified, a white precipitate forms.

3. A Group I unknown contains only Hg_2^{2+}. State what you would observe when:
   a. the unknown is treated with HCl.
   b. the product in (a) is heated with water; the resulting solution is treated with potassium chromate solution.
   c. the precipitate remaining after heating with water in (b) is treated with ammonia.
   d. the solution above the solid in (c) is acidified.

4. Calculate the solubility (moles per liter) of $PbCrO_4$ in 0.10 M NaOH. Take K_{sp} of $PbCrO_4$ to be 1×10^{-16} and K_f of $Pb(OH)_3^-$ to be 1×10^{14}.

5. Taking K_{sp} of PbI_2 to be 1×10^{-8}, determine whether or not a precipitate will form when 200 mL of 0.10 M $Pb(NO_3)_2$ is mixed with 800 mL of 0.010 M NaI.

CHAPTER 30

Answers to Problems

30.1 a. $2\ Bi_2S_3(s) + 9\ O_2(g) \longrightarrow 2\ Bi_2O_3(s) + 6\ SO_2(g)$

b. $SnO_2(s) + 2\ C(s) \longrightarrow Sn(s) + 2\ CO(g)$

c. $Sb_2S_3(s) + 3\ Fe(s) \longrightarrow 2\ Sb(s) + 3\ FeS(s)$

30.2 a. $Pb^{2+}(aq) + 2\ NH_3(aq) + 2\ H_2O \longrightarrow Pb(OH)_2(s) + 2\ NH_4^+(aq)$

b. $Bi^{3+}(aq) + 3\ NH_3(aq) + 3\ H_2O \longrightarrow Bi(OH)_3(s) + 3\ NH_4^+(aq)$

c. $Cd^{2+}(aq) + 4\ NH_3(aq) \longrightarrow Cd(NH_3)_4^{2+}(aq)$

30.3 a. $2\ Bi^{3+}(aq) + 3\ H_2S(aq) \longrightarrow Bi_2S_3(s) + 6\ H^+(aq)$

b. $Cd^{2+}(aq) + H_2S(aq) \longrightarrow CdS(s) + 2\ H^+(aq)$

c. $Sn^{4+}(aq) + 2\ H_2S(aq) \longrightarrow SnS_2(s) + 4\ H^+(aq)$

d. $2\ Sb^{3+}(aq) + 3\ H_2S(aq) \longrightarrow Sb_2S_3(s) + 6\ H^+(aq)$

30.4 a. $Cd^{2+}(aq) + 4\ Cl^-(aq) \longrightarrow CdCl_4^{2-}(aq)$

b. $Hg^{2+}(aq) + 4\ Cl^-(aq) \longrightarrow HgCl_4^{2-}(aq)$

c. $Sn^{4+}(aq) + 6\ Cl^-(aq) \longrightarrow SnCl_6^{2-}(aq)$

d. $Sb^{3+}(aq) + 6\ Cl^-(aq) \longrightarrow SbCl_6^{3-}(aq)$

30.5 a. $Cu^{2+}(aq) + 4\ NH_3(aq) \longrightarrow Cu(NH_3)_4^{2+}(aq)$

b. $2\ Bi^{3+}(aq) + 3\ Sn^{2+}(aq) \longrightarrow 2\ Bi(s) + 3\ Sn^{4+}(aq)$

c. $Hg^{2+}(aq) + 2\ OH^-(aq) \longrightarrow HgO(s) + H_2O$

d. $Sb^{3+}(aq) + Cl^-(aq) + H_2O \longrightarrow SbOCl(s) + 2\ H^+(aq)$

30.6 a. $Cu^{2+}(aq) + 2\ OH^-(aq) \longrightarrow Cu(OH)_2(s)$

b. $Sb_2S_3(s) + 8\ OH^-(aq) \longrightarrow 2\ Sb(OH)_4^-(aq) + 3\ S^{2-}(aq)$

c. $Bi^{3+}(aq) + Cl^-(aq) + H_2O \longrightarrow BiOCl(s) + 2\ H^+(aq)$

30.7 a. $CuCO_3(s) + 2\ H^+(aq) \longrightarrow Cu^{2+}(aq) + H_2CO_3(aq)$

b. $Cd(s) + 2\ H^+(aq) + 4\ Cl^-(aq) \longrightarrow CdCl_4^{2-}(aq) + H_2(g)$

c. $SbCl_6^{3-}(aq) + 4\ OH^-(aq) \longrightarrow Sb(OH)_4^-(aq) + 6\ Cl^-(aq)$

30.8 a. $Hg(l) \longrightarrow Hg^{2+}(aq) + 2\,e^-$

$$\frac{2\left[NO_3^-(aq) + 2\,H^+(aq) + e^- \longrightarrow NO_2(g) + 2\,H_2O\right]}{Hg(l) + 2NO_3^-(aq) + 4H^+(aq) \longrightarrow Hg^{2+}(aq) + 2NO_2(g) + 4H_2O}$$

b. $1.00\text{ g Hg} \times \frac{1\text{ mol Hg}}{200.6\text{ g Hg}} \times \frac{4\text{ mol }HNO_3}{1\text{ mol Hg}} = 0.0199\text{ mol }HNO_3$

$V = \frac{0.0199\text{ mol}}{6.0\text{ mol/L}} = 0.0033\text{ L} = 3.3\text{ mL}$

30.9 a. $Bi_2S_3(s) \longrightarrow 2\,Bi^{3+}(aq) + 3\,S(s) + 6\,e^-$

$$\frac{6\left[NO_3^-(aq) + 2\,H^+(aq) + e^- \longrightarrow NO_2(g) + H_2O\right]}{Bi_2S_3(s) + 6NO_3^-(aq) + 12H^+(aq) \longrightarrow 2Bi^{3+}(aq) + 3S(s) + 6NO_2(g) + 6H_2O}$$

b. $Bi(s) \longrightarrow Bi^{3+}(aq) + 3\,e^-$

$$\frac{3\left[NO_3^-(aq) + 2H^+(aq) + e^- \longrightarrow NO_2(g) + H_2O\right]}{Bi(s) + 3NO_3^-(aq) + 6H^+(aq) \longrightarrow Bi^{3+}(aq) + 3NO_2(g) + 3H_2O}$$

c. $5\left[BiO_3^-(aq) + 6\,H^+(aq) + 2\,e^- \longrightarrow Bi^{3+}(aq) + 3\,H_2O\right]$

$$\frac{2\left[Mn^{2+}(aq) + 4\,H_2O \longrightarrow MnO_4^-(aq) + 8\,H^+(aq) + 5\,e^-\right]}{5BiO_3^-(aq) + 14H^+(aq) + 2Mn^{2+}(aq) \longrightarrow 5Bi^{3+}(aq) + 7H_2O + 2MnO_4^-(aq)}$$

30.10 a. $E^o_{tot} = E^o_{red}\,Sn^{4+} + E^o_{ox}\,Sn = 0.15\text{ V} + 0.14\text{ V} = 0.29\text{ V}$; yes

b. $E^o_{tot} = E^o_{red}\,Hg^{2+} + E^o_{ox}\,Sn^{2+} = 0.92\text{ V} - 0.15\text{ V} = 0.77\text{ V}$; yes

c. $E^o_{tot} = E^o_{red}\,Sn^{4+} + E^o_{ox}\,Fe^{2+} = 0.15\text{ V} - 0.77\text{ V} = -0.62\text{ V}$; no

30.11 a. $E = -0.29\text{ V} - \frac{0.0591}{2}\log_{10}\frac{(\text{conc. }Sn^{4+})}{(\text{conc. }Sn^{2+})^2}$

$= -0.29\text{ V} - \frac{0.0591}{2}\log_{10}(1 \times 10^3) = -0.38\text{ V}$

b. $0.00 = -0.29 - \frac{0.0591}{2}\log_{10}(\text{conc. }Sn^{4+})$

$\log_{10}(\text{conc. }Sn^{4+}) = -0.58/0.0591 = -9.8$

$\text{conc. }Sn^{4+} = 2 \times 10^{-10}\text{ M}$

30.12 a. $[Cu^{2+}] = \frac{1 \times 10^{-35}}{1 \times 10^{-20}} = 1 \times 10^{-15}$ M

b. $[Cd^{2+}] = \frac{1 \times 10^{-26}}{1 \times 10^{-20}} = 1 \times 10^{-6}$ M

c. $[Bi^{3+}]^2 = \frac{1 \times 10^{-99}}{1 \times 10^{-60}} = 1 \times 10^{-39}$; $[Bi^{3+}] = 3 \times 10^{-20}$ M

30.13 conc. $Cu^{2+} = \frac{0.400 \times 0.001 \text{ M}}{1.000} = 4 \times 10^{-4}$ M

conc. $C_2O_4^{2-} = \frac{0.600 \times 0.005 \text{ M}}{1.000} = 3 \times 10^{-3}$ M

conc. Cu^{2+} x conc. $C_2O_4^{2-} = (4 \times 10^{-4})(3 \times 10^{-3})$

$= 1 \times 10^{-6} < K_{sp}$; no ppt.

30.14 a. $s^2 = 2 \times 10^{-10}$; $s = 1 \times 10^{-5}$ M

b. $s = \frac{2 \times 10^{-10}}{0.1} = 2 \times 10^{-9}$ M

c. $CuCO_3(s) + 4\ NH_3(aq) \longrightarrow Cu(NH_3)_4^{2+}(aq) + CO_3^{2-}(aq)$

$K = K_{sp}\ CuCO_3 \times K_f\ Cu(NH_3)_4^{2+} = 1 \times 10^3$

$s^2/(1 \times 10^{-1})^4 = 1 \times 10^3$; $s^2 = 0.1$; $s = 0.3$ M

30.15 a. $\frac{[Cd^{2+}]}{[CdCl_4^{2-}]} = \frac{1}{K_f \times [Cl^-]^4} = \frac{1}{(10^3)(2)^4} = 6 \times 10^{-5}$

b. $\frac{[Cd^{2+}]}{[Cd(CN)_4^{2-}]} = \frac{1}{K_f \times [CN^-]^4} = \frac{1}{(10^{19})(0.1)^4} = 1 \times 10^{-15}$

30.16 $Cu_3(PO_4)_2(s) + 3\ S^{2-}(aq) \longrightarrow 3\ CuS(s) + 2\ PO_4^{3-}(aq)$

$K = \frac{K_{sp}\ Cu_3(PO_4)_2}{(K_{sp}\ CuS)^3} = \frac{1 \times 10^{-37}}{(1 \times 10^{-35})^3} = 1 \times 10^{68}$

$\frac{[PO_4^{3-}]^2}{[S^{2-}]^3} = 1 \times 10^{68}$; $[PO_4^{3-}]^2 = (1 \times 10^{68})(1 \times 10^{-60})$

$= 1 \times 10^8$

equil. conc. $PO_4^{3-} = 1 \times 10^{4}$ M; reaction should occur

30.17 a. $K = K_{sp}\ CdS \times K_f Cd(NH_3)_4^{2+} = (1 \times 10^{-26})(1 \times 10^7)$

$= 1 \times 10^{-19}$

b. $K = \frac{1 \times 10^{-20}}{K_{sp}\ CuS} = \frac{1 \times 10^{-20}}{1 \times 10^{-35}} = 1 \times 10^{15}$

30.18 a. $[S^{2-}] = \frac{1 \times 10^{-20} \times [H_2S]}{[H^+]^2} = \frac{(1 \times 10^{-20})(1 \times 10^{-1})}{(1 \times 10^{-4})^2} = 1 \times 10^{-13}$ M

b. $[Cd^{2+}] = \frac{1 \times 10^{-26}}{1 \times 10^{-13}} = 1 \times 10^{-13}$ M

c. $[Mn^{2+}] = \frac{1 \times 10^{-15}}{1 \times 10^{-13}} = 1 \times 10^{-2}$ M

30.19 a. $CuS(s) + 2\ H^+(aq) \longrightarrow Cu^{2+}(aq) + H_2S(aq)$

$K = K_{sp}CuS/(1 \times 10^{-20}) = 1 \times 10^{-15}$

b. $s^2/6^2 = 1 \times 10^{-15}$; $s^2 = 3.6 \times 10^{-14}$; $s = 2 \times 10^{-7}$ M

c. $s \times 0.10/6^2 = 1 \times 10^{-15}$; $s = 4 \times 10^{-13}$ M

30.20 a. $Sn^{2+}(aq) + H_2O_2(aq) + 2\ H^+(aq) \longrightarrow Sn^{4+}(aq) + 2\ H_2O$

b. $Cu^{2+}(aq) + S_2O_4^{\ 2-}(aq) + 2H_2O \longrightarrow Cu(s) + 2SO_3^{\ 2-}(aq) + 4H^+(aq)$

c. $HgS(s) + 4Cl^-(aq) + 2NO_3^-(aq) + 4H^+(aq) \longrightarrow HgCl_4^{\ 2-}(aq) + 2NO_2(g) + S(s) + 2H_2O$

30.21 a. Deep blue color with NH_3; $S_2O_4^{\ 2-}$ ions reduce to Cu(s)

b. White ppt. with NH_3 or on dilution in presence of Cl^-

c. Reduced to Hg by Cu or to Hg_2Cl_2 with Sn^{2+}

30.22 a. Cd^{2+} and Cu^{2+} would precipitate rather than forming complex ions.

b. Pb^{2+} would go into solution along with Bi^{3+}; Bi^{3+} might not precipitate in Step 7.

c. no effect

30.23 Black ppt. in Step 2, partially soluble in Step 3. Remaining precipitate is completely soluble in HNO_3 in Step 4. Treatment with NH_3 (Step 5) gives a white ppt. and a colorless solution. The precipitate dissolves in Step 6; gives a white ppt. on dilution and a black solid on treatment with Sn^{2+} (Step 7).

Nothing happens in Step 9; a yellow ppt. forms in Step 10. Yellow ppt. obtained in Step 13, dissolves in 14. Grey ppt. obtained in Step 15. Nothing happens in Step 16.

30.24 Must be HgS; Hg^{2+} is only ion present

30.25 Sn^{4+}, Sb^{3+} absent; sulfides would dissolve in NaOH

Hg^{2+} absent; sulfide would not dissolve in HNO_3

Cu^{2+} present; gives blue complex with NH_3

Bi^{3+} absent; would give white ppt. with NH_3

Cd^{2+} in doubt

30.26 Sn^{4+}, Sb^{3+} absent; sulfides would dissolve in NaOH

Hg^{2+} present; only sulfide insoluble in HNO_3

Bi^{3+} present; white ppt. with NH_3

Cu^{2+} absent; would give blue color with NH_3

Cd^{2+} present; gives yellow ppt. with H_2S

30.27 a. $Cu(NH_3)_4{}^{2+}$ b. $Cd(NH_3)_4{}^{2+}$, $CdCl_4{}^{2-}$ c. $HgCl_4{}^{2-}$

d. $Sn(OH)_6{}^{2-}$, $SnCl_6{}^{2-}$ e. $SbCl_6{}^{3-}$, $Sb(OH)_4{}^{-}$

30.28 copper: +2, +1 cadmium: +2 bismuth: +3, +5

tin: +4, +2 mercury: +2, +1 antimony: +3, +5

30.29 Cu, Bi, Hg, Cd, Sb

30.30 a. color (PbS is black, CdS is yellow)

b. AgCl is soluble in NH_3

c. add OH^-; Cd^{2+} precipitates, Sb^{3+} forms complex ion

d. add NH_3; Cu^{2+} forms complex ion, Bi^{3+} precipitates

30.31 a. $Cu_2S(s) + O_2(g) \longrightarrow 2\ Cu(s) + SO_2(g)$

b. $Bi_2O_3(s) + 3\ CO(g) \longrightarrow 2\ Bi(s) + 3\ CO_2(g)$

c. $2\ CdS(s) + 3\ O_2(g) \longrightarrow 2\ CdO(s) + 2\ SO_2(g)$

30.32 a. $Pb(OH)_2(s) + 2\ H^+(aq) \longrightarrow Pb^{2+}(aq) + 2\ H_2O$

b. $Bi(OH)_3(s) + 3\ H^+(aq) \longrightarrow Bi^{3+}(aq) + 3\ H_2O$

c. $Cd(NH_3)_4{}^{2+}(aq) + 4\ H^+(aq) \longrightarrow Cd^{2+}(aq) + 4\ NH_4{}^+(aq)$

30.33 $SnS_2(s) + 6\ OH^-(aq) \longrightarrow Sn(OH)_6{}^{2-}(aq) + 2\ S^{2-}(aq)$

$Sb_2S_3(s) + 8\ OH^-(aq) \longrightarrow 2\ Sb(OH)_4{}^-(aq) + 3\ S^{2-}(aq)$

30.34 a. $Bi^{3+}(aq) + 3\ OH^-(aq) \longrightarrow Bi(OH)_3(s)$

b. $Hg^{2+}(aq) + 2\ OH^-(aq) \longrightarrow HgO(s) + H_2O$

c. $Sn^{4+}(aq) + 6\ OH^-(aq) \longrightarrow Sn(OH)_6^{\ 2-}(aq)$

d. $Sb^{3+}(aq) + 4\ OH^-(aq) \longrightarrow Sb(OH)_4^{\ -}(aq)$

30.35 a. $Cu^{2+}(aq) + S_2O_4^{\ 2-}(aq) + 2H_2O \longrightarrow Cu(s) + 2SO_3^{\ 2-}(aq) + 4H^+(aq)$

b. $Sn^{2+}(aq) + Cl^-(aq) + H_2O \longrightarrow Sn(OH)Cl(s) + H^+(aq)$

c. $CdS(s) + 2H^+(aq) + 4Cl^-(aq) \longrightarrow CdCl_4^{\ 2-}(aq) + H_2S(aq)$

d. $Sn^{2+}(aq) + \frac{1}{2}\ O_2(g) + H_2O \longrightarrow SnO_2(s) + 2H^+(aq)$

30.36 a. $Cu^{2+}(aq) + H_2S(aq) \longrightarrow CuS(s) + 2\ H^+(aq)$

b. $Sb_2S_3(s) + 6H^+(aq) + 12Cl^-(aq) \longrightarrow 2SbCl_6^{\ 3-}(aq) + 3H_2S(aq)$

c. $Bi^{3+}(aq) + 3NH_3(aq) + 3H_2O \longrightarrow Bi(OH)_3(s) + 3NH_4^{\ +}(aq)$

30.37 a. $CuCO_3(s) + 4NH_3(aq) \longrightarrow Cu(NH_3)_4^{\ 2+}(aq) + CO_3^{\ 2-}(aq)$

b. $CuO(s) + 2\ H^+(aq) \longrightarrow Cu^{2+}(aq) + H_2O$

c. $Sn(OH)Cl(s) + H^+(aq) \longrightarrow Sn^{2+}(aq) + H_2O + Cl^-(aq)$

30.38 a. $4\left[Sb(s) + 4\ OH^-(aq) \longrightarrow Sb(OH)_4^{\ -}(aq) + 3\ e^-\right]$

$3\left[O_2(g) + 2\ H_2O + 4\ e^- \longrightarrow 4\ OH^-(aq)\right]$

$4\ Sb(s) + 4\ OH^-(aq) + 3\ O_2(g) + 6\ H_2O \longrightarrow 4\ Sb(OH)_4^{\ -}(aq)$

b. $1.00\ g\ Sb \times \frac{1\ mol\ Sb}{121.8\ g\ Sb} \times \frac{1\ mol\ NaOH}{1\ mol\ Sb} = 8.21 \times 10^{-3}\ mol\ NaOH$

$V = \frac{8.21 \times 10^{-3}\ mol}{6.0\ mol/L} = 1.4 \times 10^{-3}\ L = 1.4\ mL$

30.39 a. $CuS(s) \longrightarrow Cu^{2+}(aq) + S(s) + 2\ e^-$

$2\left[NO_3^{\ -}(aq) + 2\ H^+(aq) + e^- \longrightarrow NO_2(g) + H_2O\right]$

$CuS(s) + 2NO_3^{\ -}(aq) + 4H^+(aq) \longrightarrow Cu^{2+}(aq) + 2NO_2(g) + S(s) + 2H_2O$

b. $HgS(s) + 4Cl^-(aq) \longrightarrow HgCl_4^{\ 2-}(aq) + S(s) + 2\ e^-$

$2\left[NO_3^{\ -}(aq) + 2\ H^+(aq) + e^- \longrightarrow NO_2(g) + H_2O\right]$

$HgS(s) + 4Cl^-(aq) + 2NO_3^{\ -}(aq) + 4H^+(aq) \longrightarrow HgCl_4^{\ 2-}(aq) + S(s) + 2NO_2(g) + 2H_2O$

30.39 c. $2[Bi(OH)_3(s) + 3\ e^- \longrightarrow Bi(s) + 3\ OH^-(aq)]$

$3[Sn(OH)_3^-(aq) + 3\ OH^-(aq) \longrightarrow Sn(OH)_6^{2-}(aq) + 2\ e^-]$

$2Bi(OH)_3(s) + 3Sn(OH)_3^-(aq) + 3\ OH^-(aq) \longrightarrow 2\ Bi(s) + 3Sn(OH)_6^{2-}(aq)$

30.40 a. $E^o_{tot} = E^o_{ox}Al + E^o_{red}Cu^{2+} = 1.66\ V + 0.34\ V = 2.00\ V$; yes

b. $E^o_{tot} = E^o_{ox}Cd + E^o_{red}NO_3^- = 0.40\ V + 0.96\ V = 1.36\ V$; yes

c. $E^o_{tot} = E^o_{red}Sn^{2+} + E^o_{ox}Pb = -0.14\ V + 0.13\ V = -0.01\ V$; no

30.41 a. $E = +0.37\ V - \frac{0.0591}{1} \log_{10}\frac{(1 \times 10^{-4})}{(1 \times 10^{-4})^2}$

$= +0.37\ V - 0.0591(4.0) = +0.13\ V$

b. $0.00 = 0.37 - 0.0591 \log_{10} \frac{1}{(\text{conc. } Cu^+)^2}$

$= 0.37 + 0.118 \log_{10}(\text{conc. } Cu^+)$

$\log_{10}(\text{conc. } Cu^+) = -0.37/0.118 = -3.1$

conc. $Cu^+ = 7 \times 10^{-4}$ M

30.42 a. $[Cu^{2+}] = \frac{1 \times 10^{-35}}{1 \times 10^{-18}} = 1 \times 10^{-17}$ M

b. $[Cd^{2+}] = \frac{1 \times 10^{-26}}{1 \times 10^{-18}} = 1 \times 10^{-8}$ M

c. $[Bi^{3+}]^2 = \frac{1 \times 10^{-99}}{(1 \times 10^{-18})^3} = 1 \times 10^{-45}$; $[Bi^{3+}] = 3 \times 10^{-23}$ M

30.43 conc. $Cd^{2+} = \frac{0.300 \times 0.010\ M}{0.500} = 0.0060$ M

conc. $C_2O_4^{2-} = \frac{0.200 \times 0.010\ M}{0.500} = 0.0040$ M

conc. Cd^{2+} x conc. $C_2O_4^{2-} = (6.0 \times 10^{-3})(4.0 \times 10^{-3})$

$= 2 \times 10^{-5} < K_{sp}$; no

30.44 a. $s^2 = 5 \times 10^{-12}$; $s = 2 \times 10^{-6}$ M

b. $s = (5 \times 10^{-12})/0.30$; $s = 2 \times 10^{-11}$ M

c. $CdCO_3(s) + 4\ NH_3(aq) \longrightarrow Cd(NH_3)_4^{2+}(aq) + CO_3^{2-}(aq)$

$K = K_{sp}\ CdCO_3 \times K_f\ Cd(NH_3)_4^{2+} = 5 \times 10^{-5}$

$s^2 = (2.0)^4(5 \times 10^{-5}) = 8 \times 10^{-4}$; $s = 0.03$ M

30.45 a. $\frac{[Cu(NH_3)_4^{2+}]}{[Cu^{2+}]} = 5 \times 10^{12}(6.0)^4 = 6 \times 10^{15}$

b. $(5 \times 10^{12})(0.10)^4 = 5 \times 10^8$

30.46 $Cu_3(PO_4)_2(s) + 6\ OH^-(aq) \longrightarrow 3\ Cu(OH)_2(s) + 2\ PO_4^{3-}(aq)$

$K = \frac{K_{sp}\ Cu_3(PO_4)_2}{[K_{sp}\ Cu(OH)_2]^3} = \frac{1 \times 10^{-37}}{(2 \times 10^{-19})^3} = 1 \times 10^{19}$; yes

30.47 a. $K = \frac{K_{sp}\ CdS}{1 \times 10^{-20}} = \frac{1 \times 10^{-26}}{1 \times 10^{-20}} = 1 \times 10^{-6}$

b. $K = \frac{1}{K_{sp}\ CdS \times K_f\ Cd(CN)_4^{2-}} = \frac{1}{(1 \times 10^{-26})(1 \times 10^{19})}$

$= 1 \times 10^7$

30.48 a. $[S^{2-}] = 1 \times 10^{-20} \times \frac{[H_2S]}{[H^+]^2} = 1 \times 10^{-19}$ M

b. $[Cu^{2+}] = \frac{1 \times 10^{-35}}{1 \times 10^{-19}} = 1 \times 10^{-16}$ M

c. $[Ni^{2+}] = \frac{1 \times 10^{-19}}{1 \times 10^{-19}} = 1$ M

30.49 a. $CdS(s) + 2\ H^+(aq) \longrightarrow Cd^{2+}(aq) + H_2S(aq)$

$K = \frac{K_{sp}\ CdS}{1 \times 10^{-20}} = \frac{1 \times 10^{-26}}{1 \times 10^{-20}} = 1 \times 10^{-6}$

b. $s^2 = (6.0)^2(1 \times 10^{-6})$; $s = 0.006$ M

c. $s(0.10) = (6.0)^2(1 \times 10^{-6})$; $s = 0.0004$ M

30.50 $CuS(s) + 4H^+(aq) + 2NO_3^-(aq) \longrightarrow Cu^{2+}(aq) + S(s) + 2NO_2(g) + 2H_2O$

$CdS(s) + 4H^+(aq) + 2NO_3^-(aq) \longrightarrow Cd^{2+}(aq) + S(s) + 2NO_2(g) + 2H_2O$

$Bi_2S_3(s) + 12H^+(aq) + 6NO_3^-(aq) \longrightarrow 2Bi^{3+}(aq) + 3S(s) + 6NO_2(g) + 6H_2O$

30.51 a. Cd^{2+} is precipitated as the yellow sulfide CdS

b. Sn^{4+} is reduced to Sn^{2+} and then treated with $HgCl_2$, which gives a grey precipitate of Hg_2Cl_2 + Hg

30.51 c. Sb^{3+} is converted to peach-colored Sb_2OS_2 with $S_2O_3^{2-}$

30.52 a. Possibly, some of the Group II cations would not precipitate.

b. Cu^{2+} and Cd^{2+} would precipitate, along with Bi^{3+}.

c. Would not form precipitate of BiOCl.

30.53 Black ppt. in Step 2, partially soluble in NaOH in Step 3. Remaining precipitate partially dissolves in HNO_3 (Step 4). Blue solution is formed in Step 5; deposit of Cu formed in Step 9. No precipitate in Step 10.
Precipitate of HgS dissolves in Step 11; positive tests for Hg(II) in Step 12. Orange-red ppt. forms in Step 13, dissolves in Step 14. Negative result in Step 15. Peach colored ppt. forms in Step 16.

30.54 Cu^{2+}, Bi^{3+}, Cd^{2+} and Hg^{2+} must be absent; would give precipitates with OH^- ion.

30.55 Cu^{2+}, Bi^{3+}, Hg^{2+} absent; would give dark precipitate
Cd^{2+} present; otherwise, precipitate would dissolve in NaOH
Sb^{3+} present; gives red-orange ppt.
Sn^{4+} in doubt

30.56 Hg^{2+} present; otherwise, would dissolve in HNO_3
Bi^{3+} present; white ppt. is $Bi(OH)_3$
Cu^{2+} absent; would give blue solution
Cd^{2+}, Sn^{4+}, Sb^{3+} in doubt

30.57 a. $Bi(OH)_3$, BiOCl b. Sb_2OS_2 c. Hg_2Cl_2, Hg d. none

30.58 CuF_2, CuF; BiF_3, BiF_5; HgF_2, Hg_2F_2; CdF_2; SnF_4, SnF_2; SbF_3, SbF_5

30.59 Cd, Sn

30.60 a. Add HCl; gives precipitate with $AgNO_3$

b. Add NH_3; $Cd(OH)_2$ dissolves

c. Add HCl; Pb^{2+} gives precipitate

d. Add base; Bi^{3+} gives white ppt., Hg^{2+} gives yellow ppt.

*30.61 Let $[NH_4^+] = x$; then, $[OH^-] = 0.14 + x$;

$$[NH_3] = 6.0 - x = 6.0$$

$$\frac{x(0.14 + x)}{6.0} = 1.8 \times 10^{-5}; \quad x \approx 0.0008\ M$$

$$[NH_4^+] = 0.0008\ M; \ [OH^-] = 0.14\ M$$

*30.62 no. moles Cu^{2+} = no. moles Bi^{3+} = no. moles Cd^{2+}

$$= (2.0 \times 10^{-2})(5.0 \times 10^{-3}) = 1.0 \times 10^{-4}$$

no. moles H_2S required $= 1.0 \times 10^{-4} + 1.5 \times 10^{-4} + 1.0 \times 10^{-4}$

$$= 3.5 \times 10^{-4}$$

mass thioacetamide $= 3.5 \times 10^{-4}\ \text{mol} \times \frac{75.1\ g}{1\ \text{mol}} = 0.026\ g$

*30.63 $Bi^{3+}(aq) + 3\ OH^-(aq) \rightleftharpoons Bi(OH)_3(s) \qquad K_1$

$3NH_3(aq) + 3H_2O \rightleftharpoons 3NH_4^+(aq) + 3\ OH^-(aq) \qquad K_2$

$Bi^{3+}(aq) + 3NH_3(aq) + 3H_2O \rightleftharpoons Bi(OH)_3(s) + 3NH_4^+(aq) \qquad K$

$K = K_1 \times K_2$; $K_1 = 1/(K_{sp}\ Bi(OH)_3)$; $K_2 = (K_b\ NH_3)^3$

$$K = \frac{(2 \times 10^{-5})^3}{4 \times 10^{-31}} = 2 \times 10^{16}; \text{ yes}$$

*30.64 CuS present; only colored compound

$HgCl_2$ absent; would give precipitate with base

$SnCl_4$ present; only compound other than $HgCl_2$ that dissolves in HCl

$AgNO_3$ in doubt

*30.65 $$[S^{2-}] = 1 \times 10^{-20} \times \frac{[H_2S]}{[H^+]^2} = 1 \times 10^{-20}\ M$$

$$[Cu^{2+}] = \frac{1 \times 10^{-35}}{1 \times 10^{-20}} = 1 \times 10^{-15}\ M$$

$$[Bi^{3+}]^2 = \frac{1 \times 10^{-99}}{1 \times 10^{-60}} = 1 \times 10^{-39}; \ [Bi^{3+}] = 3 \times 10^{-20}\ M$$

$$\text{ratio} = \frac{3 \times 10^{-20}}{1 \times 10^{-15}} = 3 \times 10^{-5}$$

Test Questions, Multiple Choice

1. Of the cations in Group II, ____ and ____ form stable complexes with NH_3; ____ and ____ form stable complexes with OH^-. To make this statement correct, the ions inserted in successive blanks should be:

   a. Cu^{2+}, Bi^{3+}; Cd^{2+}, Hg^{2+} b. Cu^{2+}, Cd^{2+}; Hg^{2+}, Sn^{4+}

   c. Bi^{3+}, Cd^{2+}; Sn^{4+}, Sb^{3+} d. Cu^{2+}, Cd^{2+}; Sn^{4+}, Sb^{3+}

   e. Sn^{4+}, Sb^{3+}; Cu^{2+}, Bi^{3+}

2. Which one of the following sulfides fails to dissolve in either HNO_3 or NaOH?

   a. CuS b. Bi_2S_3 c. CdS d. HgS e. SnS_2

3. How many of the Group II cations form black sulfides?

   a. 1 b. 2 c. 3 d. 4 e. 5

4. The anion used to reduce Cu^{2+} in Group II analysis is:

   a. S^{2-} b. SO_3^{2-} c. $S_2O_4^{2-}$ d. $S_2O_3^{2-}$ e. SO_4^{2-}

5. The products of the reaction between Bi^{3+} and Sn^{2+} are:

   a. Bi and Sn b. Bi^{5+} and Sn c. Bi and Sn^{4+} d. Bi^{5+} and Sn^{4+}

   c. Bi_2S_3 and N

6. When an acidic solution of $SnCl_2$ is diluted with water, a precipitate forms. The formula of the precipitate is:

   a. $SnCl_2$ b. $SnCl_4$ e. SnO_2 d. Sn(OH)Cl e. SnO

7. In Group II analysis, two sulfides are dissolved in NaOH. These are:

   a. CuS and Bi_2S_3 b. HgS and CdS c. SnS_2 and CdS

   d. SnS_2 and Sb_2S_3 e. CuS and Sb_2S_3

8. In Group II analysis, the reagent used to separate Bi^{3+} from Cu^{2+} is:

   a. NH_3 b. HNO_3 c. HCl d. NaOH e. H_2S

9. In the final step of Group II analysis, antimony is confirmed by precipitation as

a. $Sb(OH)_3$ b. $SbCl_3$ c. SbOCl d. Sb_2OS_2 e. Sb_2O_2S

10. How many of the following compounds dissolve in 1 M HCl?

CuS, $Bi(OH)_3$, HgS, $CuCO_3$, Hg_2Cl_2

a. 1 b. 2 c. 3 d. 4 e. 5

Test Problems

1. A Group II unknown forms a light-colored precipitate with H_2S, partially soluble in 6 M NaOH. When the NaOH solution is treated with HCl, Al and $HgCl_2$, no precipitate forms. Which of the Group II cations are present? absent? in doubt?

2. Write balanced net ionic equations for the reaction of

a. $Cd(NH_3)_4^{2+}$ with hydrochloric acid

b. Bi^{3+} with ammonia

c. $Sn(OH)_6^{2-}$ with nitric acid

d. Cu^{2+} with ammonia

3. Write a balanced redox equation for the reaction of HgS with aqua regia (a mixture of HCl and HNO_3), assuming the products include $HgCl_4^{2-}$ and NO(g).

4. Taking K_{sp} CdS = 1×10^{-26}, and: $\frac{[H^+]^2 \times [S^{2-}]}{[H_2S]} = 1 \times 10^{-20}$

calculate:

a. K for the reaction: $CdS(s) + 2H^+(aq) \rightleftharpoons Cd^{2+}(aq) + H_2S(aq)$

b. the solubility of CdS (moles per liter) in 6.0 M HCl.

5. Taking K_f $Cd(NH_3)_4^{2+} = 1 \times 10^7$, calculate:

a. K for the rection: $Cd(NH_3)_4^{2+}(aq) \rightleftharpoons Cd^{2+}(aq) + 4\ NH_3(aq)$

b. the concentration of NH_3 when $[Cd(NH_3)_4^{2+}] = [Cd^{2+}]$.

CHAPTER 31

Answers to Problems

31.1 a. $2\ Al_2O_3(l) \longrightarrow 4\ Al(l) + 3\ O_2(g)$

b. $ZnO(s) + C(s) \longrightarrow Zn(g) + CO(g)$

c. $2\ ZnS(s) + 3\ O_2(g) \longrightarrow 2\ ZnO(s) + 2\ SO_2(g)$

31.2 a. $Fe^{3+}(aq) + 3\ NH_3(aq) + 3\ H_2O \longrightarrow Fe(OH)_3(s) + 3\ NH_4^+(aq)$

c. $Zn^{2+}(aq) + 4\ NH_3(aq) \longrightarrow Zn(NH_3)_4^{2+}(aq)$

d. $Zn(OH)_2(s) + 4\ NH_3(aq) \longrightarrow Zn(NH_3)_4^{2+}(aq) + 2\ OH^-(aq)$

e. $Mn^{2+}(aq) + 2\ NH_3(aq) + 2\ H_2O \longrightarrow Mn(OH)_2(s) + 2\ NH_4^+(aq)$

31.3 a. $Ni(OH)_2(s) + 2\ H^+(aq) \longrightarrow Ni^{2+}(aq) + 2\ H_2O$

b. $Ni(OH)_2(s) + 6\ NH_3(aq) \longrightarrow Ni(NH_3)_6^{2+}(aq) + 2\ OH^-(aq)$

c. $Ni(OH)_2(s) + 2\ H^+(aq) \longrightarrow Ni^{2+}(aq) + 2\ H_2O$

d. $Ni(OH)_2(s) + 4\ CN^-(aq) \longrightarrow Ni(CN)_4^{2-}(aq) + 2\ OH^-(aq)$

31.4 a. $Fe(s) + 2\ H^+(aq) \longrightarrow Fe^{2+}(aq) + H_2(g)$

b. $Ni(s) + 2\ H^+(aq) \longrightarrow Ni^{2+}(aq) + H_2(g)$

c. $Al(s) + 3\ H^+(aq) \longrightarrow Al^{3+}(aq) + 3/2\ H_2(g)$

d. $Zn(s) + 2\ H^+(aq) \longrightarrow Zn^{2+}(aq) + H_2(g)$

31.5 a. $K^+(aq) + Fe^{3+}(aq) + Fe(CN)_6^{4-}(aq) \longrightarrow KFe[Fe(CN)_6](s)$

b. $2\ Fe^{3+}(aq) + 3\ S^{2-}(aq) \longrightarrow 2\ FeS(s) + S(s)$

c. $Ni(OH)_2(s) + 6\ NH_3(aq) \longrightarrow Ni(NH_3)_6^{2+}(aq) + 2\ OH^-(aq)$

d. $Al(OH)_3(s) + OH^-(aq) \longrightarrow Al(OH)_4^-(aq)$

31.6 a. $Zn(H_2O)_4^{2+}(aq) \longrightarrow Zn(H_2O)_3(OH)^+(aq) + H^+(aq)$

b. $Cr^{3+}(aq) + 3\ NH_3(aq) + 3\ H_2O \longrightarrow Cr(OH)_3(s) + 3\ NH_4^+(aq)$

c. $Fe(s) + 2\ H^+(aq) \longrightarrow Fe^{2+}(aq) + H_2(g)$

d. $Cr^{3+}(aq) + 3\ S^{2-}(aq) + 3\ H_2O \longrightarrow Cr(OH)_3(s) + 3\ HS^-(aq)$

31.7 a. $Fe^{3+}(aq) + SCN^-(aq) \longrightarrow Fe(SCN)^{2+}(aq)$

b. $ZnS(s) + 2\ H^+(aq) \longrightarrow Zn^{2+}(aq) + H_2S(aq)$

c. $Mn(s) + 2\ H_2O(l) \longrightarrow Mn(OH)_2(s) + H_2(g)$

31.8 a. $CoS(s) + 4\ H^+(aq) + 2NO_3^-(aq) \rightarrow Co^{2+}(aq) + S(s) + 2H_2O + 2NO_2(g)$

b. $Mn(OH)_2(s) + ClO^-(aq) \rightarrow MnO_2(s) + Cl^-(aq) + H_2O$

c. $2Cr(OH)_3(s) + 4\ OH^-(aq) + 3ClO^-(aq) \rightarrow 2CrO_4^{2-}(aq) + 5H_2O + 3Cl^-(aq)$

31.9 a. $Zn(s) + 2\ H^+(aq) \rightarrow Zn^{2+}(aq) + H_2(g)$

b. $2\ Cr(OH)_3(s) + 8\ OH^-(aq) \quad Cr_2O_7^{2-}(aq) + 6e^- + 7H_2O$

$3[H_2O_2(aq) + 2\ e^- \rightarrow 2\ OH^-]$

$2Cr(OH)_3(s) + 2\ OH^-(aq) + 3H_2O_2(aq) \rightarrow Cr_2O_7^{2-}(aq) + 7H_2O$

c. $4\ [Co^{2+}(aq) + 6\ NH_3(aq) \rightarrow Co(NH_3)_6^{3+}(aq) + e^-]$

$O_2(g) + 2\ H_2O + 4\ e^- \rightarrow 4\ OH^-(aq)$

$4Co^{2+}(aq) + 24NH_3(aq) + O_2(g) + 2H_2O \rightarrow 4Co(NH_3)_6^{3+}(aq) + 4\ OH^-(aq)$

d. $Cr_2O_7^{2-}(aq) + 14\ H^+(aq) + 6\ e^- \rightarrow 2Cr^{3+}(aq) + 7\ H_2O$

$3\ [H_2S(aq) \rightarrow S(s) + 2\ H^+(aq) + 2\ e^-]$

$Cr_2O_7^{2-}(aq) + 8H^+(aq) + 3H_2S(aq) \rightarrow 2Cr^{3+}(aq) + 7H_2O + 3S(s)$

31.10 a. $E^o_{tot} = E^o_{ox}Cr^{2+} + E^o_{red}Sn^{4+} = 0.41\ V + 0.15\ V = 0.56\ V$

b. $E^o_{tot} = E^o_{ox}Fe + E^o_{red}H_2O = 0.88\ V - 0.83\ V = 0.05\ V$

c. $E^o_{tot} = E^o_{ox}Fe(OH)_2 + E^o_{red}O_2 = 0.56\ V + 0.40\ V = 0.96\ V$

31.11 $E = +0.76\ V - \frac{0.0591}{2} \log_{10} \frac{1.0}{(\text{conc. } H^+)^2}$

$= +0.76\ V + 0.0591 \log_{10}(\text{conc. } H^+) = +0.76\ V - 0.0591(7.0) = +0.35\ V$

Should react with water.

31.12 a. $[OH^-]^2 = \frac{2 \times 10^{-16}}{1 \times 10^{-2}} = 2 \times 10^{-14}$; $[OH^-] \approx 1 \times 10^{-7}$ M

$[H^+] = 1 \times 10^{-7}$ M; pH = 7

b. $[Co^{2+}] = \frac{2 \times 10^{-16}}{(1 \times 10^{-5})^2} = 2 \times 10^{-6}$ M

$f = (2 \times 10^{-6})/(1 \times 10^{-2}) = 2 \times 10^{-4}$

31.13 a. $Ni(OH)_2(s) + 6\ NH_3(aq) \longrightarrow Ni(NH_3)_6^{2+}(aq) + 2\ OH^-(aq)$

$K = K_{sp}\ Ni(OH)_2 \times K_f\ Ni(NH_3)_6^{2+}$

$= (1 \times 10^{-16})(5 \times 10^{8}) = 5 \times 10^{-8}$

b. $4s^3 = (6)^6 \times (5 \times 10^{-8})$; $4s^3 = 2 \times 10^{-3}$; $s = 0.08$ M

31.14 $\frac{[Zn^{2+}]}{[Zn(OH)_4^{2-}]} = \frac{1}{[OH^-]^4 \times (3 \times 10^{15})}$

a. $1 = \frac{1}{[OH^-]^4 \times (3 \times 10^{15})}$; $[OH^-]^4 = \frac{1}{3 \times 10^{15}} = 3 \times 10^{-16}$

$[OH^-] \approx 1 \times 10^{-4}$ M; $[H^+] = 1 \times 10^{-10}$ M; pH = 10

b. $10^{-4} = \frac{1}{[OH^-]^4 \times (3 \times 10^{15})}$; $[OH^-]^4 = \frac{1}{3 \times 10^{11}} = 3 \times 10^{-12}$

$[OH^-] \approx 1 \times 10^{-3}$ M; $[H^+] = 1 \times 10^{-11}$ M; pH = 11

c. $[OH^-] \approx 1 \times 10^{-2}$ M; $[H^+] = 1 \times 10^{-12}$ M; pH = 12

31.15 $s = 5 \times [OH^-]$; $[H^+] = 5 \times 10^{-11}$ M; $[OH^-] = 2 \times 10^{-4}$ M

$s = 1 \times 10^{-3}$ M; $1 \times 10^{-3}\ \frac{mol}{L} \times \frac{78.0\ g}{1\ mol} = 0.08$ g/L

31.16 $s = [H^+] \times (1 \times 10^{-3})$

$[H^+] = 0.01/0.001 = 10$ M; pH = -1

31.17 $4s^3 = 4 \times 10^{-14}$; $s^3 = 1 \times 10^{-14}$; $s = 2 \times 10^{-5}$ M

$[OH^-] = 2s = 4 \times 10^{-5}$ M; $[H^+] = 2 \times 10^{-10}$ M; pH = 9.7

31.18 a. $\frac{[NH_4^+]}{[NH_3]} = \frac{[H^+]}{K_a} = \frac{1 \times 10^{-9}}{5.6 \times 10^{-10}} = 2$

b. no. moles $NH_4^+ = 1.0 \times 10^{-3}$; no. moles $NH_3 = \frac{1.0 \times 10^{-3}}{2}$

$= 5 \times 10^{-4}$

$V = \frac{5 \times 10^{-4}}{0.6 \times 6.0} = 1 \times 10^{-4}$ L = 0.1 mL

31.19 $Fe^{2+}(aq) + S^{2-}(aq) \longrightarrow FeS(s)$

$2\ Fe^{3+}(aq) + 3\ S^{2-}(aq) \longrightarrow 2\ FeS(s) + S(s)$

$Co^{2+}(aq) + S^{2-}(aq) \longrightarrow CoS(s)$

31.19 (cont.)

$Ni^{2+}(aq) + S^{2-}(aq) \longrightarrow NiS(s)$

$Al^{3+}(aq) + 3\ S^{2-}(aq) + 3\ H_2O \longrightarrow Al(OH)_3(s) + 3\ HS^-(aq)$

$Mn^{2+}(aq) + S^{2-}(aq) \longrightarrow MnS(s)$

$Cr^{3+}(aq) + 3\ S^{2-}(aq) + 3\ H_2O \longrightarrow Cr(OH)_3(s) + 3\ HS^-(aq)$

$Zn^{2+}(aq) + S^{2-}(aq) \longrightarrow ZnS(s)$

31.20 a. Step 2   b. Steps 1, 3   c. Step 8

31.21 Colored precipitate formed in Step 1, partially soluble in Step 2. Remainder of precipitate dissolves in Step 3. Negative test in Step 4; red ppt. in Step 5. May get precipitate in Step 6 if Ni^{2+} carries over (precipitate would dissolve in Step 7). Addition of NH_3 in Step 10 would give deep blue color; red ppt. in Step 11.
Solution from Step 6 fails to give precipitate in Step 13. Yellow precipitate is obtained in Step 15, blue color in Step 16. Green precipitate observed in Step 17.

31.22 a. To remove H_2S, dissolve precipitate, convert Fe^{2+} to Fe^{3+}

b. If NaOH were used, Al^{3+} would stay in solution as $Al(OH)_4^-$

c. Need oxidizing agent to oxidize Mn^{2+}, Cr^{3+}

31.23 Add H_2S at pH 9 to give CoS, MnS, $Cr(OH)_3$. Treat with HCl; black residue shows Co^{2+} is present. Treat HCl solution with NaOH, NaClO, NH_3. If precipitate (MnO_2) is obtained, Mn^{2+} must be present. If solution is yellow (CrO_4^{2-}), Cr^{3+} is present.

31.24 Could only be $Al(OH)_3$ and/or ZnS. All other ions would give colored precipitates so must be absent.

31.25 Fe^{3+} present; gives red ppt. with NH_3, black sulfide
Co^{2+} and Ni^{2+} absent; sulfides would not dissolve in HCl
Cr^{3+} absent; would give yellow color with ClO^-
Mn^{2+} absent; MnO_2 would not dissolve in H_2SO_4
Al^{3+}, Zn^{2+} in doubt

31.26 Ag^+ present; AgCl soluble in NH_3
Pb^{2+}, Hg_2^{2+} absent; chlorides insoluble in NH_3
All Group II ions absent; would precipitate with H_2S
Mn^{2+} absent; MnO_2 would not dissolve in H_2SO_4
Ni^{2+} absent; would give red ppt. with DMG
Co^{2+} present; must be present to account for residue with HCl
Fe^{3+} present; accounts for precipitate with NaOH
Cr^{3+}, Al^{3+}, Zn^{2+} in doubt

31.27 $Pb(OH)_2$, $Zn(OH)_2$ and $Al(OH)_3$ are absent; would dissolve in NaOH. $Cu(OH)_2$ present; dissolves in NH_3. $Fe(OH)_3$ present; insoluble in NH_3

31.28 a. Add excess OH^-; Mn^{2+} will give precipitate

b. Add excess OH^-; Fe^{3+} gives precipitate

c. Note color; $Fe(OH)_3$ is red

d. Add NH_3; Cu^{2+} gives deep blue color

31.29 a. Al^{3+}, Zn^{2+} b. Fe^{2+}, Co^{2+}, Ni^{2+}, Mn^{2+}, Zn^{2+}

c. Al^{3+}, Cr^{3+}, Zn^{2+} d. all

31.30 a. Fe^{2+}, Fe^{3+}, Mn^{2+}, Cr^{3+} b. Fe^{3+}, Co^{2+}, Ni^{2+}, Mn^{2+}, Cr^{3+}

c. Co^{2+}, Ni^{2+}, Al^{3+}, Zn^{2+}, Fe^{3+}

31.31 a. $Fe_2O_3(s) + 3\ CO(g) \longrightarrow 2\ Fe(s) + 3\ CO_2(g)$

b. $Ni(s) + 4\ CO(g) \longrightarrow Ni(CO)_4(g)$

c. $2\ NiS(s) + 3\ O_2(g) \longrightarrow 2\ NiO(s) + 2\ SO_2(g)$

31.32 a. $Fe^{3+}(aq) + 3\ OH^-(aq) \longrightarrow Fe(OH)_3(s)$

b. $Al^{3+}(aq) + 4\ OH^-(aq) \longrightarrow Al(OH)_4{}^-(aq)$

c. $Zn(OH)_2(s) + 2\ OH^-(aq) \longrightarrow Zn(OH)_4{}^{2-}(aq)$

d. $Al(OH)_3(s) + OH^-(aq) \longrightarrow Al(OH)_4{}^-(aq)$

e. $Ni^{2+}(aq) + 2\ OH^-(aq) \longrightarrow Ni(OH)_2(s)$

31.33 a. $ZnCO_3(s) + 2\ H^+(aq) \longrightarrow Zn^{2+}(aq) + H_2CO_3(aq)$

b. $ZnCO_3(s) + 4\ NH_3(aq) \longrightarrow Zn(NH_3)_4{}^{2+}(aq) + CO_3{}^{2-}(aq)$

c. $ZnCO_3(s) + 4\ OH^-(aq) \longrightarrow Zn(OH)_4{}^{2-}(aq) + CO_3{}^{2-}(aq)$

31.33 d. $ZnCO_3(s) + 4\ CN^-(aq) \longrightarrow Zn(CN)_4^{2-}(aq) + CO_3^{2-}(aq)$

31.34 a. $Al(s) + 3\ H_2O + OH^-(aq) \longrightarrow Al(OH)_4^-(aq) + 3/2\ H_2(g)$

b. $Sn(s) + 4\ H_2O + 2\ OH^-(aq) \longrightarrow Sn(OH)_6^{2-}(aq) + 2\ H_2(g)$

d. $Zn(s) + 2\ H_2O + 2\ OH^-(aq) \longrightarrow Zn(OH)_4^{2-}(aq) + H_2(g)$

31.35 a. $Fe(s) + 2\ H^+(aq) \longrightarrow Fe^{2+}(aq) + H_2(g)$

b. $Co^{2+}(aq) + 4\ SCN^-(aq) \longrightarrow Co(SCN)_4^{2-}(aq)$

c. $3K^+(aq) + Co^{2+}(aq) + 7NO_2^-(aq) + 2H^+(aq) \longrightarrow$

$K_3[Co(NO_2)_6](s) + NO(g) + H_2O$

31.36 a. $Al(H_2O)_6^{3+}(aq) \longrightarrow Al(H_2O)_5(OH)^{2+}(aq) + H^+(aq)$

b. $Fe^{2+}(aq) + 2\ NH_3(aq) + 2\ H_2O \longrightarrow Fe(OH)_2(s) + 2NH_4^+(aq)$

c. $Al^{3+}(aq) + 3\ S^{2-}(aq) + 3\ H_2O \longrightarrow Al(OH)_3(s) + 3\ HS^-(aq)$

d. $MnS(s) + 2\ H^+(aq) \longrightarrow Mn^{2+}(aq) + H_2S(aq)$

31.37 a. $Zn^{2+}(aq) + 2\ OH^-(aq) \longrightarrow Zn(OH)_2(s)$

$Zn(OH)_2(s) + 2\ OH^-(aq) \longrightarrow Zn(OH)_4^{2-}(aq)$

b. $Co^{2+}(aq) + 4\ SCN^-(aq) \longrightarrow Co(SCN)_4^{2-}(aq)$

31.38 a. $Co(NO_2)_6^{4-}(aq) + NO_2^-(aq) + 2H^+(aq) \longrightarrow Co(NO_2)_6^{3-}(aq) +$

$NO(g) + H_2O$

b. $MnO_2(s) + H_2O_2(aq) + 2H^+(aq) \longrightarrow Mn^{2+}(aq) + O_2(g) + 2H_2O$

c. $Cr_2O_7^{2-}(aq) + 4H_2O_2(aq) + 2H^+(aq) \longrightarrow 2CrO_5(aq) + 5H_2O$

31.39 a. $Fe(s) + 6\ H^+(aq) + 3NO_3^-(aq) \longrightarrow Fe^{3+}(aq) + 3NO_2(g) + 3H_2O$

b. $CoS(s) + 4\ Cl^-(aq) \longrightarrow CoCl_4^{2-}(aq) + S(s) + 2\ e^-$

$2\ [NO_3^-(aq) + 2\ H^+(aq) + e^- \longrightarrow NO_2(g) + H_2O]$

$CoS(s) + 4Cl^-(aq) + 2NO_3^-(aq) + 4H^+(aq) \longrightarrow$

$CoCl_4^{2-}(aq) + S(s) + 2NO_2(g) + 2H_2O$

c. $2\ [Mn^{2+}(aq) + 4H_2O \longrightarrow MnO_4^-(aq) + 8H^+(aq) + 5\ e^-]$

$5\ [BiO_3^-(aq) + 6H^+(aq) + 2\ e^- \longrightarrow Bi^{3+}(aq) + 3\ H_2O]$

$2Mn^{2+}(aq) + 5BiO_3^-(aq) + 14H^+(aq) \longrightarrow 2MnO_4^-(aq) + 7H_2O$

$+ 5Bi^{3+}(aq)$

31.39 d. $2\,[MnO_2(s) + 2H_2O \rightarrow MnO_4^-(aq) + 4\,H^+(aq) + 3\,e^-]$

$3\,[ClO^-(aq) + 2\,H^+(aq) + 2\,e^- \rightarrow Cl^-(aq) + H_2O]$

$2MnO_2(s) + H_2O + 3ClO^-(aq) \rightarrow 2MnO_4^-(aq) + 2H^+(aq) + 3Cl^-$
$+\,2\,OH^- \qquad\qquad + 2\,OH^-$

$2MnO_2(s) + 3ClO^-(aq) + 2OH^-(aq) \rightarrow 2MnO_4^-(aq) + 3Cl^-(aq) + H_2O$

31.40 a. $E^o_{tot} = E^o_{ox}Fe + E^o_{red}Fe^{3+} = 0.44\text{ V} + 0.77\text{ V} = 1.21\text{ V}$

b. $E^o_{tot} = E^o_{ox}Mn^{2+} + E^o_{red}H_2O_2 = -1.23\text{ V} + 1.77\text{ V} = 0.54\text{ V}$

c. $E^o_{tot} = E^o_{ox}Cr(OH)_3 + E^o_{red}O_2 = +0.12\text{ V} + 0.40\text{ V} = 0.52\text{ V}$

31.41 $Al(s) + 3\,H^+(aq) \rightarrow Al^{3+}(aq) + 3/2\,H_2(g)$

$$E = +1.66\text{ V} - \frac{0.0591}{3}\log_{10}\frac{1}{(\text{conc. }H^+)^3}$$

$$= +1.66\text{ V} + 0.0591\log_{10}(\text{conc. }H^+)$$

$$= +1.66\text{ V} - 0.0591(7.0) = +1.25\text{ V; yes}$$

31.42 a. $[OH^-]^3 = \dfrac{4 \times 10^{-38}}{1 \times 10^{-2}} = 4 \times 10^{-36}$; $[OH^-] = 1.6 \times 10^{-12}$ M

$$[H^+] = \frac{1.0 \times 10^{-14}}{1.6 \times 10^{-12}} = 6 \times 10^{-3}\text{ M; pH} = 2.2$$

b. $[H^+] = 1 \times 10^{-9}$ M; $[OH^-] = 1 \times 10^{-5}$ M

$$[Cr^{3+}] = \frac{4 \times 10^{-38}}{(1 \times 10^{-5})^3} = 4 \times 10^{-23}\text{ M}$$

$$f = \frac{4 \times 10^{-23}}{1 \times 10^{-2}} = 4 \times 10^{-21}$$

31.43 a. $ZnS(s) + 4\,NH_3(aq) \rightarrow Zn(NH_3)_4^{2+}(aq) + S^{2-}(aq)$

$$K = K_{sp}\,ZnS \times K_f\,Zn(NH_3)_4^{2+} = (1 \times 10^{-20})(1 \times 10^{9}) = 1 \times 10^{-11}$$

b. $s^2 = (1 \times 10^{-11}) \times [NH_3]^4 = (6)^4(1 \times 10^{-11}) = 1 \times 10^{-8}$

$s = 1 \times 10^{-4}$ M

31.44 $[Zn^{2+}] = \frac{[Zn(NH_3)_4^{2+}]}{K_f \times [NH_3]^4} = \frac{1 \times 10^{-1}}{(1 \times 10^9) \times [NH_3]^4} = \frac{1 \times 10^{-10}}{[NH_3]^4}$

a. $\frac{1 \times 10^{-10}}{(1 \times 10^{-2})^4} = 1 \times 10^{-2}$ M   b. $\frac{1 \times 10^{-10}}{(1 \times 10^{-1})^4} = 1 \times 10^{-6}$ M

c. 1×10^{-10} M

31.45 $s = (5 \times 10^4) \times [H^+]^3$

$= (5 \times 10^4)(1 \times 10^{-4})^3 = 5 \times 10^{-8}$ mol/L

$= 5 \times 10^{-8} \frac{mol}{L} \times \frac{107\ g}{1\ mol} = 5 \times 10^{-6}$ g/L

31.46 $K = (1 \times 10^{-20})/(1 \times 10^{-20}) = 1$

$s = [H^+]$; $[H^+] = 0.01$ M; pH = 2.0

31.47 $Fe(OH)_2(s) \rightleftharpoons Fe^{2+}(aq) + 2\ OH^-(aq)$

$4s^3 = 1 \times 10^{-15}$; $s = (1 \times 10^{-15}/4)^{1/3} = 6 \times 10^{-6}$ M

$[OH^-] \approx 1 \times 10^{-5}$ M; $[H^+] = 1 \times 10^{-9}$ M; pH = 9

31.48 a. $\frac{[NH_4^+]}{[NH_3]} = \frac{1.0}{6.0} = 0.17$

b. $[H^+] = (5.6 \times 10^{-10}) \times 0.17 = 9.5 \times 10^{-11}$; pH = 10.02

31.49 $Fe^{3+}(aq) + 3\ OH^-(aq) \longrightarrow Fe(OH)_3(s)$

$Al^{3+}(aq) + 4\ OH^-(aq) \longrightarrow Al(OH)_4^-(aq)$

$Mn^{2+}(aq) + 2\ OH^-(aq) + ClO^-(aq) \longrightarrow MnO_2(s) + Cl^-(aq) + H_2O$

$2Cr^{3+}(aq) + 10\ OH^-(aq) + 3ClO^-(aq) \longrightarrow 2CrO_4^{2-}(aq) + 3Cl^-(aq) + 5H_2O$

$Zn^{2+}(aq) + 4\ OH^-(aq) \longrightarrow Zn(OH)_4^{2-}(aq)$

$Ni^{2+}(aq) + 2\ OH^-(aq) \longrightarrow Ni(OH)_2(s)$

31.50 a. Steps 1, 5, 6, 10, 11   b. Steps 1, 6, 12   c. Steps 1, 13

31.51 Black ppt. in Step 1, soluble in Step 2. Dark ppt., colorless solution in Step 6. Precipitate partially dissolves in Step 7. Remaining precipitate dissolves in Step 8, gives purple color in Step 9.

Solution from Step 7 gives deep red solution, red ppt. in Step 10; results in Steps 11, 12 are negative. Solution from 6 gives white ppt. in Step 13, blue color in Step 14. Steps 15 and 17 give negative results.

31.52 a. Buffer solution at pH 9; precipitate Al^{3+}

b. Reduce MnO_2 to Mn^{2+}

c. Ensure oxidation of Mn^{2+} to MnO_4^-

31.53 Step 1: add excess OH^- to form $Al(OH)_4^-$, $Zn(OH)_4^{2-}$ and $Fe(OH)_3$. Red ppt. indicates Fe^{3+}

Step 2: add $HC_2H_3O_2$, NH_3 to form $Al(OH)_3$, $Zn(NH_3)_4^{2+}$. White ppt. indicates Al^{3+}

Step 3: To solution from Step 2, add H^+, $K_4Fe(CN)_6$; green ppt. indicates Zn^{2+}

31.54 Either Co^{2+} or Ni^{2+} or both must be present; CoS and NiS do not dissolve in HCl. Fe^{3+}, Mn^{2+} must be absent; would give precipitates with NaOH, NaClO.

31.55 Co^{2+}, Ni^{2+} absent; CoS, NiS insoluble in HCl
Cr^{3+} present; yellow color is that of CrO_4^{2-}
Mn^{2+} present; MnO_2 fails to dissolve in H_2SO_4
Fe^{3+} present; $Fe(OH)_3$ dissolves in H_2SO_4
Al^{3+}, Zn^{2+} in doubt

31.56 Sb^{3+}, Sn^{4+} absent; sulfides would dissolve in NaOH
Hg^{2+} absent; HgS insoluble in HNO_3
Cu^{2+} present; gives colored solution with NH_3
Bi^{3+} absent; would give precipitate with NH_3
Cd^{2+} in doubt
Co^{2+}, Ni^{2+} absent; sulfides insoluble in HCl
Fe^{3+}, Mn^{2+}, Cr^{3+} absent; would give precipitates with OH^-
Al^{3+}, Zn^{2+} in doubt

31.57 Sb_2S_3, $Al(OH)_3$ absent; would dissolve in NaOH
NiS present; only NiS would dissolve in HNO_3 and not give precipitate with NH_3
Bi_2S_3 absent; would give precipitate with NH_3
FeS in doubt

31.58 a. Add excess OH^-; Fe^{3+} gives precipitate

b. Add NH_3; Ni^{2+} gives deep blue solution

c. Add HCl; ZnS dissolves

d. Add HCl; FeS dissolves completely

31.59 a. Co, Ni, Fe, Zn, Cr, Mn b. Co, Fe, Cr c. Cr, Mn

d. Co, Ni, Zn

31.60 a. none b. CoS, NiS, FeS, ZnS, MnS, $Cr(OH)_3$, $Al(OH)_3$

c. $Co(OH)_2$, $Fe(OH)_3$, $Mn(OH)_2$, $Cr(OH)_3$, $Al(OH)_3$ d. none

*31.61

$$Mn(OH)_2(s) \rightleftharpoons Mn^{2+}(aq) + 2\ OH^-(aq) \qquad K_1$$
$$2\ NH_4^+(aq) \rightleftharpoons 2\ NH_3(aq) + 2\ H^+(aq) \qquad K_2$$
$$2\ OH^-(aq) + 2\ H^+(aq) \rightleftharpoons 2\ H_2O \qquad K_3$$
$$Mn(OH)_2(s) + 2NH_4^+(aq) \rightleftharpoons Mn^{2+}(aq) + 2NH_3(aq) + 2H_2O; \quad K$$
$$K = K_1 \times K_2 \times K_3 = K_{sp}\ Mn(OH)_2 \times (K_a\ NH_4^+)^2 \times \frac{1}{K_w^2}$$
$$= 4 \times 10^{-14} \times \frac{(5.6 \times 10^{-10})^2}{(1.0 \times 10^{-14})^2} = 1 \times 10^{-4}$$

*31.62 1, 2, 3, 9, 10, 11, 12, 13, 14, 16, 18, 19, 25, 26, 36, 37, 38, 42, 43, 44, 45, 46, 47, 48, 49, 54; 26 in all

*31.63 For CoS: $[S^{2-}] = \frac{1 \times 10^{-20}}{1 \times 10^{-2}} = 1 \times 10^{-18}\ M$

$$[H^+]^2 = 1 \times 10^{-20} \times \frac{1 \times 10^{-1}}{1 \times 10^{-18}} = 1 \times 10^{-3}$$
$$[H^+] = (1 \times 10^{-3})^{\frac{1}{2}} = 3 \times 10^{-2}\ M;\ pH = 1.5$$

For $Al(OH)_3$: $[OH^-]^3 = \frac{5 \times 10^{-33}}{1 \times 10^{-2}} = 5 \times 10^{-31}$

$$[OH^-] = 8 \times 10^{-11}\ M;\ [H^+] = \frac{1 \times 10^{-14}}{8 \times 10^{-11}}$$
$$= 1 \times 10^{-4}\ M;\ pH = 4.0$$

CoS precipitates first; black

*31.64 $5.00 \times 10^{-3}\ g\ CoS \times \frac{1\ mol\ CoS}{91.0\ g\ CoS} \times \frac{4\ mol\ SCN^-}{1\ mol\ CoS}$

$= 2.20 \times 10^{-4}$ mol SCN^- required; use 3.30×10^{-4} mol

Volume $= \frac{3.30 \times 10^{-4}\ mol}{0.200\ mol/L} = 1.65 \times 10^{-3}\ L = 1.65\ mL$

*31.65 formula = $NiC_8H_{14}N_4O_4$; molar mass = 288.91 g/mol

$$\text{no. atoms} = 1.00 \times 10^{-3}\ \text{g} \times \frac{1\ \text{mol}}{288.91\ \text{g}} \times \frac{6.022 \times 10^{23}\ \text{atoms}}{1\ \text{mol}}$$
$$= 2.08 \times 10^{18}$$

Test Questions, Multiple Choice

1. Which of the Group III cations form stable complexes with both NH_3 and OH^-?

   a. Co^{2+} and Ni^{2+} b. Al^{3+} and Mn^{2+} c. Al^{3+} and Ni^{2+}
   d. only Zn^{2+} e. none

2. How many of the following hydroxides are soluble in NH_3?

   $Fe(OH)_3$, $Ni(OH)_2$, $Al(OH)_3$, $Mn(OH)_2$, $Zn(OH)_2$

   a. 1 b. 2 c. 3 d. 4 e. 5

3. How many of the following reagents will dissolve $Ni(OH)_2$?

   water, HCl, NaOH, NH_3

   a. 0 b. 1 c. 2 d. 3 e. 4

4. Which two Group III aations form colored complexes with the SCN^- ion in the qualitative analysis scheme?

   a. Ni^{2+} and Co^{2+} b. Al^{3+} and Zn^{2+} c. Fe^{3+} and Al^{3+}

   d. Co^{2+} and Fe^{3+} e. Mn^{2+} and Zn^{2+}

5. How many of the seven metals in Group III (Fe, Co, Ni, Al, Cr, Zn, Mn) form stable +1 cations?

   a. 0 b. 1 c. 2 d. 3 e. 4

6. In the qualitative analysis scheme, which two Group III cations originally precipitate as hydroxides rather than sulfides?

   a. Co^{2+} and Ni^{2+} b. Fe^{2+} and Mn^{2+} c. Al^{3+} and Zn^{2+}

   d. Zn^{2+} and Cr^{3+} e. Al^{3+} and Cr^{3+}

7. Which two Group III sulfides fail to dissolve in HCl?

   a. CoS and NiS b. NiS and FeS c. CoS and ZnS

   d. MnS and FeS e. MnS and ZnS

8. Which one of the following compounds would be referred to as an "alum"?

a. $Al_2(SO_4)_3$ b. $NaAlF_3$ c. Al_2O_3 d. $NaAl(SO_4)_2 \cdot 12H_2O$

e. all of the above

9. Which two of the Group III metals readily form oxyanions?

a. Co and Ni b. Fe and Zn c. Al and Cr d. Mn and Cr

e. Mn and Ni

10. When MnO_2 is treated with hydrogen peroxide in the qualitative analysis of Group III, it is converted to

a. $Mn(OH)_2$ b. MnO_5 c. Mn^{2+} d. MnO_4^- e. Mn^{3+}

<u>Test Problems</u>

1. Write net ionic equations for the following reactions, all of which take place during the qualitative analysis of a Group III known.

a. Fe^{3+} with S^{2-}

b. Fe^{3+} with NH_3

c. Al^{3+} with sodium hydroxide

d. Cr^{3+} with hydrogen sulfide at pH 9

2. Consider the half-reaction for the reduction of MnO_2:

$$MnO_2(s) + 4\ H^+(aq) + 2\ e^- \longrightarrow Mn^{2+}(aq) + 2\ H_2O$$

Taking E^o_{red} to be +1.23 V, calculate the reduction voltage at pH 5.3 if conc. Mn^{2+} = 0.10 M.

3. Taking K_{sp} $Al(OH)_3 = 5 \times 10^{-33}$, K_f $Al(OH)_4^- = 1 \times 10^{33}$, calculate:

a. K for the dissolving of $Al(OH)_3$ in sodium hydroxide

b. the solubility of aluminum hydroxide in 0.20 M NaOH.

4. Explain in your own words how the following reagents are used in Group III analysis.

a. NH_3 b. SCN^- c. OCl^- d. H_2SO_4

5. A certain Group III unknown is treated with H_2S at pH 9. It forms a black precipitate completely insoluble in HCl. On that basis, what can you say about the identity of the unknown? How would you proceed to identify the cations present in the unknown?

CHAPTER 32

Answers to Problems

32.1 a. $Ba(s) + \frac{1}{2} O_2(g) \longrightarrow BaO(s)$

$Ba(s) + O_2(g) \longrightarrow BaO_2(s)$

b. $Ca(s) + \frac{1}{2} O_2(g) \longrightarrow CaO(s)$

c. $Mg(s) + \frac{1}{2} O_2(g) \longrightarrow MgO(s)$

32.2 a. $3\ BaO(s) + 2\ Al(s) \longrightarrow 3\ Ba(s) + Al_2O_3(s)$

b. $2\ NaCl(l) \longrightarrow 2\ Na(l) + Cl_2(g)$

c. $KCl(l) + Na(l) \longrightarrow K(g) + NaCl(l)$

32.3 a. $Ba(s) + 2\ H_2O(l) \longrightarrow H_2(g) + Ba^{2+}(aq) + 2\ OH^-(aq)$

b. $Na(s) + H_2O(l) \longrightarrow \frac{1}{2} H_2(g) + Na^+(aq) + OH^-(aq)$

c. $Mg(s) + H_2O(g) \longrightarrow H_2(g) + MgO(s)$

32.4 a. $Mg^{2+}(aq) + CO_3^{2-}(aq) \longrightarrow MgCO_3(s)$

b. $Ba^{2+}(aq) + CO_3^{2-}(aq) \longrightarrow BaCO_3(s)$

32.5 a. $Mg^{2+}(aq) + 2\ OH^-(aq) \longrightarrow Mg(OH)_2(s)$

d. $NH_4^+(aq) + OH^-(aq) \longrightarrow NH_3(aq) + H_2O$

32.6 a. $Na(s) + H^+(aq) \longrightarrow Na^+(aq) + \frac{1}{2} H_2(g)$

b. $2\ BaCrO_4(s) + 2\ H^+(aq) \longrightarrow 2\ Ba^{2+}(aq) + Cr_2O_7^{2-}(aq) + H_2O$

c. $MgO(s) + 2\ H^+(aq) \longrightarrow Mg^{2+}(aq) + H_2O$

d. $MgC_2O_4(s) + 2\ H^+(aq) \longrightarrow Mg^{2+}(aq) + H_2C_2O_4(aq)$

32.7 a. $BaCO_3(s) + 2\ H^+(aq) \longrightarrow Ba^{2+}(aq) + H_2CO_3(aq)$

b. $Mg^{2+}(aq) + NH_3(aq) + HPO_4^{2-}(aq) \longrightarrow MgNH_4PO_4(s)$

32.8 $\dfrac{[H^+]^2}{0.40} = 5.6 \times 10^{-10}$; $[H^+]^2 = 2.2 \times 10^{-10}$

$[H^+] = 1.5 \times 10^{-5}$ M; pH = 4.82

32.9 $[H^+] = 5.6 \times 10^{-10} \times \dfrac{[NH_4^+]}{[NH_3]} = 5.6 \times 10^{-11}$; pH = 10.25

32.10 a. $K = \dfrac{K_a\ HPO_4^{2-}}{K_a\ NH_4^+} = \dfrac{1.7 \times 10^{-12}}{5.6 \times 10^{-10}} = 3.0 \times 10^{-3}$

32.10 b. $\frac{x^2}{(0.10 - x)^2} = 3.0 \times 10^{-3}$; $\frac{x}{0.10 - x} = 0.055$

$x = \frac{(0.10)(0.055)}{1.055} = 0.0052$; $[NH_4^+] = [PO_4^{3-}] = 0.0052$ M

32.11 $MgNH_4PO_4(s) \rightleftharpoons Mg^{2+}(aq) + NH_4^+(aq) + PO_4^{3-}(aq)$

$s^3 = 2.5 \times 10^{-13}$; $s = 6.3 \times 10^{-5}$ mol/L $= 8.6 \times 10^{-3}$ mg/mL

32.12 a. $K = \frac{K_{sp}\ Mg(OH)_2}{K_{sp}\ MgCO_3} = \frac{1 \times 10^{-11}}{2 \times 10^{-8}} = 5 \times 10^{-4}$

b. $[OH^-]^2 = 0.1 \times 5 \times 10^{-4} = 5 \times 10^{-5}$; $[OH^-] = 7 \times 10^{-3}$ M

32.13 $[H^+] = 1 \times 10^{-10}$ M; $[OH^-] = 1 \times 10^{-4}$ M

$[Mg^{2+}] = \frac{K_{sp}\ Mg(OH)_2}{(1 \times 10^{-4})^2} = \frac{1 \times 10^{-11}}{1 \times 10^{-8}} = 1 \times 10^{-3}$ M

$f = \frac{1 \times 10^{-3}}{1 \times 10^{-2}} = 0.1$

32.14 a. $K = 1/K_a\ NH_4^+ = 1/(5.6 \times 10^{-10}) = 1.8 \times 10^9$

b. $K = 1/K_a\ H_2CO_3 = 1/(4.2 \times 10^{-7}) = 2.4 \times 10^6$

c. $K = \frac{K_a\ HCO_3^-}{K_a\ H_2CO_3} = \frac{4.8 \times 10^{-11}}{4.2 \times 10^{-7}} = 1.1 \times 10^{-4}$

32.15 a. $Ba^{2+}(aq) + CO_3^{2-}(aq) \longrightarrow BaCO_3(s)$

$Ca^{2+}(aq) + CO_3^{2-}(aq) \longrightarrow CaCO_3(s)$

$Mg^{2+}(aq) + CO_3^{2-}(aq) \longrightarrow MgCO_3(s)$

b. $Mg^{2+}(aq) + NH_3(aq) + HPO_4^{2-}(aq) \longrightarrow MgNH_4PO_4(s)$

c. $Ba^{2+}(aq) + CrO_4^{2-}(aq) \longrightarrow BaCrO_4(s)$

$Ba^{2+}(aq) + SO_4^{2-}(aq) \longrightarrow BaSO_4(s)$

32.16 a. precipitation as $MgNH_4PO_4$ b. yellow flame test

c. conversion to NH_3 with OH^-

32.17 a. precipitation of Ba^{2+} b. precipitation of Mg^{2+}

c. precipitation of Ba^{2+}, Ca^{2+}, Mg^{2+}

32.18 a. Reagents used in other tests may contain NH_4^+, Na^+

b. Make solution sufficiently basic to precipitate CaC_2O_4

c. Adjust pH so that $BaCrO_4$ will precipitate

32.19 Ba^{2+}, Ca^{2+}, Mg^{2+} absent; other ions in doubt. (NH_3 comes from $(NH_4)_2CO_3$ used)

32.20 Cr^{3+} present; reddish-violet color, colored ppt. with NaOH

NH_4^+ present; NH_3 given off with NaOH

Pb^{2+} present; $PbSO_4$ precipitates

Sb^{3+} in doubt

32.21 NH_4^+ present; odor of NH_3 with NaOH

Ba^{2+} absent; would give $BaCrO_4$ precipitate

Ca^{2+} absent; would give CaC_2O_4 precipitate

Mg^{2+} present; white precipitate must be $MgCO_3$

Na^+, K^+ in doubt

32.22 Bi^{3+} present; sulfide insoluble in NaOH

Sn^{2+} absent; would give sulfide soluble in NaOH

Cr^{3+} absent; would give precipitate with NH_3

Ca^{2+} present; gives precipitate with $(NH_4)_2CO_3$

NH_4^+ in doubt

32.23 a. Add H_2SO_4; Ba^{2+} precipitates

b. Add NH_3; Mg^{2+} precipitates

c. $MgSO_4$ soluble in water d. flame test

32.24 a. heat with Na_2CO_3 solution

b. treat with HCl c. treat with HNO_3, evaporate

d. add $BaCl_2$, filter off $BaCrO_4$, evaporate

32.25 a. H_2O; $BaS(s) \longrightarrow Ba^{2+}(aq) + S^{2-}(aq)$

b. H^+; $Cu(OH)_2(s) + 2\ H^+(aq) \longrightarrow Cu^{2+}(aq) + 2\ H_2O$

NH_3; $Cu(OH)_2(s) + 4\ NH_3(aq) \longrightarrow Cu(NH_3)_4^{2+}(aq) + 2\ OH^-(aq)$

3225 c. HNO_3; $CdS(s) + 4H^+(aq) + 2NO_3^-(aq) \longrightarrow Cd^{2+}(aq) + S(s) + 2NO_2(g) + 2H_2O$

d. OH^-: $PbCl_2(s) + 3\ OH^-(aq) \longrightarrow Pb(OH)_3^-(aq) + 2\ Cl^-(aq)$

32.26 a. $Ag^+(aq) + 2\ NH_3(aq) \longrightarrow Ag(NH_3)_2^+(aq)$

b. $Bi^{3+}(aq) + 3\ NH_3(aq) + 3H_2O \longrightarrow Bi(OH)_3(s) + 3NH_4^+(aq)$

c. $Zn^{2+}(aq) + 4\ NH_3(aq) \longrightarrow Zn(NH_3)_4^{2+}(aq)$

d. $Mn^{2+}(aq) + 2NH_3(aq) + 2H_2O \longrightarrow Mn(OH)_2(s) + 2NH_4^+(aq)$

e. $Mg^{2+}(aq) + 2\ NH_3(aq) + 2H_2O \longrightarrow Mg(OH)_2(s) + 2NH_4^+(aq)$

32.27 a. $Cu(OH)_2(s) + 2\ H^+(aq) \longrightarrow Cu^{2+}(aq) + 2\ H_2O$

b. $Zn(NH_3)_4^{2+}(aq) + 4\ H^+(aq) \longrightarrow Zn^{2+}(aq) + 4\ NH_4^+(aq)$

d. $ZnS(s) + 2\ H^+(aq) \longrightarrow Zn^{2+}(aq) + H_2S(aq)$

e. $NH_3(aq) + H^+(aq) \longrightarrow NH_4^+(aq)$

32.28 a. yellow b. yellow c. red d-f white g. red

32.29 Stir with water; brown residue indicates Ag_2O. Test a portion of the water solution with Ag^+; white ppt. indicates NaCl. Add excess NaOH to another portion of the solution; white ppt. indicates $CdSO_4$. Carefully neutralize NaOH solution with acetic acid; white ppt. of $Zn(OH)_2$ indicates $Zn(NO_3)_2$.

32.30 a. Add HCl to precipitate Hg_2Cl_2; add H_2S to precipitate CdS.

b. Add NH_3 to precipitate $Al(OH)_3$; deep blue color indicates Cu^{2+}.

c. Add NaOH to precipitate $Co(OH)_2$; heat NaOH solution to drive off NH_3.

d. Add HCl to precipitate AgCl; add excess NaOH to acidic solution to precipitate $Mg(OH)_2$.

e. Add NH_3 to precipitate $Fe(OH)_3$; add H_2SO_4 to precipitate $BaSO_4$.

f. Add excess NaOH to precipitate $Bi(OH)_3$; acidify NaOH solution and add H_2S to precipitate Sb_2S_3.

g. Add H_2SO_4 to precipitate $PbSO_4$; add H_2S, H_2O_2 to solution to precipitate SnS_2.

32.31 a. $2\ Na(s) + O_2(g) \longrightarrow Na_2O_2(s)$

$2\ Na(s) + \frac{1}{2}\ O_2(g) \longrightarrow Na_2O(s)$

b. $K(s) + O_2(g) \longrightarrow KO_2(s)$

32.32 a. $2\ NaCl(l) \longrightarrow 2\ Na(l) + Cl_2(g)$

b. $MgCl_2(l) \longrightarrow Mg(l) + Cl_2(g)$

c. $2\ LiCl(l) \longrightarrow 2\ Li(l) + Cl_2(g)$

32.33 a. $K(s) + H_2O(l) \longrightarrow \frac{1}{2}\ H_2(g) + K^+(aq) + OH^-(aq)$

b. $Ca(s) + 2\ H_2O(l) \longrightarrow H_2(g) + Ca^{2+}(aq) + 2\ OH^-(aq)$

32.34 b. $Ba^{2+}(aq) + CrO_4^{\ 2-}(aq) \longrightarrow BaCrO_4(s)$

32.35 a. $Mg^{2+}(aq) + 2\ NH_3(aq) + 2\ H_2O \longrightarrow Mg(OH)_2(s) + 2NH_4^{\ +}(aq)$

d. $NH_3(aq) + HPO_4^{\ 2-}(aq) \longrightarrow NH_4^{\ +}(aq) + PO_4^{\ 3-}(aq)$

32.36 a. $Mg(s) + 2\ H^+(aq) \longrightarrow Mg^{2+}(aq) + H_2(g)$

b. $Mg(OH)_2(s) + 2\ H^+(aq) \longrightarrow Mg^{2+}(aq) + 2\ H_2O$

c. $NH_3(aq) + H^+(aq) \longrightarrow NH_4^{\ +}(aq)$

d. $CaC_2O_4(s) + 2\ H^+(aq) \longrightarrow Ca^{2+}(aq) + H_2C_2O_4(aq)$

32.37 a. $BaC_2O_4(s) + 2\ H^+(aq) \longrightarrow Ba^{2+}(aq) + H_2C_2O_4(aq)$

b. $NH_4^{\ +}(aq) + OH^-(aq) \longrightarrow NH_3(aq) + H_2O$

c. $S^{2-}(aq) + H_2O \longrightarrow OH^-(aq) + HS^-(aq)$

32.38 $\dfrac{[H^+]^2}{0.15} = 5.6 \times 10^{-10}$; $[H^+]^2 = 8.4 \times 10^{-11}$

$[H^+] = 9.2 \times 10^{-6}$ M; pH = 5.04

32.39 $[H^+] = 5.6 \times 10^{-10} \times \dfrac{[NH_4^{\ +}]}{[NH_3]} = 5.6 \times 10^{-9}$ M

32.40 a. $K = \dfrac{K_a\ H_2PO_4^{\ -}}{K_a\ NH_4^{\ +}} = \dfrac{6.2 \times 10^{-8}}{5.6 \times 10^{-10}} = 1.1 \times 10^2$

b. $\dfrac{x^2}{(0.10 - x)^2} = 1.1 \times 10^2$; $\dfrac{x}{0.10 - x} = 10$; $x = 0.091$

$[NH_4^{\ +}] = [H_2PO_4^{\ -}] = 0.091$ M

32.41 conc. $Mg^{2+} = \frac{0.500 \times 0.0010\ M}{1.000} = 0.00050\ M$

conc. $PO_4^{3-} = 0.00050\ M$; conc. $NH_4^+ = 0.0015\ M$

conc. product $= (5.0 \times 10^{-4})(5.0 \times 10^{-4})(1.5 \times 10^{-3})$

$= 3.8 \times 10^{-10} > K_{sp}$; precipitate forms

32.42 a. $K = \frac{K_{sp}\ CaCO_3}{K_{sp}\ CaC_2O_4} = \frac{5 \times 10^{-9}}{1.3 \times 10^{-9}} = 4$

b. $[CO_3^{2-}] = 4 \times 0.5\ M = 2\ M$

32.43 $[OH^-]^2 = \frac{1 \times 10^{-11}}{2 \times 10^{-2}} = 5 \times 10^{-10}$; $[OH^-] \approx 2 \times 10^{-5}\ M$

$[H^+] = (1 \times 10^{-14})/(2 \times 10^{-5}) = 5 \times 10^{-10}\ M$; pH = 9.3

32.44 a. $K = K_a\ NH_4^+ = 5.6 \times 10^{-10}$

b. $K = 1/K_a\ HCO_3^- = 1/(4.8 \times 10^{-11}) = 2.1 \times 10^{10}$

c. $K = \frac{K_a\ NH_4^+}{K_a\ HCO_3^-} = \frac{5.6 \times 10^{-10}}{4.8 \times 10^{-11}} = 12$

32.45 a. $2\ BaCrO_4(s) + 2\ H^+(aq) \rightarrow 2\ Ba^{2+}(aq) + Cr_2O_7^{2-}(aq) + H_2O$

b. $CaC_2O_4(s) + 2\ H^+(aq) \rightarrow Ca^{2+}(aq) + H_2C_2O_4(aq)$

c. $NH_4^+(aq) + OH^-(aq) \rightarrow NH_3(aq) + H_2O$

32.46 a. flame test b. precipitated as CaC_2O_4

c. precipitated as $BaCrO_4$, $BaSO_4$

32.47 a. test for NH_4^+

b. dissolving $BaCrO_4$, CaC_2O_4, $MgNH_4PO_4$

32.48 a. forms buffer to control pH

b. $BaCrO_4$ would not precipitate from strong acid

c. masks yellow color of Na^+

32.49 Ba^{2+} absent; would give yellow ppt. Other ions in doubt.

32.50 Ni^{2+} absent; would give colored solution

Hg^{2+} absent; would give precipitate with NH_3

Ba^{2+} absent; would give precipitate with H_2SO_4

Ag^+ present; gives brown ppt. of Ag_2O with NaOH

32.51 Ba^{2+} present; yellow ppt. is $BaCrO_4$
Ca^{2+} absent; would give ppt. with $K_2C_2O_4$
Na^+ absent; gives strong yellow flame test
K^+ probably absent; violet flame test
Mg^{2+}, NH_4^+ in doubt

32.52 Cu^{2+} present; black ppt. is CuS
Cd^{2+} in doubt
Fe^{3+} absent; would give red ppt. with NH_3
Al^{3+} present; white ppt. is $Al(OH)_3$
Ba^{2+} present; gives precipitate with H_2SO_4
Ca^{2+} in doubt

32.53 a. Add H_2SO_4; Ba^{2+} gives precipitate

b. Add OH^-; Mg^{2+} gives precipitate

c. Add CrO_4^{2-}; $BaCl_2$ gives precipitate

d. Test pH; NH_3 is basic, NH_4^+ acidic

32.54 a. Add H_2SO_4 to precipitate $BaSO_4$

b. Heat with base c. Heat to form CaO; dissolve in water

d. Treat with strong acid

32.55 a. HCl, H_2SO_4, - - b. HCl, HNO_3, - - c. HCl, HNO_3, - -

d. NaOH, HNO_3, - - e. water, HCl, - - f. HCl, NaOH, NH_3

32.56 a. $Pb(OH)_3^-$ b. $Sn(OH)_6^{2-}$ c. $Cd(OH)_2$ d. $Zn(OH)_4^{2-}$

e. NH_3

32.57 a. $Ag(NH_3)_2^+(aq) + 2\ H^+(aq) \longrightarrow Ag^+(aq) + 2\ NH_4^+(aq)$

b. $Pb(OH)_3^-(aq) + 3\ H^+(aq) \longrightarrow Pb^{2+}(aq) + 3\ H_2O$

c. $BaCO_3(s) + 2\ H^+(aq) \longrightarrow Ba^{2+}(aq) + H_2CO_3(aq)$

32.58 a. deep blue b. pink c. white d. red-orange

e. blue f. black g. yellow

32.59 Step 1: extract with water, test for Ba^{2+}

Step 2: extract residue from (1) with hot water, test for Pb^{2+}

Step 3: extract residue from (2) with NaOH, test for Al^{3+}

Step 4: extract residue from (3) with a weak acid such as $HC_2H_3O_2$. If there is a red residue, $Fe(OH)_3$ is present. Test solution for Mg^{2+}.

32.60 a. Add HCl to precipitate $PbCl_2$; add H_2S to HCl solution to precipitate CdS.

b. Add NH_3 to precipitate $Bi(OH)_3$; to NH_3 solution, add $C_2O_4^{2-}$ to precipitate CaC_2O_4.

c. Add excess OH^- to precipitate Mg^{2+}; neutralize basic solution to precipitate $Zn(OH)_2$.

d. Add H_2SO_4 to precipitate $BaSO_4$; make H_2SO_4 solution basic to precipitate $Mn(OH)_2$.

e. Add OH^-, OCl^-; green ppt. indicates Ni^{2+}, yellow solution indicates Cr^{3+}.

f. Add OH^- to precipitate HgO; use flame test for K^+

g. Add $C_2O_4^{2-}$ to test for Ca^{2+}; use flame test for Na^+

h. Add excess OH^- to precipitate $Fe(OH)_3$; test basic solution for Al^{3+}.

*32.61 $CoCl_2$ absent; would give colored solution

$AgNO_3$ present; white ppt. with HNO_3 is AgCl

$AlCl_3$ present; source of Cl^- ions

$Mg(NO_3)_2$ present; white ppt. with NaOH is $Mg(OH)_2$

*32.62 $[OH^-]^2 = (1 \times 10^{-11})/(1 \times 10^{-2}) = 1 \times 10^{-9}$

$[OH^-] = 3 \times 10^{-5}$ M; $[H^+] = 3 \times 10^{-10}$ M

$$\frac{[NH_4^+]}{[NH_3]} = \frac{[H^+]}{K_a\ NH_4^+} = \frac{3 \times 10^{-10}}{5.6 \times 10^{-10}} = 0.5$$

*32.63 $Mg(OH)_2(s) \rightleftharpoons Mg^{2+}(aq) + 2\ OH^-(aq)$ K_1

$2NH_4^+(aq) + 2\ OH^-(aq) \rightleftharpoons 2NH_3(aq) + 2H_2O$ K_2

$Mg(OH)_2(s) + 2NH_4^+(aq) \rightleftharpoons Mg^{2+}(aq) + 2NH_3(aq) + 2H_2O$; K

$$K = K_1 \times K_2 = \frac{K_{sp}\ Mg(OH)_2}{(K_b\ NH_3)^2} = \frac{1 \times 10^{-11}}{(1.8 \times 10^{-5})^2} = 0.03$$

*32.64 $CaCO_3(s) \rightleftharpoons Ca^{2+}(aq) + CO_3^{2-}(aq)$ K_1

$CO_3^{2-}(aq) + H^+(aq) \rightleftharpoons HCO_3^-(aq)$ K_2

$H_2CO_3(aq) \rightleftharpoons H^+(aq) + HCO_3^-(aq)$ K_3

$CaCO_3(s) + H_2CO_3(aq) \rightleftharpoons Ca^{2+}(aq) + 2HCO_3^-(aq)$ K

$$K = K_1 \times K_2 \times K_3 = \frac{K_{sp}\ CaCO_3}{K_a\ HCO_3^-} \times K_a\ H_2CO_3$$

$$= \frac{(5 \times 10^{-9})(4.2 \times 10^{-7})}{4.8 \times 10^{-11}} = 4 \times 10^{-5}$$

*32.65 no. moles $NH_3 = (1.0 \times 10^{-3}\ L)(1.0 \times 10^{-1}\ mol/L)$

$= 1.0 \times 10^{-4}$ mol

$P\ NH_3 = 721$ mm Hg

$$V = \frac{(1.0 \times 10^{-4}\ mol)(0.0821\ L{\cdot}atm/mol{\cdot}K)(298\ K)}{(721/760\ atm)}$$

$= 0.0026\ L = 2.6\ mL$

Test Questions, Multiple Choice

1. Of the cations in Group IV (Ba^{2+}, Ca^{2+}, Mg^{2+}, Na^+, K^+, NH_4^+) how many form insoluble carbonates?

   a. 1 b. 2 c. 3 d. 4 e. 5

2. Which one of the following compounds is brightly colored?

   a. $BaSO_4$ b. $Ba(OH)_2$ c. $BaCrO_4$ d. CaC_2O_4 e. $CaCO_3$

3. Calcium occurs in nature as the

   a. fluoride b. carbonate c. sulfate d. all of these

4. In Group IV analysis, Ca^{2+} is confirmed by precipitation as the

   a. hydroxide b. oxalate c. chromate d. sulfate e. oxide

5. The central cation in chlorophyll is

a. Ca^{2+} b. Ba^{2+} c. Na^{+} d. K^{+} e. none of these

6. Which one of the following sodium compounds occurs in nature?

a. NaI b. Na_2S c. NaOH d. Na_2CO_3 e. Na_2O_2

7. Potassium, in Group IV analysis, is detected by its _____ flame.

a. yellow b. violet c. green d. red e. blue

8. Which one of the following ions gives off a gas when heated with base?

a. Na^{+} b. K^{+} c. NH_4^{+} d. Ca^{2+} e. Ba^{2+}

9. In Group IV analysis, Mg^{2+} is confirmed by precipitation as

a. $Mg(OH)_2$ b. MgC_2O_4 c. $Mg_3(PO_4)_2$ d. $MgNH_4PO_4$

e. $Mg(NH_3)_2^{2+}$

10. Addition of sodium sulfate solution to a Group IV unknown gives a precipitate. The unknown must contain:

a. Mg^{2+} b. Ca^{2+} c. Ba^{2+} d. Na^{+} e. K^{+}

Test Problems

1. Write balanced net ionic equations for the reaction of

a. Mg^{2+} with NH_3 and HPO_4^{2-} ions.

b. CaC_2O_4 with hydrochloric acid.

c. ammonium sulfate with potassium hydroxide.

d. Ba^{2+} with CrO_4^{2-} ions.

2. In Group IV analysis, what ion(s) is (are)

a. only partially precipitated as the carbonate?

b. identified by flame tests?

c. precipitated as the chromate?

d. tested for on the original solution?

3. A Group IV unknown gives a white precipitate with NH_3 and $(NH_4)_2CO_3$. When this precipitate is dissolved in acetic acid and treated with potassium chromate, the solution turns yellow, but no precipitate forms. The unknown gives a yellow flame test. What ions are present? absent? in doubt?

4. What volume of 0.10 M NH_4Cl solution should be added to 25 mL of 0.10 M NH_3 to give a buffer with a pH of 9.00? K_a NH_4^+ = 5.6×10^{-10}

5. Consider the reaction:

$$CaC_2O_4(s) + 2\ H^+(aq) \longrightarrow Ca^{2+}(aq) + H_2C_2O_4(aq)$$

Taking K_{sp} $CaC_2O_4 = 1.3 \times 10^{-9}$, K_a $H_2C_2O_4 = 4 \times 10^{-2}$, K_a $HC_2O_4^- = 5 \times 10^{-5}$, calculate:

a. K for this reaction

b. the solubility of CaC_2O_4 in 0.10 M HCl

CHAPTER 33

Answers to Problems

33.1 a. $2\ Sc(s) + 3\ Cl_2(g) \longrightarrow 2\ ScCl_3(s)$

b. $Sc_2O_3(s) + 6H^+(aq) + 6Cl^-(aq) \longrightarrow 2Sc^{3+}(aq) + 6Cl^-(aq) + 3H_2O$

c. $Sc(OH)_3(s) + 3H^+(aq) + 3Cl^-(aq) \longrightarrow Sc^{3+}(aq) + 3Cl^-(aq) + 3H_2O$

d. $Sc_2(CO_3)_3(s) + 6H^+(aq) + 6Cl^-(aq) \longrightarrow 2Sc^{3+}(aq) + 6Cl^-(aq) + 3CO_2(g) + 3H_2O$

33.2 a. $Cl_2(g) + 2Br^-(aq) \longrightarrow 2Cl^-(aq) + Br_2(l)$

b. $Ag^+(aq) + Br^-(aq) \longrightarrow AgBr(s)$

c. $Hg^{2+}(aq) + 4\ Br^-(aq) \longrightarrow HgBr_4{}^{2-}(aq)$

d. $Hg_2{}^{2+}(aq) + 2\ Br^-(aq) \longrightarrow Hg_2Br_2(s)$

33.3 a. $Hg_2{}^{2+}(aq) + 2\ SCN^-(aq) \longrightarrow Hg_2(SCN)_2(s)$

b. $Fe^{3+}(aq) + SCN^-(aq) \longrightarrow Fe(SCN)^{2+}(aq)$

c. $3H_2O_2(aq) + SCN^-(aq) \longrightarrow SO_4{}^{2-}(aq) + HCN(aq) + H^+(aq) + 2H_2O$

d. $Pd^{2+}(aq) + 4\ SCN^-(aq) \longrightarrow Pd(SCN)_4{}^{2-}(aq)$

33.4 a. $Fe_2O_3(s) + 6H^+(aq) \longrightarrow 2\ Fe^{3+}(aq) + 3H_2O$

b. $Hg^{2+}(aq) + 4\ I^-(aq) \longrightarrow HgI_4{}^{2-}(aq)$

c. $Ag^+(aq) + SCN^-(aq) \longrightarrow AgSCN(s)$

33.5 a. $HNCS(aq) \longrightarrow H^+(aq) + NCS^-(aq)$

b. $MnO_2(s) + 4H^+(aq) + 2Cl^-(aq) \longrightarrow Mn^{2+}(aq) + Cl_2(g) + 2H_2O$

c. $Ba^{2+}(aq) + SO_4{}^{2-}(aq) \longrightarrow BaSO_4(s)$

33.6 a. $H_2(g) + Cl_2(g) \longrightarrow 2\ HCl(g)$

b. $SCN^-(aq) + H_3PO_4(l) \longrightarrow HNCS(l) + H_2PO_4{}^-(aq)$

c. $Pb(SCN)_2(s) + Br_2 \longrightarrow PbBr_2(s) + (SCN)_2$

33.7 a. $K_2CO_3(s) + 2H^+(aq) + 2Cl^-(aq) \longrightarrow 2K^+(aq) + 2Cl^-(aq) + 2H_2O + 2CO_2(g)$

b. $NaCl(s) + H_2SO_4(l) \longrightarrow NaHSO_4(s) + HCl(g)$

33.7 c. $NaI(s) + H_3PO_4(l) \longrightarrow NaH_2PO_4(s) + HI(g)$

33.8 a. $AgCl(s) + 2\ NH_3(aq) \longrightarrow Ag(NH_3)_2^+(aq) + Cl^-(aq)$

b. $Ag(NH_3)_2^+(aq) + Cl^-(aq) + 2H^+(aq) \longrightarrow AgCl(s) + 2NH_4^+(aq)$

33.9 a.

$$5[SCN^-(aq) + 4H_2O \rightarrow SO_4^{2-}(aq) + HCN(aq) + 7H^+(aq) + 6e^-]$$
$$6[MnO_4^-(aq) + 8H^+(aq) + 5e^- \rightarrow Mn^{2+}(aq) + 4H_2O]$$
$$5SCN^-(aq) + 6MnO_4^-(aq) + 13H^+(aq) \rightarrow 5SO_4^{2-}(aq) + 5HCN(aq) + 6Mn^{2+}(aq) + 4H_2O$$

b. $\text{no. moles } SCN^- = (5.0 \times 10^{-3}\ L)(0.10\ mol/L) = 5.0 \times 10^{-4}\ mol$

$$\text{no. moles } MnO_4^- = 5.0 \times 10^{-4}\ mol\ SCN^- \times \frac{6\ mol\ MnO_4^-}{5\ mol\ SCN^-} = 6.0 \times 10^{-4}\ mol\ MnO_4^-$$

$$V = \frac{6.0 \times 10^{-4}\ mol}{1.0\ mol/L} = 6.0 \times 10^{-4}\ L = 0.60\ mL$$

33.10 a.

$$I_2(s) + 2\ e^- \rightarrow 2\ I^-(aq)$$
$$I_2(s) + 2\ H_2O \rightarrow 2\ IO^-(aq) + 4\ H^+(aq) + 2\ e^-$$
$$2I_2(s) + 2H_2O \rightarrow 2I^-(aq) + 2IO^-(aq) + 4H^+(aq)$$
$$I_2(s) + H_2O \rightarrow I^-(aq) + IO^-(aq) + 2\ H^+(aq)$$
$$I_2(s) + 2\ OH^-(aq) \rightarrow I^-(aq) + IO^-(aq) + H_2O$$

b.

$$2\ I^-(aq) \rightarrow I_2(s) + 2\ e^-$$
$$SO_4^{2-}(aq) + 4\ H^+(aq) + 2\ e^- \rightarrow SO_2(g) + 2\ H_2O$$
$$2\ I^-(aq) + SO_4^{2-}(aq) + 4\ H^+(aq) \rightarrow I_2(s) + SO_2(g) + 2H_2O$$

c.

$$SCN^-(aq) + 4H_2O \rightarrow SO_4^{2-}(aq) + HCN(aq) + 7H^+(aq) + 6e^-$$
$$3[MnO_2(s) + 4H^+(aq) + 2e^- \rightarrow Mn^{2+}(aq) + 2H_2O]$$
$$SCN^-(aq) + 3MnO_2(s) + 5H^+(aq) \rightarrow SO_4^{2-}(aq) + HCN(aq) + 3Mn^{2+}(aq) + 2H_2O$$

33.11 a. $E^o_{tot} = E^o_{red}Br_2 + E^o_{ox}Cl^- = 1.07\ V - 1.36\ V = -0.29\ V$; no

b. $E^o_{tot} = E^o_{red}Br_2 + E^o_{ox}SCN^- = 1.07\ V - 0.77\ V = 0.30\ V$; yes

c. $E^o_{tot} = E^o_{red}Br_2 + E^o_{ox}Fe^{2+} = 1.07\ V - 0.77\ V = 0.30\ V$; yes

33.12 a. $E = +0.70\text{ V} - \frac{0.0591}{4} \log_{10} \frac{1}{(\text{conc. } H^+)^4}$

$= +0.70\text{ V} + 0.0591 \log_{10}(\text{conc. } H^+) = 0.70\text{ V}$

b. $E = +0.70\text{ V} + 0.0591(-7.0) = +0.29\text{ V}$

c. $E = +0.70\text{ V} + 0.0591(-14.0) = -0.13\text{ V}$

33.13 a. $[I^-] = \frac{K_{sp}\text{ AgI}}{[Ag^+]} = \frac{1 \times 10^{-16}}{2 \times 10^{-5}} = 5 \times 10^{-12}\text{ M}$

b. $[I^-]^2 = \frac{K_{sp}\text{ }Hg_2I_2}{[Hg_2^{\ 2+}]} = \frac{5 \times 10^{-29}}{3 \times 10^{-6}} = 1.7 \times 10^{-23}$

$[I^-] = 4 \times 10^{-12}\text{ M}$

c. $[I^-]^2 = \frac{K_{sp}\text{ }PbI_2}{[Pb^{2+}]} = \frac{1 \times 10^{-8}}{1.2 \times 10^{-2}} = 8 \times 10^{-7}$; $[I^-] = 9 \times 10^{-4}\text{ M}$

33.14 $\frac{[PtBr_4^{\ 2-}]}{[Pt^{2+}]} = \frac{0.999}{0.001} = [Br^-]^4 \times 7 \times 10^{17}$

$[Br^-]^4 = \frac{1 \times 10^3}{7 \times 10^{17}} = 1.4 \times 10^{-15}$; $[Br^-] = 2 \times 10^{-4}\text{ M}$

33.15 a. $K = 1/K_f\text{ }Ag(SCN)_2^{\ -} = 1/(2 \times 10^8) = 5 \times 10^{-9}$

b. $K = \frac{K_{sp}\text{ AgSCN}}{K_{sp}\text{ AgBr}} = \frac{1 \times 10^{-12}}{5 \times 10^{-13}} = 2$

c. $K = K_{sp}\text{ AgSCN} \times K_f Ag(SCN)_2^{\ -} = (1 \times 10^{-12})(2 \times 10^8)$

$= 2 \times 10^{-4}$

33.16 a. $K = K_{sp}\text{ AgI} \times K_f\text{ }Ag(NH_3)_2^{\ +} = 2 \times 10^{-9}$

b. $s^2 = [NH_3]^2 \times 2 \times 10^{-9}$; $s = [NH_3] \times 4 \times 10^{-5}$

c. 4×10^{-5} M, 4×10^{-4} M

33.17 a. $\frac{[Ag^+]}{[Ag(SCN)_2^{\ -}]} = \frac{1}{K_f \times [SCN^-]^2} = \frac{1}{(2 \times 10^8)(1 \times 10^{-2})} = 5 \times 10^{-7}$

b. $\frac{1}{(2 \times 10^8)(1 \times 10^{-10})} = 50$

c. $\frac{1}{(2 \times 10^8)(1 \times 10^{-20})} = 5 \times 10^{11}$

33.18 $[Cl^-] = \frac{1.6 \times 10^{-10}}{2 \times 10^{-5}} = 8 \times 10^{-6}$ M; $f = \frac{8 \times 10^{-6}}{0.01} = 8 \times 10^{-4}$

33.19 a. $Zn(s) + 2AgCl(s) \rightarrow Zn^{2+}(aq) + 2Cl^-(aq) + 2Ag(s)$

$Zn(s) + 2AgBr(s) \rightarrow Zn^{2+}(aq) + 2Br^-(aq) + 2Ag(s)$

$Zn(s) + 2AgI(s) \rightarrow Zn^{2+}(aq) + 2I^-(aq) + 2Ag(s)$

$Zn(s) + 2AgSCN(s) \rightarrow Zn^{2+}(aq) + 2SCN^-(aq) + 2Ag(s)$

b. $2I^-(aq) + 2NO_2^-(aq) + 4H^+(aq) \rightarrow I_2(s) + 2NO(g) + 2H_2O$

c. $NH_2SO_3H(aq) + NO_2^-(aq) \rightarrow N_2(g) + SO_4^{2-}(aq) + H^+(aq) + H_2O$

33.20 a. converted to red complex with Fe^{3+}

b. oxidized with NO_2^-

c. oxidized with MnO_4^- to Br_2; gives orange color in TCE

33.21 a. MnO_4^- b. Fe^{3+} c. Zn, NH_3

33.22 White ppt. in Step 1, brought into solution in Step 2. Red color, followed by white ppt. in Step 3. Negative test in Step 4; orange color in Step 6. Negative test in Step 8.

33.23 Test one portion of solution with Fe^{3+}; red color indicates SCN^-. Test another portion with KNO_2, followed by extraction with TCE; purple color indicates I^-.

33.24 SCN^- present; red color due to $Fe(SCN)^{2+}$

Br^-, I^- absent; would give colored TCE layer on oxidation

Cl^- in doubt

33.25 SCN^- absent; would give red color with Fe^{3+}

I^- present; gives brown color with KNO_2

Br^- absent; would give orange color

Cl^- present; white ppt. with $AgNO_3$

33.26 Br^-, I^- absent; would give colored layer

Cl^- present; white ppt. is AgCl

SCN^- in doubt

33.27 a. purple (solid is black) b. white c. red d. colorless

e. white

33.28 a. Cl^- b. I^- c. SCN^- d. Cl^-

33.29 $1.00 \times 10^3 \text{ g NaBr} \times \frac{80.9 \text{ g HBr}}{102.9 \text{ g NaBr}} = 786 \text{ g HBr}$

33.30 N-C-S angle is 180°; C-S-S angle is 109°

33.31 a. $Zn(s) + I_2(s) \rightarrow ZnI_2(s)$

b. $ZnO(s) + 2H^+(aq) + 2I^-(aq) \rightarrow Zn^{2+}(aq) + 2I^-(aq) + H_2O$

c. $Zn(OH)_2(s) + 2H^+(aq) + 2I^-(aq) \rightarrow Zn^{2+}(aq) + 2I^-(aq) + 2H_2O$

d. $ZnCO_3(s) + 2H^+(aq) + 2I^-(aq) \rightarrow Zn^{2+}(aq) + 2I^-(aq) + H_2O + CO_2(g)$

33.32 a. $Cl_2(g) + 2I^-(aq) \rightarrow 2\,Cl^-(aq) + I_2(s)$

b. $Ag^+(aq) + I^-(aq) \rightarrow AgI(s)$

c. $Pd^{2+}(aq) + 4\,I^-(aq) \rightarrow PdI_4{}^{2-}(aq)$

d. $Pb^{2+}(aq) + 2\,I^-(aq) \rightarrow PbI_2(s)$

33.33 a. $Hg_2{}^{2+}(aq) + 2Cl^-(aq) \rightarrow Hg_2Cl_2(s)$

b. $MnO_2(s) + 4H^+(aq) + 2Cl^-(aq) \rightarrow Mn^{2+}(aq) + Cl_2(g) + 2H_2O$

c. $Sb^{3+}(aq) + Cl^-(aq) + H_2O \rightarrow SbOCl(s) + 2\,H^+(aq)$

d. $Ag^+(aq) + Cl^-(aq) \rightarrow AgCl(s)$

33.34 a. $ZnO(s) + 2\,H^+(aq) \rightarrow Zn^{2+}(aq) + H_2O$

b. $Zn(s) + 2AgCl(s) \rightarrow Zn^{2+}(aq) + 2Cl^-(aq) + 2Ag(s)$

c. $Br_2(l) + 2I^-(aq) \rightarrow 2Br^-(aq) + I_2(s)$

33.35 a. $HCl(aq) \rightarrow H^+(aq) + Cl^-(aq)$

b. $I_2(s) + I^-(aq) \rightarrow I_3{}^-(aq)$

c. $Fe^{3+}(aq) + SCN^-(aq) \rightarrow Fe(SCN)^{2+}(aq)$

33.36 a. $H_2(g) + Br_2(l) \rightarrow 2\,HBr(g)$

b. $2\,Fe(s) + 3\,Cl_2(g) \rightarrow 2\,FeCl_3(s)$

c. $KCN(l) + S(l) \rightarrow KSCN(l)$

33.37 a. $Co(s) + I_2(s) \longrightarrow CoI_2(s)$

b. $ZnCO_3(s) + 2H^+(aq) + 2Cl^-(aq) \longrightarrow Zn^{2+}(aq) + 2Cl^-(aq) + CO_2(g) + H_2O$

c. $NaBr(s) + H_3PO_4(l) \longrightarrow HBr(g) + NaH_2PO_4(s)$

33.38 a. $Ag^+(aq) + Br^-(aq) \longrightarrow AgBr(s)$

$AgBr(s) + 2\ CN^-(aq) \longrightarrow Ag(CN)_2^-(aq) + Br^-(aq)$

b. $H_2SO_4(l) + KCl(s) \longrightarrow HCl(g) + KHSO_4(s)$

33.39 a. $3\left[2\ Cl^-(aq) \longrightarrow Cl_2(g) + 2\ e^-\right]$

$Cr_2O_7^{2-}(aq) + 14\ H^+(aq) + 6\ e^- \longrightarrow 2\ Cr^{3+}(aq) + 7H_2O$

$Cr_2O_7^{2-}(aq) + 14H^+(aq) + 6Cl^-(aq) \longrightarrow 2Cr^{3+}(aq) + 3Cl_2(g) + 7\ H_2O$

b. no. moles $Cl^- = 5.0 \times 10^{-3}\ L \times 0.10\ \frac{mol}{L} = 5.0 \times 10^{-4}$ mol

no. moles $Cr_2O_7^{2-} = 5.0 \times 10^{-4}\ mol\ Cl^- \times \frac{1\ mol\ Cr_2O_7^{2-}}{6\ mol\ Cl^-}$

$= 8.3 \times 10^{-5}\ mol\ Cr_2O_7^{2-}$

$V = \frac{8.3 \times 10^{-5} mol}{0.50\ mol/L} = 1.7 \times 10^{-4}\ L = 0.17\ mL$

33.40 a. $Cl_2(g) + 2\ Br^-(aq) \longrightarrow 2\ Cl^-(aq) + Br_2(l)$

b. $MnO_2(s) + 4H^+(aq) + 2Br^-(aq) \longrightarrow Mn^{2+}(aq) + 2H_2O + Br_2(l)$

c. $SCN^-(aq) + 4H_2O \longrightarrow SO_4^{2-}(aq) + HCN(aq) + 7H^+(aq) + 6e^-$

$6\left[NO_3^-(aq) + 2\ H^+(aq) + e^- \longrightarrow NO_2(g) + H_2O\right]$

$SCN^-(aq) + 6NO_3^-(aq) + 5H^+(aq) \longrightarrow SO_4^{2-}(aq) + HCN(aq) + 6NO_2(g) + 2H_2O$

33.41 a. $E^o_{tot} = E^o_{red} H_2O_2 + E^o_{ox} Cl^- = 1.77\ V - 1.36\ V = 0.41\ V$; yes

b. $E^o_{tot} = E^o_{red} H_2O_2 + E^o_{ox} Br^- = 1.77\ V - 1.07\ V = 0.70\ V$; yes

c. $E^o_{tot} = E^o_{red} H_2O_2 + E^o_{ox} I^- = 1.77\ V - 0.53\ V = 1.24\ V$; yes

33.42 $E^o_{tot} = E^o_{ox} I^- + E^o_{red} O_2 = -0.53\ V + 0.40\ V = -0.13\ V$

a. $E = -0.13\ V - \frac{0.0591}{4} \log_{10}(\text{conc. } OH^-)^4$

$= -0.13\ V - 0.0591 \log_{10}(\text{conc. } OH^-) = -0.13\ V$

33.42 b. $E = -0.13\ V - 0.0591(-7.0) = +0.28\ V$

c. $E = -0.13\ V - 0.0591(-14.0) = +0.70\ V$

33.43 a. $[Br^-] = \frac{K_{sp}\ AgBr}{[Ag^+]} = \frac{5 \times 10^{-13}}{3 \times 10^{-4}} = 2 \times 10^{-9}\ M$

b. $[Br^-]^2 = \frac{K_{sp}\ Hg_2Br_2}{[Hg_2^{\ 2+}]} = \frac{6 \times 10^{-23}}{2 \times 10^{-7}} = 3 \times 10^{-16}$

$[Br^-] = 2 \times 10^{-8}\ M$

c. $[Br^-]^2 = \frac{K_{sp}\ PbBr_2}{[Pb^{2+}]} = \frac{5 \times 10^{-6}}{1 \times 10^{-5}} = 5 \times 10^{-1}$; $[Br^-] = 0.7\ M$

33.44 $\frac{[BiBr_4^{\ -}]}{[Bi^{3+}]} = \frac{0.999}{0.001} = [Br^-]^4(6 \times 10^7)$

$[Br^-]^4 = \frac{1 \times 10^3}{6 \times 10^7} = 1.7 \times 10^{-5}$; $[Br^-] = 0.06\ M$

33.45 a. $K = K_{sp}PbI_2 \times K_fPbI_4^{\ 2-} = (1 \times 10^{-8})(1.7 \times 10^4) = 2 \times 10^{-4}$

b. $K = 1/K_{sp}\ PbI_2 = 1 \times 10^8$

c. $K = \frac{K_f\ BiI_4^{\ -}}{K_f\ PbI_4^{\ 2-}} = \frac{9 \times 10^{14}}{1.7 \times 10^4} = 5 \times 10^{10}$

33.46 a. $K = K_{sp}AgSCN \times K_fAg(NH_3)_2^{\ +} = (1 \times 10^{-12})(2 \times 10^7)$

$= 2 \times 10^{-5}$

b. $s^2 = (2 \times 10^{-5}) \times [NH_3]^2$; $s = (4 \times 10^{-3}) \times [NH_3]$

c. 0.004 M; 0.04 M

33.47 a. $\frac{[Pd^{2+}]}{[Pd(SCN)_4^{\ 2-}]} = \frac{1}{(1 \times 10^{24}) \times [SCN^-]^4} = \frac{1}{1 \times 10^{20}} = 1 \times 10^{-20}$

b. $1/(1 \times 10^4) = 1 \times 10^{-4}$

c. $1/(1 \times 10^{-16}) = 1 \times 10^{16}$

33.48 $[SCN^-] = (1 \times 10^{-12})/(2 \times 10^{-5}) = 5 \times 10^{-8}$

$f = (5 \times 10^{-8})/(1 \times 10^{-2}) = 5 \times 10^{-6}$

33.49 a. $Fe^{3+}(aq) + SCN^-(aq) \longrightarrow Fe(SCN)^{2+}(aq)$

$Fe(SCN)^{2+}(aq) + Ba^{2+}(aq) + 3H_2O_2(aq) \longrightarrow BaSO_4(s) + HCN(aq) + Fe^{3+}(aq) + H^+(aq) + 2H_2O$

$H_2O_2(aq) + 2\ H^+(aq) + 2\ I^-(aq) \longrightarrow I_2(s) + 2\ H_2O$

b. $Ag^+(aq) + Cl^-(aq) \longrightarrow AgCl(s)$

$AgCl(s) + 2\ NH_3(aq) \longrightarrow Ag(NH_3)_2^+(aq) + Cl^-(aq)$

$Ag(NH_3)_2^+(aq) + Cl^-(aq) + 2\ H^+(aq) \longrightarrow AgCl(s) + 2NH_4^+(aq)$

33.50 a. Br^-, I^-, SCN^- b. none   c. Cl^- d. SCN^-, I^-, Br^-

33.51 a. Br^-, I^-, SCN^- b. I^- c. MnO_4^-

33.52 White ppt. formed in Step 1, dissolves in Step 2. Negative test for SCN^- in Step 3. Purple organic layer in Step 4. Negative test in Step 6. White ppt. in Step 8.

33.53 Treat with $KMnO_4$, extract with TCE; orange color indicates Br^-. To remaining water solution, add $AgNO_3$; white ppt. indicates Cl^-.

33.54 I^- present; purple color due to I_2

Br^- absent; would give orange color with MnO_4^-

SCN^-, Cl^- in doubt

33.55 I^- absent; would give color with KNO_2

SCN^- present; white ppt. is $BaSO_4$

Br^- in doubt; reaction with $KMnO_4$ could be due to SCN^-

Cl^- in doubt; white ppt. could be AgSCN

33.56 SCN^- present; red color is $Fe(SCN)^{2+}$

I^- present; brown color with KNO_2

Br^-, Cl^- in doubt

33.57 a. white   b. red   c. pale yellow   d. white   e. white

33.58 a. Br^-, I^-, Cl^-, SCN^- b. none   c. none

33.59 $1.00 \times 10^6\ g\ I_2 \times \dfrac{70.9\ g\ Cl_2}{253.8\ g\ I_2} = 2.79 \times 10^5\ g$

33.60 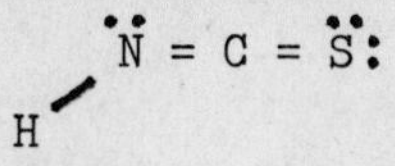

; bent molecule, 120°, 180° bond angles

*33.61 Step 1: extract with water, test solution with $AgNO_3$; white ppt. indicates KBr

Step 2: extract remaining solid with HCl, test solution with OH^-; white ppt. indicates $Bi(OH)_3$

Step 3: react remaining solid with Zn, $HC_2H_3O_2$; test solution with Fe^{3+}. Red color indicates AgSCN

Step 4: if residue remains from (3), it must be CoS (black)

*33.62 no. moles NaBr $= \dfrac{10.0\ g}{102.9\ g/mol} = 0.0972$ mol NaBr

no. moles $H_3PO_4 = 0.100\ L \times 6.00\ \dfrac{mol}{L} = 0.600$ mol H_3PO_4

NaBr is limiting. Theor. yield $= 10.0\ g\ NaBr \times \dfrac{80.9\ g\ HBr}{102.9\ g\ NaBr}$

$= 7.86$ g HBr

% yield $= \dfrac{6.21}{7.86} \times 100 = 79.0\%$

*33.63 $Ag(s) + H^+(aq) \longrightarrow Ag^+(aq) + \frac{1}{2} H_2(g)$

$$E = -0.80\ V - 0.0591 \log_{10}\frac{(\text{conc. } Ag^+)}{(\text{conc. } H^+)}$$

$$= -0.80\ V - 0.0591 \log_{10}\frac{(1.6 \times 10^{-9})}{(1.0 \times 10^{-1})}$$

$$= -0.80\ V - 0.0591(-7.80) = -0.34\ V$$

*33.64

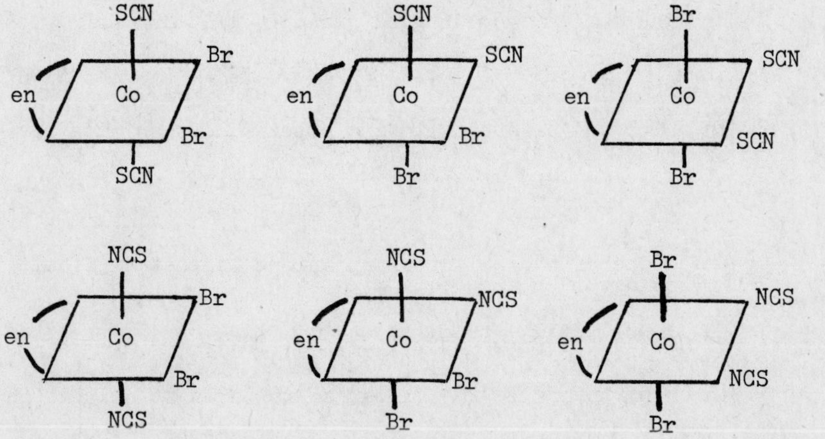

*33.64 (cont.)

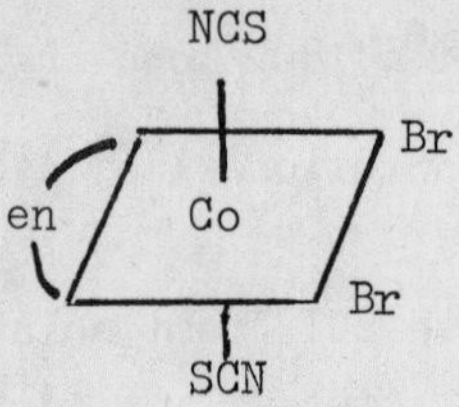

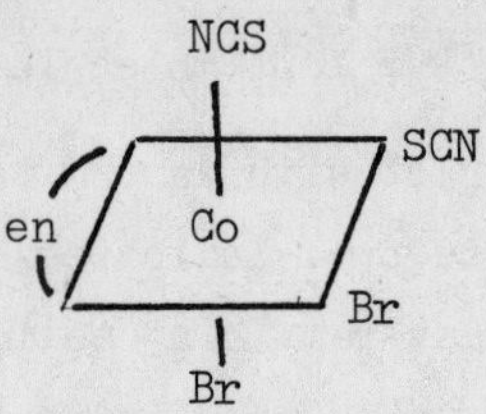

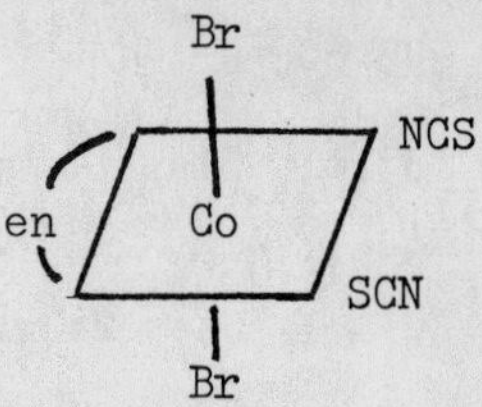

(10 in all)

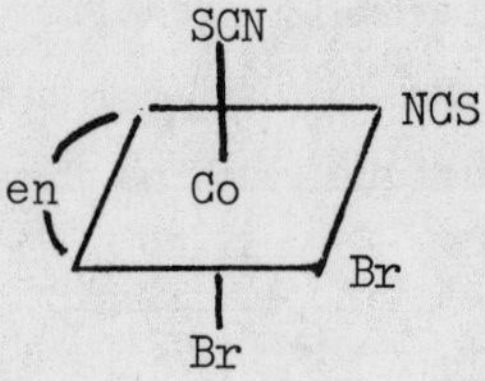

*33.65

	orig.	change	equil.
conc. Fe^{3+}	0.100	-x	0.100 - x
conc. SCN^-	0.100	-x	0.100 - x
conc. $FeSCN^{2+}$	0.000	x	x

$$\frac{x}{(0.100 - x)^2} = 900; \quad x = 0.090$$

$$[FeSCN^{2+}] = 0.090\ M$$

<u>Test Questions, Multiple Choice</u>

1. What is the approximate pH of a 0.10 M KCl solution?

   a. 1 b. 3 c. 5 d. 7 e. 9

2. Bromine can be made in good yield by treating NaBr with

   a. H_2 b. H_2O c. H_2SO_4 d. H_3PO_4 e. NaOH

3. In which one of the following species does an atom have an expanded octet?

   a. I_2 b. I_3^- c. I^- d. SCN^- e. Br^-

4. Which one of the Silver Group anions is most difficult to oxidize?

   a. Cl^- b. Br^- c. I^- d. SCN^-

5. Which of the following silver salts is least soluble in NH_3?

   a. AgCl b. AgBr c. AgI d. AgSCN

6. In the qualitative analysis of the Silver Group anions, what reagent is used to bring all the silver salts into solution?

a. HNO_3 b. NH_3 c. Zn d. HCl e. fe

7. How many of the following substances are brightly colored?

$NaCl$, Br_2, $NaBr$, I_2, NaI

a. 0 b. 1 c. 2 d. 3 e. 4

8. When the SCN^- ion is oxidized in qualitative analysis, the sulfur atom ends up as

a. S(s) b. H_2S c. SO_2 d. $S_2O_3^{2-}$ e. SO_4^{2-}

9. Which one of the following anions is oxidized by NO_2^- in qualitative analysis?

a. Cl^- b. Br^- c. I^- d. SCN^-

10. Which one of the following anions gives a blood-red color with $Fe(NO_3)_3$ solution?

a. Cl^- b. Br^- c. I^- d. SCN^-

Test Problems

1. Write a balanced net ionic equation to describe the reaction of

a. $Ag(NH_3)_2^+$ with nitric acid

b. Zn and silver thiocyanate

c. $Cl_2(g)$ and a solution of sodium iodide

d. solutions of $AgNO_3$ and HCl

2. Explain briefly how each of the following reagents is used in the analysis of the Silver Group anions.

a. Zn b. NO_2^- c. MnO_4^-

3. A Silver Group unknown gives a deep red color when treated with $Fe(NO_3)_3$. Treatment with Cl_2 water followed by extraction with trichloroethane does not give a color change. On this basis, state which ions are present, absent, and in doubt. Explain your reasoning.

4. An unknown containing Br^- ions can be oxidized by an acidic solution of potassium permanganate; products include Br_2 and Mn^{2+}.

   a. Write a balanced equation for the reaction.

   b. What volume of 0.10 M $KMnO_4$ is required to react with 5.0 mL of an unknown which is 0.020 M in Br^-?

5. K_f for the $Fe(SCN)^{2+}$ ion is 9.0×10^2. What concentration of SCN^- is required to convert 99.0% of a sample of Fe^{3+} to $Fe(SCN)^{2+}$?

CHAPTER 34

Answers to Problems

34.1 a. $CaCO_3(s) + 2H^+(aq) + SO_4^{2-}(aq) \rightarrow Ca^{2+}(aq) + SO_4^{2-}(aq) + CO_2(g) + H_2O$

b. $CaO(s) + 2H^+(aq) + SO_4^{2-}(aq) \rightarrow Ca^{2+}(aq) + SO_4^{2-}(aq) + H_2O$

c. $Ca(OH)_2(s) + 2H^+(aq) + SO_4^{2-}(aq) \rightarrow Ca^{2+}(aq) + SO_4^{2-}(aq) + 2\ H_2O$

d. $Ca(s) + 2H^+(aq) + SO_4^{2-}(aq) \rightarrow Ca^{2+}(aq) + SO_4^{2-}(aq) + H_2(g)$

34.2 a. $2BaCrO_4(s) + 2H^+(aq) \rightarrow 2Ba^{2+}(aq) + Cr_2O_7^{2-}(aq) + H_2O$

b. $BaC_2O_4(s) + 2H^+(aq) \rightarrow Ba^{2+}(aq) + H_2C_2O_4(aq)$

c. $FePO_4(s) + 3H^+(aq) \rightarrow Fe^{3+}(aq) + H_3PO_4(aq)$

34.3 a. $Ca_3(PO_4)_2(s) + 6HC_2H_3O_2(aq) \rightarrow 3Ca^{2+}(aq) + 6C_2H_3O_2^-(aq) + 2H_3PO_4(aq)$

b. $ZnC_2O_4(s) + 2HC_2H_3O_2(aq) \rightarrow Zn^{2+}(aq) + 2C_2H_3O_2^-(aq) + H_2C_2O_4(aq)$

c. $2Ag_2CrO_4(s) + 2HC_2H_3O_2(aq) \rightarrow 4Ag^+(aq) + 2C_2H_3O_2^-(aq) + Cr_2O_7^{2-}(aq) + H_2O$

34.4 a. $Pb^{2+}(aq) + SO_4^{2-}(aq) \rightarrow PbSO_4(s)$

b. $Ca^{2+}(aq) + C_2O_4^{2-}(aq) \rightarrow CaC_2O_4(s)$

c. $PO_4^{3-}(aq) + 12MoO_4^{2-}(aq) + 3NH_4^+(aq) + 24H^+(aq) \rightarrow (NH_4)_3PO_4\cdot 12MoO_3(s) + 12H_2O$

34.5 a. $H_2SO_4(aq) \rightarrow H^+(aq) + HSO_4^-(aq)$

b. $HC_2O_4^-(aq) \rightleftharpoons H^+(aq) + C_2O_4^{2-}(aq)$

c. $HPO_4^{2-}(aq) \rightleftharpoons H^+(aq) + PO_4^{3-}(aq)$

d. $Cu(C_2O_4)_2^{2-}(aq) \rightleftharpoons Cu^{2+}(aq) + 2C_2O_4^{2-}(aq)$

34.6 a. $CuCO_3(s) + 2H^+(aq) \rightarrow Cu^{2+}(aq) + H_2CO_3(aq)$

b. $NH_3(aq) + H^+(aq) \rightarrow NH_4^+(aq)$

c. $CaC_2O_4\cdot H_2O(s) + 2H^+(aq) \rightarrow Ca^{2+}(aq) + H_2C_2O_4(aq) + H_2O$

d. $H_2PO_4^-(aq) + OH^-(aq) \rightarrow HPO_4^{2-}(aq) + H_2O$

34.7 a. $Cu(s) + 4H^+(aq) + SO_4^{2-}(aq) \rightarrow Cu^{2+}(aq) + SO_2(g) + 2H_2O$

b. $Ca_3(PO_4)_2(s) + 4H_3PO_4(l) \rightarrow 3Ca(H_2PO_4)_2(s)$

c. $2Ca_3(PO_4)_2(s) + 6SiO_2(s) + 10C(s) \rightarrow P_4(g) + 10CO(g) + 6CaSiO_3(l)$

d. $FePO_4(s) + 3H^+(aq) \rightarrow Fe^{3+}(aq) + H_3PO_4(aq)$

34.8 $Cr_2O_7^{2-}(aq) + 6Fe^{2+}(aq) + 14H^+(aq) \rightarrow 2Cr^{3+}(aq) + 6Fe^{3+}(aq) + 7H_2O$

$$\text{no. moles } Cr_2O_7^{2-} = 1.00 \text{ g } FeSO_4 \times \frac{1 \text{ mol } FeSO_4}{151.9 \text{ g } FeSO_4} \times \frac{1 \text{ mol } Cr_2O_7^{2-}}{6 \text{ mol } Fe^{2+}}$$

$$= 0.00110 \text{ mol } Cr_2O_7^{2-}$$

$$V = \frac{0.00110 \text{ mol}}{0.100 \text{ mol/L}} = 0.0110 \text{ L} = 11.0 \text{ mL}$$

34.9 a. $Cr_2O_7^{2-}(aq) + 14H^+(aq) + 6e^- \rightarrow 2Cr^{3+}(aq) + 7H_2O$

$3[C_2O_4^{2-}(aq) \rightarrow 2CO_2(g) + 2e^-]$

$Cr_2O_7^{2-}(aq) + 14H^+(aq) + 3C_2O_4^{2-}(aq) \rightarrow 2Cr^{3+}(aq) + 7H_2O + 6CO_2(g)$

b. $2[CrO_4^{2-}(aq) + 4H_2O + 3e^- \rightarrow Cr(OH)_3(s) + 5\ OH^-(aq)]$

$3[S^{2-}(aq) \rightarrow S(s) + 2e^-]$

$2CrO_4^{2-}(aq) + 8H_2O + 3S^{2-}(aq) \rightarrow 2Cr(OH)_3(s) + 10\ OH^-(aq) + 3S(s)$

34.10 a. $E^o_{tot} = E^o_{red}\,CrO_4^{2-} + E^o_{ox}\,S^{2-} = -0.12 \text{ V} + 0.43 \text{ V} = 0.31 \text{ V}$; yes

b. $E^o_{tot} = E^o_{red}\,CrO_4^{2-} + E^o_{ox}\,Cl^- = -0.12 \text{ V} - 0.89 \text{ V} = -1.01 \text{ V}$; no

c. $E^o_{tot} = E^o_{red}\,CrO_4^{2-} + E^o_{ox}\,H_2 = -0.12 \text{ V} + 0.83 \text{ V} = 0.71 \text{ V}$; yes

d. $E^o_{tot} = E^o_{red}\,CrO_4^{2-} + E^o_{ox}\,OH^- = -0.12 \text{ V} - 0.40 \text{ V} = -0.52 \text{V}$; no

34.11 $$E_{red} = +0.20 \text{ V} - \frac{0.0591}{2} \log \frac{1}{(\text{conc. } H^+)^4}$$

$$= +0.20 \text{ V} + 0.1182 \log_{10}(\text{conc. } H^+)$$

a. $E_{red} = +0.20 \text{ V}$

b. $E_{red} = +0.20 \text{ V} - 0.1182(2.0) = -0.04 \text{ V}$

c. $E_{red} = +0.20 \text{ V} - 0.1182(4.0) = -0.27 \text{ V}$

34.11 d. $E_{red} = +0.20\ V - 0.1182(6.0) = -0.51\ V$

34.12 a. $s^2 = 1.3 \times 10^{-9}$; $s = 3.6 \times 10^{-5}$ M

b. conc. $Ca^{2+} = \dfrac{0.400 \times 0.010}{0.800} = 0.0050$ M

conc. $C_2O_4^{2-} = \dfrac{0.400 \times 0.0010}{0.800} = 0.00050$ M

conc. Ca^{2+} x conc. $C_2O_4^{2-} = (5.0 \times 10^{-3})(5.0 \times 10^{-4})$

$= 2.5 \times 10^{-6} > K_{sp}$; yes

34.13 a. $K = K_{sp}\ CdC_2O_4 \times K_f\ Cd(C_2O_4)_2^{2-}$

$= (9 \times 10^{-5})(6 \times 10^5) = 5 \times 10^1$

b. $s = K \times [C_2O_4^{2-}] = 0.5$ M

34.14 a. $\dfrac{[H_2C_2O_4]}{[HC_2O_4^-]} = \dfrac{[H^+]}{K_a} = \dfrac{1 \times 10^{-1}}{4 \times 10^{-2}} = 2$

$\dfrac{[HC_2O_4^-]}{[C_2O_4^{2-}]} = \dfrac{[H^+]}{K_a} = \dfrac{1 \times 10^{-1}}{5 \times 10^{-5}} = 2 \times 10^3$

$H_2C_2O_4$ is the principal species

b. $\dfrac{[H_2C_2O_4]}{[HC_2O_4^-]} = \dfrac{1 \times 10^{-5}}{4 \times 10^{-2}} = 2 \times 10^{-4}$

$\dfrac{[HC_2O_4^-]}{[C_2O_4^{2-}]} = \dfrac{1 \times 10^{-5}}{5 \times 10^{-5}} = 0.2$; $C_2O_4^{2-}$ is the principal species

c. $\dfrac{[H_2C_2O_4]}{[HC_2O_4^-]} = \dfrac{1 \times 10^{-9}}{4 \times 10^{-2}} = 2 \times 10^{-8}$

$\dfrac{[HC_2O_4^-]}{[C_2O_4^{2-}]} = \dfrac{1 \times 10^{-9}}{5 \times 10^{-5}} = 2 \times 10^{-5}$; principal species = $C_2O_4^{2-}$

34.15 $[OH^-]^2 = 1.6 \times 10^{-13}$; $[OH^-] = 4 \times 10^{-7}$ M

$[H^+] = 2 \times 10^{-8}$ M; pH = 7.7

34.16 $\dfrac{[PO_4^{3-}]}{[HPO_4^{2-}]} \dfrac{K_a}{[H^+]} = \dfrac{1.7 \times 10^{-12}}{1.0 \times 10^{-11}} = 0.17$

$\dfrac{[PO_4^{3-}]}{[HPO_4^{2-}]} = \dfrac{1.7 \times 10^{-12}}{1.0 \times 10^{-12}} = 1.7$

34.17 $\frac{[H^+]^2}{0.40} \approx 7.5 \times 10^{-3}$; $[H^+]^2 = 3.0 \times 10^{-3}$; $[H^+] = 0.055$ M

$\frac{[H^+]^2}{0.34} = 7.5 \times 10^{-3}$; $[H^+]^2 = 2.6 \times 10^{-3}$; $[H^+] = 0.051$ M

pH = $-\log_{10}(0.051) = 1.29$

34.18 a. $K = K_a\ H_3PO_4 \times K_a\ H_2PO_4^- \times K_a\ HPO_4^{2-}$

$= (7.5 \times 10^{-3})(6.2 \times 10^{-8})(1.7 \times 10^{-12}) = 7.9 \times 10^{-22}$

b. $K = \frac{K_{sp}\ Ag_3PO_4}{7.9 \times 10^{-22}} = \frac{1.8 \times 10^{-18}}{7.9 \times 10^{-22}} = 2.3 \times 10^3$

34.19 Step 1:

$2CrO_4^{2-}(aq) + 2HC_2H_3O_2(aq) \rightarrow Cr_2O_7^{2-}(aq) + 2C_2H_3O_2^-(aq) + H_2O$

$C_2O_4^{2-}(aq) + HC_2H_3O_2(aq) \rightarrow HC_2O_4^-(aq) + C_2H_3O_2^-(aq)$

$PO_4^{3-}(aq) + 2HC_2H_3O_2(aq) \rightarrow H_2PO_4^-(aq) + 2C_2H_3O_2^-(aq)$

Step 2:

$Ca^{2+}(aq) + HC_2O_4^-(aq) + H_2O \rightarrow CaC_2O_4 \cdot H_2O(s) + H^+(aq)$

$CaC_2O_4 \cdot H_2O(s) + 2H^+(aq) \rightarrow Ca^{2+}(aq) + H_2C_2O_4(aq) + H_2O$

$5H_2C_2O_4(aq) + 2MnO_4^-(aq) + 6H^+(aq) \rightarrow 2Mn^{2+}(aq) + 8H_2O + 10CO_2(g)$

34.20 a. $C_2O_4^{2-}$ b. SO_4^{2-}, CrO_4^{2-}, $C_2O_4^{2-}$, PO_4^{3-}

c. CrO_4^{2-}, PO_4^{3-}

34.21 Step 1: add $CaCl_2$ in $HC_2H_3O_2$ solution. White ppt. shows oxalate is present.

Step 2: to solution from Step 1, add $Fe(NO_3)_3$; light tan ppt. shows phosphate is present.

34.22 Extract with water; test water solution with $AgNO_3$. White ppt. shows NaBr is present. Extract remaining solid with HCl; white residue indicates $BaSO_4$. Treat acidic solution with MnO_4^-; if decolorized, CaC_2O_4 must be present.

34.23 Chromate is present; other ions in doubt

34.24 SO_4^{2-} present (only $BaSO_4$ insoluble in acid)

$C_2O_4^{2-}$ absent (would give precipitate with Ca^{2+})

CrO_4^{2-} present (yellow solution)

PO_4^{3-} in doubt

34.25 $C_2O_4^{2-}$ absent (would give precipitate with Ca^{2+})

SO_4^{2-} present (precipitate is $BaSO_4$)

PO_4^{3-} present (precipitate is $FePO_4$)

CrO_4^{2-} absent; would give yellow precipitate with Ba^{2+}

34.26 $C_2O_4^{2-}$, PO_4^{3-}

34.27 a. 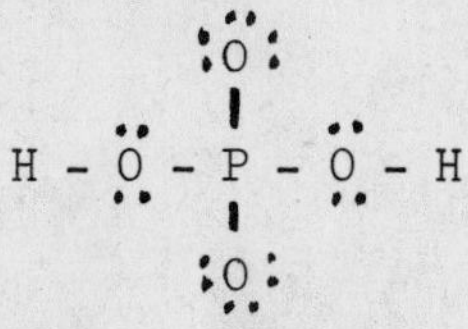 b. 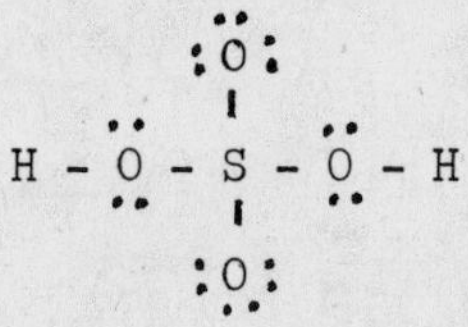

c. 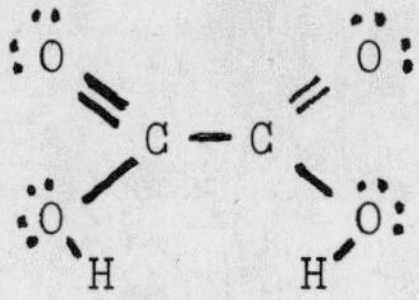

34.28 a. color; CrO_4^{2-} is yellow

b. color; $Cr_2O_7^{2-}$ is red

c. water solubility; CaC_2O_4 insoluble

d. water solubility; $FePO_4$ insoluble

34.29 a. SO_4^{2-}, CrO_4^{2-} b. SO_4^{2-} c. CrO_4^{2-}

34.30 a. $1.00 \text{ g } Ca_3(PO_4)_2 \times \frac{578 \text{ g SP}}{310 \text{ g } Ca_3(PO_4)_2} = 1.86 \text{ g}$

b. $\frac{61.94}{578} \times 100 = 10.7\%$

34.31 a. $Fe_2O_3(s) + 6H^+(aq) + 3SO_4^{2-}(aq) \longrightarrow 2Fe^{3+}(aq) + 3SO_4^{2-}(aq) + 3H_2O$

b. $2Fe(OH)_3(s) + 6H^+(aq) + 3SO_4^{2-}(aq) \longrightarrow 2Fe^{3+}(aq) + 6H_2O + 3SO_4^{2-}(aq)$

c. see (a) or (b)

34.32 a. $2SrCrO_4(s) + 2H^+(aq) \rightarrow 2Sr^{2+}(aq) + Cr_2O_7^{2-}(aq) + H_2O$

b. $CaC_2O_4(s) + 2H^+(aq) \rightarrow Ca^{2+}(aq) + H_2C_2O_4(aq)$

c. $Ag_3PO_4(s) + 3H^+(aq) \rightarrow 3Ag^+(aq) + H_3PO_4(aq)$

34.33 a. $Ca_3(PO_4)_2(s) + 6HSO_4^-(aq) \rightarrow 3Ca^{2+}(aq) + 2H_3PO_4(aq) + 6SO_4^{2-}(aq)$

b. $ZnC_2O_4(s) + 2HSO_4^-(aq) \rightarrow Zn^{2+}(aq) + H_2C_2O_4(aq) + 2SO_4^{2-}(aq)$

c. $2Ag_2CrO_4(s) + 2HSO_4^-(aq) \rightarrow 4Ag^+(aq) + Cr_2O_7^{2-}(aq) + 2SO_4^{2-}(aq) + H_2O$

34.34 a. $Ba^{2+}(aq) + SO_4^{2-}(aq) \rightarrow BaSO_4(s)$

b. $2Ag^+(aq) + CrO_4^{2-}(aq) \rightarrow Ag_2CrO_4(s)$

c. $Fe^{3+}(aq) + PO_4^{3-}(aq) \rightarrow FePO_4(s)$

34.35 a. $HSO_4^-(aq) \rightleftharpoons H^+(aq) + SO_4^{2-}(aq)$

b. $H_2C_2O_4(aq) \rightleftharpoons H^+(aq) + HC_2O_4^-(aq)$

c. $H_2PO_4^-(aq) \rightleftharpoons H^+(aq) + HPO_4^{2-}(aq)$

d. $Fe(C_2O_4)_3^{3-}(aq) \rightleftharpoons Fe^{3+}(aq) + 3\,C_2O_4^{2-}(aq)$

34.36 a. $CuO(s) + 2H^+(aq) \rightarrow Cu^{2+}(aq) + H_2O$

b. $C_2O_4^{2-}(aq) + H_2O \rightleftharpoons HC_2O_4^-(aq) + OH^-(aq)$

c. $H_3PO_4(aq) + OH^-(aq) \rightarrow H_2PO_4^-(aq) + H_2O$

d. $PO_4^{3-}(aq) + H_2O \rightleftharpoons HPO_4^{2-}(aq) + OH^-(aq)$

34.37 a. $Zn(s) + 2H^+(aq) \rightarrow Zn^{2+}(aq) + H_2(g)$

b. $Ca_3(PO_4)_2(s) + 2H_2SO_4(l) + 4H_2O(l) \rightarrow Ca(H_2PO_4)_2(s) + 2[CaSO_4 \cdot 2H_2O(s)]$

c. $2NaCHO_2(s) \rightarrow Na_2C_2O_4(s) + H_2(g)$

d. $P_4(s) + 5\,O_2(g) \rightarrow P_4O_{10}(s)$

$P_4O_{10}(s) + 6H_2O(l) \rightarrow 4H_3PO_4(l)$

34.38 a. $3[C_2O_4^{2-}(aq) \longrightarrow 2CO_2(g) + 2e^-]$

$ClO_3^-(aq) + 6H^+(aq) + 6e^- \longrightarrow Cl^-(aq) + 3H_2O$

$3C_2O_4^{2-}(aq) + ClO_3^-(aq) + 6H^+(aq) \longrightarrow 6CO_2(g) + Cl^-(aq) + 3H_2O$

b. $1.00 \text{ g } Na_2C_2O_4 \times \frac{1 \text{ mol } Na_2C_2O_4}{134 \text{ g } Na_2C_2O_4} \times \frac{1 \text{ mol } KClO_3}{3 \text{ mol } Na_2C_2O_4} \times \frac{123 \text{ g } KClO_3}{1 \text{ mol } KClO_3}$

$= 0.305 \text{ g}$

34.39 a. $Cr_2O_7^{2-}(aq) + 14\,H^+(aq) + 6e^- \longrightarrow 2Cr^{3+}(aq) + 7H_2O$

$3[H_2S(aq) \longrightarrow S(s) + 2H^+(aq) + 2\,e^-]$

$Cr_2O_7^{2-}(aq) + 8H^+(aq) + 3H_2S(aq) \longrightarrow 2Cr^{3+}(aq) + 7H_2O + 3S(s)$

b. $5H_2C_2O_4(aq) + 2MnO_4^-(aq) + 6H^+(aq) \longrightarrow 10CO_2(aq) + 8H_2O + 2Mn^{2+}(aq)$

34.40 a. $E^o_{tot} = E^o_{red}Cr_2O_7^{2-} + E^o_{ox}Cl^- = 1.33\text{ V} - 1.36\text{ V} = -0.03\text{ V}$

b. $E^o_{tot} = E^o_{red}Cr_2O_7^{2-} + E^o_{ox}Br^- = 1.33\text{ V} - 1.07\text{ V} = 0.26\text{ V}$

c. $E^o_{tot} = E^o_{red}Cr_2O_7^{2-} + E^o_{ox}I^- = 1.33\text{ V} - 0.53\text{ V} = 0.80\text{ V}$

d. $E^o_{tot} = E^o_{red}Cr_2O_7^{2-} + E^o_{ox}Fe^{2+} = 1.33\text{ V} - 0.77\text{ V} = 0.56\text{ V}$

34.41 a. $E = E^o_{tot} - \frac{0.0591}{n} \log_{10} \frac{1}{(\text{conc. } H^+)^4}$

$= -0.14\text{ V} + 0.1182 \log_{10}(\text{conc. } H^+)$

$= -0.14\text{ V} - 0.1182\text{ pH}$

b. -0.14 V

$-0.14\text{ V} - 0.1182(4.0) = -0.61\text{ V}$

$-0.14\text{ V} - 0.1182(7.0) = -0.97\text{ V}$

34.42 a. $s^2 = 1.5 \times 10^{-8}$; $s = 1.2 \times 10^{-4}$ M

b. $s = \frac{1.5 \times 10^{-8}}{1.0 \times 10^{-1}} = 1.5 \times 10^{-7}$ M

c. $s = \frac{1.5 \times 10^{-8}}{2.5 \times 10^{-2}} = 6.0 \times 10^{-7}$ M

34.43 a. $K = K_{sp}CoC_2O_4 \times K_fCo(C_2O_4)_2^{2-} = (2 \times 10^{-5})(1 \times 10^7)$
$= 2 \times 10^2$

b. $s = 2 \times 10^2 \times 0.01 = 2$ M

34.44 a. $\frac{[H_2C_2O_4]}{[HC_2O_4^-]} = \frac{1.0 \times 10^{-2}}{4 \times 10^{-2}} = 0.2$

$\frac{[HC_2O_4^-]}{[C_2O_4^{2-}]} = \frac{1 \times 10^{-2}}{5 \times 10^{-5}} = 2 \times 10^2$; $HC_2O_4^-$ is principal species

b. $\frac{[H_2C_2O_4]}{[HC_2O_4^-]} = \frac{1.0 \times 10^{-6}}{4 \times 10^{-2}} = 2 \times 10^{-5}$

$\frac{[HC_2O_4^-]}{[C_2O_4^{2-}]} = \frac{1 \times 10^{-6}}{5 \times 10^{-5}} = 2 \times 10^{-2}$; $C_2O_4^{2-}$ principal species

c. $\frac{[H_2C_2O_4]}{[HC_2O_4^-]} = \frac{1.0 \times 10^{-10}}{4 \times 10^{-2}} = 2 \times 10^{-9}$

$\frac{[HC_2O_4^-]}{[C_2O_4^{2-}]} = \frac{1 \times 10^{-10}}{5 \times 10^{-5}} = 2 \times 10^{-6}$; $C_2O_4^{2-}$ principal species

34.45 $\frac{[OH^-]^2}{0.200} \approx 5.9 \times 10^{-3}$; $[OH^-]^2 = 1.2 \times 10^{-3}$; $[OH^-] = 0.035$ M

$\frac{[OH^-]^2}{0.165} = 5.9 \times 10^{-3}$; $[OH^-]^2 = 0.97 \times 10^{-3}$; $[OH^-] = 0.031$ M

$[H^+] = \frac{1.0 \times 10^{-14}}{0.031} = 3.2 \times 10^{-13}$ M; pH = 12.49

34.46 $[H^+] = 6.2 \times 10^{-8} \times \frac{[H_2PO_4^-]}{[HPO_4^{2-}]} = 6.2 \times 10^{-8}$; pH = 7.21

34.47 $\frac{[H^+]^2}{1.20} \approx 4.0 \times 10^{-2}$; $[H^+]^2 = 4.8 \times 10^{-2}$; $[H^+] = 0.22$ M

$\frac{[H^+]^2}{0.98} = 4.0 \times 10^{-2}$; $[H^+]^2 = 4.0 \times 10^{-2}$; $[H^+] = 0.20$ M

pH = 0.70

34.48 a. $K = K_aH_2C_2O_4 \times K_aHC_2O_4^- = (4 \times 10^{-2})(5 \times 10^{-5})$
$= 2 \times 10^{-6}$

b. $K = \frac{K_{sp}\ CaC_2O_4}{K_aH_2C_2O_4 \times K_aHC_2O_4^-} = \frac{1.3 \times 10^{-9}}{(4 \times 10^{-2})(5 \times 10^{-5})}$
$= 6 \times 10^{-4}$

34.49 $Ba^{2+}(aq) + SO_4^{2-}(aq) \longrightarrow BaSO_4(s)$

$2Ba^{2+}(aq) + Cr_2O_7^{2-}(aq) + H_2O \longrightarrow 2BaCrO_4(s) + 2H^+(aq)$

$2BaCrO_4(s) + 2H^+(aq) \longrightarrow 2Ba^{2+}(aq) + Cr_2O_7^{2-}(aq) + H_2O$

34.50 SO_4^{2-} forms white ppt. of $BaSO_4$

$C_2O_4^{2-}$ forms white ppt. of $CaC_2O_4 \cdot H_2O$

CrO_4^{2-} forms yellow ppt. of $BaCrO_4$

PO_4^{3-} forms light tan ppt. of $FePO_4$

34.51 Observe color; yellow if CrO_4^{2-} is present. Acidify with HCl, add $BaCl_2$; white ppt. of $BaSO_4$ shows presence of SO_4^{2-}.

34.52 Stir with water; yellow residue indicates $BaCrO_4$. To the solution, add $HC_2H_3O_2$ and $CaCl_2$; white ppt. indicates $Na_2C_2O_4$. Centrifuge and add $Fe(NO_3)_3$ to solution; light tan ppt. indicates Na_3PO_4.

34.53 SO_4^{2-} must be present, all other anions absent

34.54 $C_2O_4^{2-}$ absent; would give precipitate with $CaCl_2$

CrO_4^{2-} present; $BaCrO_4$ dissolves in HNO_3

SO_4^{2-} absent; $BaSO_4$ would not dissolve in acid

PO_4^{3-} in doubt

34.55 $C_2O_4^{2-}$ present; gives ppt. with Ca^{2+}

CrO_4^{2-} absent; would give precipitate with Ba^{2+}

SO_4^{2-} absent; would give precipitate with Ba^{2+}

PO_4^{3-} present; tan ppt. is $FePO_4$

34.56 PO_4^{3-}; HPO_4^{2-} is weakest acid

34.57 a. 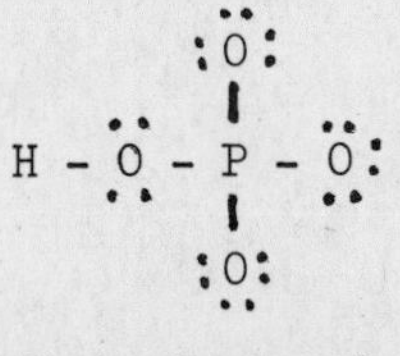

b. 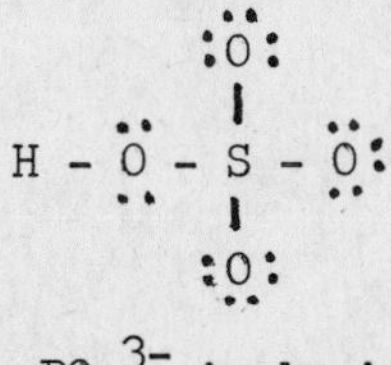

c. 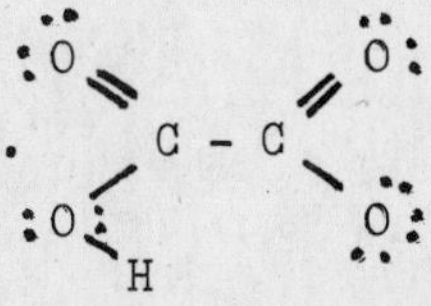

34.58 a. Test pH of solution; PO_4^{3-} is basic

b. Observe color; CrO_4^{2-} is yellow

34.58 c. Add Fe^{3+} to solution; Na_3PO_4 gives precipitate

d. Dissolve in acid, add MnO_4^-; decolorized by $C_2O_4^{2-}$

34.59 a. all b. CrO_4^{2-} c. $C_2O_4^{2-}$

34.60 a. $1.00 \times 10^3 \text{ g TSP} \times \frac{1 \text{ mol TSP}}{234 \text{ g TSP}} \times \frac{4 \text{ mol } H_3PO_4}{3 \text{ mol TSP}} \times \frac{98.0 \text{ g } H_3PO_4}{1 \text{ mol } H_3PO_4}$

$= 558 \text{ g } H_3PO_4$

b. $\frac{61.94}{234.0} \times 100 = 26.47\%$

*34.61 required ratio $\frac{[H_2C_2O_4]}{[HC_2O_4^-]} \approx \frac{0.20}{0.80} = 0.25$

Mix $H_2C_2O_4$, KHC_2O_4 in a 1:4 volume ratio

*34.62 Molecular formula = $C_{10}H_{12}O_{13}N_5P_3$

$\% \text{ P} = \frac{3(30.97)}{503.2} \times 100 = 18.46$

*34.63 Let $[Cr_2O_7^{2-}] = x$; $4 \times 10^{14} = \frac{x}{[H^+]^2 \times (0.10 - 2x)^2}$

$\frac{x}{(0.10 - 2x)^2} = 4 \times 10^{14}(1 \times 10^{-7})^2 = 4; \; x = 0.018$

$[CrO_4^{2-}] = 0.064 \text{ M}$; $[Cr_2O_7^{2-}] = 0.018 \text{ M}$

*34.64 a. $[C_2O_4^{2-}] = \frac{K_{sp}CdC_2O_4}{1.0} = \frac{9 \times 10^{-5}}{1.0} = 9 \times 10^{-5} \text{ M}$

b. $CdC_2O_4(s) + C_2O_4^{2-}(aq) \longrightarrow Cd(C_2O_4)_2^{2-}(aq)$

$K = K_{sp}CdC_2O_4 \times K_f Cd(C_2O_4)_2^{2-}$

$= (9 \times 10^{-5})(6 \times 10^5) = 54$

$10 \frac{\text{mg}}{\text{mL}} CdC_2O_4 \times \frac{0.001 \text{ g}}{1 \text{ mg}} \times \frac{1 \text{ mL}}{0.001 \text{ L}} \times \frac{1 \text{ mol } CdC_2O_4}{200 \text{ g } CdC_2O_4}$

$= 0.05 \text{ mol/L}$

$54 = \frac{0.05}{[C_2O_4^{2-}]}$; $[C_2O_4^{2-}] = 1 \times 10^{-3} \text{ M}$

Will form precipitate of CdC_2O_4 first, since required concentration of $C_2O_4^{2-}$ is lower. Precipitate will later dissolve as concentration of $C_2O_4^{2-}$ increases.

*34.65 $K = \frac{K_{sp}\ CaC_2O_4 \times K_a\ HC_2H_3O_2}{K_a\ HC_2O_4^-} = \frac{(1.3 \times 10^{-9})(1.8 \times 10^{-5})}{5 \times 10^{-5}}$

$= 5 \times 10^{-10}$; no

Test Questions, Multiple Choice

1. Of the four anions in the Ca-Ba-Fe group, which one forms a precipitate with Ba^{2+} in strongly acidic solution?

   a. SO_4^{2-} b. CrO_4^{2-} c. $C_2O_4^{2-}$ d. PO_4^{3-}

2. Of the four anions in the Ca-Ba-Fe group, which one is most easily oxidized?

   a. SO_4^{2-} b. CrO_4^{2-} c. $C_2O_4^{2-}$ d. PO_4^{3-}

3. In strongly acidic solution, which one of the following is the strongest oxidizing agent?

   a. SO_4^{2-} b. $Cr_2O_7^{2-}$ c. $H_2C_2O_4$ d. H_3PO_4

4. Which of the following is the strongest base?

   a. SO_4^{2-} b. $HC_2O_4^-$ c. $H_2PO_4^-$ d. HPO_4^{2-} e. PO_4^{3-}

5. The chromate ion can be converted to dichromate by treatment with:

   a. a strong acid
   b. a strong base
   c. a strong oxidizing agent
   d. a strong reducing agent
   e. none of the above

6. How many of the following compounds are brightly colored?

   $BaSO_4$, $NiSO_4$, K_2CrO_4, $K_2Cr_2O_7$, $Na_2C_2O_4$

   a. 1 b. 2 c. 3 d. 4 e. 5

7. Which of the following anions acts as a chelating agent?

   a. SO_4^{2-} b. CrO_4^{2-} c. $C_2O_4^{2-}$ d. PO_4^{3-}

8. In the analysis of the Ca-Ba-Fe group, which ion is precipitated as the iron(III) salt?

   a. SO_4^{2-} b. CrO_4^{2-} c. $C_2O_4^{2-}$ d. PO_4^{3-}

9. In the analysis of the Ca-Ba-Fe group, which ion is detected by its reaction with $KMnO_4$?

a. SO_4^{2-} b. CrO_4^{2-} c. $C_2O_4^{2-}$ d. PO_4^{3-}

10. In the analysis of the Ca-Ba-Fe group, which ion is detected by its ability to form a blue peroxide?

a. SO_4^{2-} b. CrO_4^{2-} c. $C_2O_4^{2-}$ d. PO_4^{3-}

Test Problems

1. Write a balanced net ionic equation to explain why

a. a white precipitate forms when solutions of $BaCl_2$ and Na_2SO_4 are mixed.

b. barium chromate dissolves in acid.

c. a precipitate forms when solutions of $Fe_2(SO_4)_3$ and $(NH_4)_3PO_4$ are mixed.

d. a solution of sodium phosphate is basic.

2. Consider the half-reaction for the reduction of CrO_4^{2-} in basic solution:

$$CrO_4^{2-}(aq) + 4H_2O + 3e^- \longrightarrow Cr(OH)_3(s) + 5\ OH^-(aq)$$

for which E^o_{red} = -0.12 V. Calculate the pH at which E_{red} = 0.00 V, taking conc. CrO_4^{2-} = 1.0 M.

3. An unknown which may contain any anion in the Silver Group (Cl^-, Br^-, I^-, SCN^-) or the Ba-Ca-Fe Group (SO_4^{2-}, CrO_4^{2-}, $C_2O_4^{2-}$, PO_4^{3-}) is colorless and has a pH very close to 7. On this basis, what can you say about the identity of the unknown? Explain your reasoning.

4. Taking K_a $H_2C_2O_4 = 4 \times 10^{-2}$, K_a $HC_2O_4^- = 5 \times 10^{-5}$, determine which of the three species, $H_2C_2O_4$, $HC_2O_4^-$, or $C_2O_4^{2-}$ is present at the highest concentration at pH 3.3.

5. Calculate K for the reaction:

$$HPO_4^{2-}(aq) + HC_2O_4^-(aq) \longrightarrow H_2PO_4^-(aq) + C_2O_4^{2-}(aq)$$

($K_aH_2C_2O_4 = 4 \times 10^{-2}$, $HC_2O_4^- = 5 \times 10^{-5}$, $H_3PO_4 = 7.5 \times 10^{-3}$, $H_2PO_4^- = 6.2 \times 10^{-8}$, $HPO_4^{2-} = 1.7 \times 10^{-12}$)

CHAPTER 35

Answers to Problems

35.1 $H^+(aq) + NO_2^-(aq) \longrightarrow HNO_2(aq)$

$H^+(aq) + C_2H_3O_2^-(aq) \longrightarrow HC_2H_3O_2(aq)$

35.2 a. $AgNO_2(s) + H^+(aq) \longrightarrow HNO_2(aq) + Ag^+(aq)$

b. $AgC_2H_3O_2(s) + H^+(aq) \longrightarrow HC_2H_3O_2(aq) + Ag^+(aq)$

c. $Hg_2(C_2H_3O_2)_2(s) + 2H^+(aq) \longrightarrow Hg_2^{2+}(aq) + 2HC_2H_3O_2(aq)$

35.3 a. $CuO(s) + 2H^+(aq) + 2NO_3^-(aq) \longrightarrow Cu^{2+}(aq) + 2NO_3^-(aq) + H_2O$

b. $4Zn(s) + NO_3^-(aq) + 10\ H^+(aq) \longrightarrow 4Zn^{2+}(aq) + NH_4^+(aq) + 3H_2O$

c. $Cr_2O_7^{2-}(aq) + 8H^+(aq) + 3NO_2^-(aq) \longrightarrow 2Cr^{3+}(aq) + 3NO_3^-(aq) + 4H_2O$

35.4 a. $NH_4NO_3(s) \longrightarrow N_2O(g) + 2H_2O(l)$

b. $2HC_2H_3O_2(aq) + Ca^{2+}(aq) + 2\ OH^-(aq) \longrightarrow Ca^{2+}(aq) + 2H_2O + 2C_2H_3O_2^-(aq)$

c. $Ba(ClO_3)_2(s) + 2H^+(aq) + SO_4^{2-}(aq) \longrightarrow BaSO_4(s) + 2H^+(aq) + 2ClO_3^-(aq)$

35.5 a. $NO_2^-(aq) + NH_2SO_3H(aq) \longrightarrow N_2(g) + SO_4^{2-}(aq) + H^+(aq) + H_2O$

b. $3Cl_2(g) + 6\ OH^-(aq) \longrightarrow ClO_3^-(aq) + 5Cl^-(aq) + 3H_2O$

c. $2H^+(aq) + 2ClO_3^-(aq) \longrightarrow 2ClO_2(g) + \frac{1}{2}O_2(g) + H_2O$

d. $H^+(aq) + C_2H_3O_2^-(aq) \longrightarrow HC_2H_3O_2(aq)$

35.6 a. $HNO_3(aq) \longrightarrow H^+(aq) + NO_3^-(aq)$

b. $NO_2^-(aq) + H_2O \rightleftharpoons HNO_2(aq) + OH^-(aq)$

c. $HC_2H_3O_2(aq) \rightleftharpoons H^+(aq) + C_2H_3O_2^-(aq)$

35.7 a. $ZnO(s) + 2H^+(aq) + 2NO_3^-(aq) \longrightarrow Zn^{2+}(aq) + 2NO_3^-(aq) + H_2O$

b. $2NaNO_3(s) \longrightarrow 2NaNO_2(s) + O_2(g)$

c. $Ba(ClO_3)_2(s) + 2H^+(aq) + SO_4^{2-}(aq) \longrightarrow BaSO_4(s) + 2H^+(aq) + 2ClO_3^-(aq)$

35.8 a. $2KNO_3(s) \longrightarrow 2KNO_2(s) + O_2(g)$

b. $NH_4NO_3(s) \longrightarrow N_2O(g) + 2H_2O(l)$

c. $C_2H_5OH(aq) + O_2(g) \longrightarrow HC_2H_3O_2(aq) + H_2O$

35.9 a.

$$3\left[2I^-(aq) \longrightarrow I_2(s) + 2e^-\right]$$

$$\underline{ClO_3^-(aq) + 6H^+(aq) + 6e^- \longrightarrow Cl^-(aq) + 3H_2O}$$

$$6I^-(aq) + 6H^+(aq) + ClO_3^-(aq) \longrightarrow 3I_2(s) + Cl^-(aq) + 3H_2O$$

b. $$\text{moles } ClO_3^- = 0.100\ \frac{\text{mol}}{\text{L}} \times 0.0250\ \text{L} = 2.50 \times 10^{-3}\ \text{mol}$$

$$\text{moles } I^- = 2.50 \times 10^{-3}\ \text{mol } ClO_3^- \times \frac{6\ \text{mol } I^-}{1\ \text{mol } ClO_3^-}$$

$$= 1.50 \times 10^{-2}\ \text{mol } I^-$$

$$\text{conc. } I^- = \frac{1.50 \times 10^{-2}\ \text{mol}}{5.00 \times 10^{-2}\ \text{L}} = 0.300\ \text{M}$$

35.10 a.

$$3\left[Cu(s) \longrightarrow Cu^{2+}(aq) + 2e^-\right]$$

$$\underline{2\left[NO_3^-(aq) + 4H^+(aq) + 3e^- \longrightarrow NO(g) + 2H_2O\right]}$$

$$3Cu(s) + 2NO_3^-(aq) + 8H^+(aq) \longrightarrow 3Cu^{2+}(aq) + 2NO(g) + 4H_2O$$

b.

$$3\left[Fe^{2+}(aq) \longrightarrow Fe^{3+}(aq) + e^-\right]$$

$$\underline{NO_3^-(aq) + 4H^+(aq) + 3e^- \longrightarrow NO(g) + 2H_2O}$$

$$3Fe^{2+}(aq) + NO_3^-(aq) + 4H^+(aq) \longrightarrow 3Fe^{3+}(aq) + NO(g) + 2H_2O$$

c.

$$2\left[CrO_4^{2-}(aq) + 5H^+(aq) + 3e^- \longrightarrow Cr(OH)_3(s) + H_2O\right]$$

$$\underline{3\left[NO_2^-(aq) + H_2O \longrightarrow NO_3^-(aq) + 2H^+(aq) + 2e^-\right]}$$

$$2CrO_4^{2-}(aq) + 4H^+(aq) + 3NO_2^-(aq) + H_2O \longrightarrow 2Cr(OH)_3(s) + 3NO_3^-$$

$$2CrO_4^{2-}(aq) + 5H_2O + 3NO_2^-(aq) \longrightarrow 2Cr(OH)_3(s) + 4OH^-(aq) + 3NO_3^-(aq)$$

35.11 a. $E^o_{tot} = 1.25\ V + E^o_{ox}\ ClO_3^- = 1.25\ V - 1.19\ V = +0.06\ V$

b. $E^o_{tot} = 1.25\ V + E^o_{ox}\ HClO_2 = 1.25\ V - 1.21\ V = +0.04\ V$

c. $E^o_{tot} = 1.25\ V + E^o_{ox}\ HClO = 1.25\ V - 1.43\ V = -0.18\ V$

35.12 $NO_3^-(aq) + 6H^+(aq) + 5e^- \longrightarrow \frac{1}{2} N_2(g) + 3H_2O$

$$E = +1.25\ V - \frac{0.0591}{5} \log_{10} \frac{1}{(\text{conc. } H^+)^6}$$

$$= +1.25\ V + \frac{0.0591(6)}{5} \log_{10}(\text{conc. } H^+)$$

$$= +1.25\ V + 0.0709 \log_{10}(\text{conc. } H^+)$$

a. +1.25 V   b. +1.25 V - 0.50 V = +0.75 V

c. +1.25 V - 0.99 V = +0.26 V

35.13 a. 4.7   b. 4.0   c. 5.4

35.14 a. $\frac{[H^+]^2}{1.0} \approx 4.5 \times 10^{-4}$; $[H^+] = 0.021$ M

b. $\frac{[NO_2^-]}{[HNO_2]} = \frac{K_a}{[H^+]} = \frac{4.5 \times 10^{-4}}{1.0 \times 10^{-3}} = 0.45$

c. $\frac{1.0 \times 10^{-14}}{4.5 \times 10^{-4}} = 2.2 \times 10^{-11}$

35.15 $\frac{x^2}{0.10} \approx 5.6 \times 10^{-10}$; $x = 7.5 \times 10^{-6}$ M $= [OH^-]$

$[H^+] = \frac{1.0 \times 10^{-14}}{7.5 \times 10^{-6}} = 1.3 \times 10^{-9}$ M; pH = 8.89

35.16 a. conc. $HNO_2 = \frac{0.0500 \times 0.100\ M}{0.0800} = 0.0625$ M

conc. $NO_2^- = \frac{0.0300 \times 0.200\ M}{0.0800} = 0.0750$ M

b. $[H^+] = 4.5 \times 10^{-4} \times \frac{0.0625}{0.0750} = 3.8 \times 10^{-4}$ M; pH = 3.42

c. 3.42

35.17 a. $\frac{[HC_2H_3O_2]}{[C_2H_3O_2^-]} = \frac{[H^+]}{1.8 \times 10^{-5}} = \frac{3.2 \times 10^{-5}}{1.8 \times 10^{-5}} = 1.8$

b. $\frac{[HC_2H_3O_2]}{[C_2H_3O_2^-]} = \frac{1.0 \times 10^{-5}}{1.8 \times 10^{-5}} = 0.56$

c. $\frac{[HC_2H_3O_2]}{[C_2H_3O_2^-]} = \frac{3.2 \times 10^{-6}}{1.8 \times 10^{-5}} = 0.18$

35.18 a. $K = \frac{K_a\ HC_2H_3O_2}{K_a\ HNO_2} = \frac{1.8 \times 10^{-5}}{4.5 \times 10^{-4}} = 0.040$

b. $K = 1/K_a\ HC_2H_3O_2 = 1/(1.8 \times 10^{-5}) = 5.6 \times 10^4$

c. $K = 1/K_b\ NO_2^- = 1/(2.2 \times 10^{-11}) = 4.5 \times 10^{10}$

35.19 $2Fe^{2+}(aq) + NO_2^-(aq) + 2H^+(aq) \longrightarrow Fe^{3+}(aq) + Fe(NO)^{2+}(aq) + H_2O$

$4Fe^{2+}(aq) + NO_3^-(aq) + 4H^+(aq) \longrightarrow 3Fe^{3+}(aq) + Fe(NO)^{2+}(aq) + 2H_2O$

35.20 a. Test for ClO_3^- involves precipitating AgCl

b. NO_2^-, like NO_3^-, gives brown ring test

35.21 a. $Ag^+(aq) + SCN^-(aq) \longrightarrow AgSCN(s)$

b. $Ba^{2+}(aq) + SO_4^{2-}(aq) \longrightarrow BaSO_4(s)$

c. $NO_2^-(aq) + NH_2SO_3H(aq) \longrightarrow N_2(g) + SO_4^{2-}(aq) + H^+(aq) + H_2O$

35.22 NO_3^- present; brown color in conc. acid

NO_2^- absent; no color in dilute acid

$C_2H_3O_2^-$ present; blue color

ClO_3^- in doubt

35.23 CrO_4^{2-} present; converted to orange $Cr_2O_7^{2-}$ by acid

$C_2H_3O_2^-$ present; $HC_2H_3O_2$ has odor of vinegar

Cl^-, NO_3^- in doubt

35.24 a. 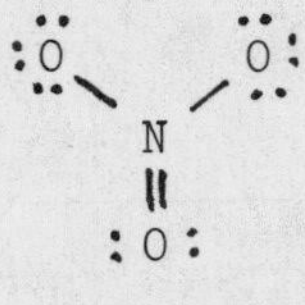 equilateral triangle b. 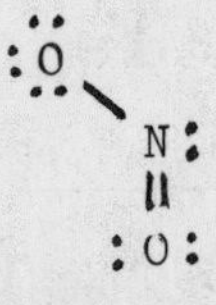 bent

35.25 a. Reduce, test with Ag^+; ClO_3^- would give white ppt.

b. Add Fe^{3+}; PO_4^{3-} would give precipitate

c. Add Fe^{3+}; SCN^- would give red color

d. Observe color; CrO_4^{2-} is yellow

35.26 a. Oxidize with NO_2^- or precipitate with Ag^+

b. Precipitate with Ba^{2+} c. Precipitate with Ba^{2+}

d. Precipitate with H_2S

35.27 a. R b. R c. O d. O, R

35.28 a. $C_2H_3O_2^-$, NO_3^- b. $C_2H_3O_2^-$, NO_2^- c. $C_2H_3O_2^-$

35.29 Step 1: add $AgNO_3$ in acid; white ppt. indicates Cl^-

Step 2: to solution from (1) add NO_2^-; white ppt. shows ClO_3^-

Step 3: add $La(NO_3)_3$ and KI_3 to sample; dark blue color indicates $C_2H_3O_2^-$

35.30 a. moles $Cr_2O_7^{2-} = 2.0 \times 10^{-3}$ L $\times 1.0 \frac{mol}{L} = 2.0 \times 10^{-3}$ mol

moles $NO_2^- = 6.0 \times 10^{-3}$

$$V = \frac{6.0 \times 10^{-3} \text{ mol}}{2.0 \times 10^{-2} \text{ mol/L}} = 0.30 \text{ L} = 300 \text{ mL}$$

b. 4.0×10^{-3} mol $\times \frac{52.0 \text{ g}}{1 \text{ mol}} = 0.21$ g

35.31 $NO_2^-(aq) + H_2O \rightleftharpoons HNO_2(aq) + OH^-(aq)$

$C_2H_3O_2^-(aq) + H_2O \rightleftharpoons HC_2H_3O_2(aq) + OH^-(aq)$

35.32 a. $PbSO_4(s) + 4C_2H_3O_2^-(aq) \rightarrow Pb(C_2H_3O_2)_4^{2-}(aq) + SO_4^{2-}(aq)$

b. $Co^{3+}(aq) + 6NO_2^-(aq) \rightarrow Co(NO_2)_6^{3-}(aq)$

c. $Cu(s) + 4H^+(aq) + 2NO_3^-(aq) \rightarrow Cu^{2+}(aq) + 2NO_2(g) + 2H_2O$

35.33 a. $Fe_2O_3(s) + 6NO_3^-(aq) + 6H^+(aq) \rightarrow 2Fe^{3+}(aq) + 6NO_3^-(aq) + 3H_2O$

b. $3Cu(s) + 8H^+(aq) + 2NO_3^-(aq) \rightarrow 3Cu^{2+}(aq) + 2NO(g) + 4H_2O$

c. $NO_2^-(aq) + Fe^{2+}(aq) + 2H^+(aq) \rightarrow NO(g) + Fe^{3+}(aq) + H_2O$

35.34 a. $4HNO_3(aq) \rightarrow 4NO_2(g) + 2H_2O + O_2(g)$

b. $3HNO_2(aq) \rightarrow 2NO(g) + H^+(aq) + NO_3^-(aq) + H_2O$

c. $2H^+(aq) + 2ClO_3^-(aq) \rightarrow 2ClO_2(aq) + \frac{1}{2} O_2(g) + H_2O$

35.35 a. $H^+(aq) + NO_2^-(aq) \longrightarrow HNO_2(aq)$

b. $Fe^{2+}(aq) + NO(g) \longrightarrow Fe(NO)^{2+}(aq)$

c. $ClO_3^-(aq) + 3NO_2^-(aq) \longrightarrow Cl^-(aq) + 3NO_3^-(aq)$

d. $HC_2H_3O_2(aq) + OH^-(aq) \longrightarrow C_2H_3O_2^-(aq) + H_2O$

35.36 a. $HNO_2(aq) \rightleftharpoons H^+(aq) + NO_2^-(aq)$

b. $HClO_3(aq) \longrightarrow H^+(aq) + ClO_3^-(aq)$

c. $C_2H_3O_2^-(aq) + H_2O \rightleftharpoons HC_2H_3O_2(aq) + OH^-(aq)$

35.37 a. $NiO(s) + 2H^+(aq) + 2NO_3^-(aq) \longrightarrow Ni^{2+}(aq) + 2NO_3^-(aq) + H_2O$

b. $3Cu(s) + 8H^+(aq) + 2NO_3^-(aq) \longrightarrow 3Cu^{2+}(aq) + 2NO(g) + 4H_2O$

c. $Cu^{2+}(aq) + 2NO_3^-(aq) + 2\ OH^-(aq) \longrightarrow Cu(OH)_2(s) + 2NO_3^-(aq)$

$Cu(OH)_2(s) \longrightarrow CuO(s) + H_2O(g)$

35.38 a. $2NaNO_3(s) \longrightarrow 2NaNO_2(s) = O_2(g)$

b. $2Pb(NO_3)_2(s) \longrightarrow 2PbO(s) + 4NO_2(g) + O_2(g)$

c. $CH_3CHO(l) + \frac{1}{2}\ O_2(g) \longrightarrow CH_3COOH(l)$

35.39 a. ClO_4^-

b. $2[ClO_3^-(aq) + 6H^+(aq) + 5e^- \longrightarrow \frac{1}{2}\ Cl_2(g) + 3H_2O]$

$\underline{5[ClO_3^-(aq) + H_2O \longrightarrow ClO_4^-(aq) + 2H^+(aq) + 2e^-]}$

$7ClO_3^-(aq) + 2H^+(aq) \longrightarrow Cl_2(g) + H_2O + 5ClO_4^-(aq)$

$7ClO_3^-(aq) + H_2O \longrightarrow 5ClO_4^-(aq) + Cl_2(g) + 2\ OH^-(aq)$

c. $1.00\ g\ ClO_3^- \times \frac{1\ mol\ ClO_3^-}{83.4\ g\ ClO_3^-} \times \frac{5\ mol\ ClO_4^-}{7\ mol\ ClO_3^-} \times \frac{99.4\ g\ ClO_4^-}{1\ mol\ ClO_4^-}$

$= 0.851\ g\ ClO_4^-$

35.40 a. $4[Zn(s) \longrightarrow Zn^{2+}(aq) + 2e^-]$

$\underline{NO_3^-(aq) + 10\ H^+(aq) + 8e^- \longrightarrow NH_4^+(aq) + 3H_2O}$

$4Zn(s) + NO_3^-(aq) + 10H^+(aq) \longrightarrow 4Zn^{2+}(aq) + NH_4^+(aq) + 3H_2O$

b. $2I^-(aq) \longrightarrow I_2(s) + 2e^-$

$\underline{2[NO_3^-(aq) + 2H^+(aq) + e^- \longrightarrow NO_2(g) + H_2O]}$

$2I^-(aq) + 2NO_3^-(aq) + 4H^+(aq) \longrightarrow I_2(s) + 2NO_2(g) + 2H_2O$

35.40 c. $S^{2-}(aq) \longrightarrow S(s) + 2\ e^-$

$$\frac{2[NO_2^-(aq) + e^- + H_2O \longrightarrow NO(g) + 2\ OH^-(aq)]}{S^{2-}(aq) + 2NO_2^-(aq) + 2H_2O \longrightarrow S(s) + 2NO(g) + 4\ OH^-(aq)}$$

35.41 a. $E^o_{tot} = 1.45\ V + E^o_{ox}NO = 1.45\ V - 0.96\ V = 0.49\ V$

b. $E^o_{tot} = 1.45\ V + E^o_{ox}HNO_2 = 1.45\ V - 0.94\ V = 0.51\ V$

c. $E^o_{tot} = 1.45\ V + E^o_{ox}HNO_2 = 1.45\ V - 1.10\ V = 0.35\ V$

35.42 $ClO_3^-(aq) + 6H^+(aq) + 6e^- \longrightarrow Cl^-(aq) + 3H_2O$

$$E = +1.45\ V - \frac{0.0591}{6} \log_{10} \frac{1}{(\text{conc. } H^+)^6}$$

$$= +1.45\ V - 0.0591\ pH$$

a. +1.45 V b. +1.04 V c. +0.62 V

35.43 a. 3.3 b. 4.0 c. 2.7

35.44 a. $\frac{[H^+]^2}{0.20} \approx 1.8 \times 10^{-5}$; $[H^+] = 1.9 \times 10^{-3}$ M

b. $\frac{[HC_2H_3O_2]}{[C_2H_3O_2^-]} = \frac{[H^+]}{K_a} = \frac{5.0 \times 10^{-5}}{1.8 \times 10^{-5}} = 2.8$

c. $\frac{1.0 \times 10^{-14}}{1.8 \times 10^{-5}} = 5.6 \times 10^{-10}$

35.45 $\frac{[OH^-]^2}{0.10} \approx 2.2 \times 10^{-11}$; $[OH^-] = 1.5 \times 10^{-6}$ M

$[H^+] = 6.7 \times 10^{-9}$ M; pH = 8.17

35.46 a. conc. $HC_2H_3O_2 = \frac{0.080 \times 0.200\ M}{0.100} = 0.160$ M

conc. $C_2H_3O_2^- = \frac{0.020 \times 1.00\ M}{0.100} = 0.200$ M

b. $[H^+] = 1.8 \times 10^{-5} \times \frac{0.160}{0.200} = 1.4 \times 10^{-5}$ M; pH = 4.85

c. moles $HC_2H_3O_2 = 0.0160 - 0.0040 = 0.0120$

moles $C_2H_3O_2^- = 0.0200 + 0.0040 = 0.0240$

$[H^+] = 1.8 \times 10^{-5} \times \frac{0.0120}{0.0240} = 9.0 \times 10^{-6}$; pH = 5.05

35.47 a. $\frac{[HNO_2]}{[NO_2^-]} = \frac{[H^+]}{K_a} = \frac{1.0 \times 10^{-3}}{4.5 \times 10^{-4}} = 2.2$

b. $\frac{3.2 \times 10^{-4}}{4.5 \times 10^{-4}} = 0.71$ c. $\frac{1.0 \times 10^{-4}}{4.5 \times 10^{-4}} = 0.22$

35.48 a. $K = \frac{K_a HNO_2}{K_a HC_2H_3O_2} = \frac{4.5 \times 10^{-4}}{1.8 \times 10^{-5}} = 25$

b. $K = 1/K_a HNO_2 = 1/(4.5 \times 10^{-4}) = 2.2 \times 10^3$

c. $K = 1/K_b C_2H_3O_2^- = 1/(5.6 \times 10^{-10}) = 1.8 \times 10^9$

35.49 $ClO_3^-(aq) + 3NO_2^-(aq) \longrightarrow Cl^-(aq) + 3NO_3^-(aq)$

$Ag^+(aq) + Cl^-(aq) \longrightarrow AgCl(s)$

35.50 a. oxidized to I_2, which interferes with color

b. gives precipitate with Ag^+

35.51 a. $Ag^+(aq) + Br^-(aq) \longrightarrow AgBr(s)$

b. $Ag^+(aq) + Br^-(aq) \longrightarrow AgBr(s)$

c. $Ba^{2+}(aq) + CrO_4^{2-}(aq) \longrightarrow BaCrO_4(s)$

35.52 NO_2^- absent; would give N_2 with sulfamic acid

$C_2H_3O_2^-$ present; odor is that of $HC_2H_3O_2$

ClO_3^- absent; would give precipitate

NO_3^- in doubt

35.53 $C_2H_3O_2^-$ absent; would give basic solution

NO_2^- absent; would give basic solution

ClO_3^- present; gives precipitate of AgCl

NO_3^- in doubt

35.54 a. 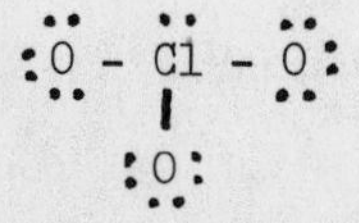 (109°) b. 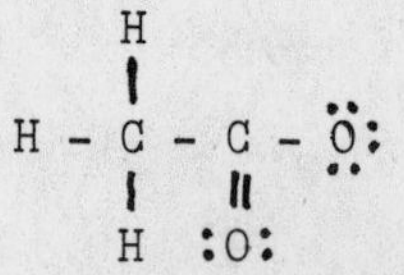 109°, 120°

35.55 a. Test pH; $NaNO_2$ is basic b. Add Ag^+; NaCl gives ppt.

c. Add $La(NO_3)_3$ + KI_3; blue color indicates $NaC_2H_3O_2$

35.55 d. Add Ag^+; NaBr gives precipitate

35.56 a. Heat with acid b. Reduce with NO_2^-; precipitate as AgCl

c. Precipitate with Ag^+ d. Neutralize with H^+

35.57 a. O b. O c. R d. O

35.58 a. all except $C_2H_3O_2^-$ b. NO_2^- c. NO_2^-, ClO_3^-, NO_3^-

35.59 Add Ag^+ in acid; I^- forms white ppt. of AgI. To another sample of unknown, add Fe^{3+}; light tan ppt. indicates PO_4^{3-}. To test for NO_3^-, first remove I^- with Ag^+, then carry out brown ring test.

35.60 a. $1.00 \text{ g Cu} \times \frac{1 \text{ mol Cu}}{63.5 \text{ g Cu}} \times \frac{4 \text{ mol HNO}_3}{1 \text{ mol Cu}} = 0.0630 \text{ mol HNO}_3$

$V = \frac{0.0630 \text{ mol}}{6.0 \text{ mol/L}} = 0.010 \text{ L} = 10 \text{ mL}$

b. $1.00 \text{ g Cu} \times \frac{1 \text{ mol Cu}}{63.5 \text{ g Cu}} \times \frac{2 \text{ mol NO}_2}{1 \text{ mol Cu}} \times \frac{46.0 \text{ g NO}_2}{1 \text{ mol NO}_2} = 1.45 \text{ g NO}_2$

*35.61 $FeSO_4 \cdot 4H_2O$ absent; colored

$KClO_3$, $ZnBr_2$ present; react to give Br_2

$NaC_2H_3O_2$ present; odor is that of $HC_2H_3O_2$

*35.62 3.7 - 5.7

*35.63 $E^o_{tot} = E^o_{ox}\,Ag + E^o_{red}\,NO_3^- = -0.80 \text{ V} + 0.78 \text{ V} = -0.02 \text{ V}$

$\log_{10}K = \frac{1(-0.02)}{0.0591} = -0.3;\ K = 0.5$

$K = \frac{[Ag^+] \times 1.0}{(0.10)^2(0.10)} = 0.5;\ [Ag^+] = 5 \times 10^{-4} \text{ M}$

*35.64 moles $NO_2^- = 0.020 \frac{\text{mol}}{\text{L}} \times 5.0 \times 10^{-3} \text{ L} = 1.0 \times 10^{-4} \text{ mol}$

moles NH_2SO_3H required $= 1.0 \times 10^{-4}$ mol

mass required $= 1.0 \times 10^{-4} \text{ mol} \times \frac{97.1 \text{ g}}{1 \text{ mol}} = 9.7 \times 10^{-3} \text{ g}$

mass used = 0.0097 g/0.80 = 0.012 g

*35.65 orig. conc. $I_2 = \frac{0.25/253.8}{0.010}$ M = 0.10 M

	orig. conc.	change	equil. conc.
I_2	0.10 M	-x	0.10 - x
I^-	0.10 M	-x	0.10 - x
I_3^-	0.00 M	+x	x

$$\frac{x}{(0.10 - x)^2} = 700; \; x = 0.09 \text{ M}$$

Test Questions, Multiple Choice

1. Of the anions in the soluble group (NO_3^-, NO_2^-, ClO_3^-, $C_2H_3O_2^-$) how many are derived from weak acids?

   a. 0 b. 1 c. 2 d. 3 e. 4

2. The nitrate ion in acidic solution is:

   a. a strong oxidizing agent and a poor complexing agent

   b. a weak oxidizing agent and a poor complexing agent

   c. a strong oxidizing agent and a good complexing agent

   d. a weak oxidizing agent and a good complexing agent

3. The brown ring test involves a complex between

   a. Fe^{3+} and NO_2 b. Fe^{3+} and NO c. Fe^{2+} and NO_2

   d. Fe^{2+} and NO e. Fe^{2+} and NO_2^-

4. How many of the following acids decompose spontaneously by redox reactions in water solution?

   HNO_3, HNO_2, $HClO_3$, $HC_2H_3O_2$

   a. 0 b. 1 c. 2 d. 3 e. 4

5. Which one of the following oxyanions can act as either an oxidizing or reducing agent?

   a. Cl^- b. NO_3^- c. NO_2^- d. $Cr_2O_7^{2-}$ e. CrO_4^{2-}

6. Which Soluble Group anion is detected by precipitating AgCl?

   a. NO_3^- b. NO_2^- c. ClO_3^- d. $C_2H_3O_2^-$

7. A water solution of $NaClO_3$ is commonly made by the reaction of

a. Na with Cl_2 and O_2 b. NaOH with ClO_2

c. NaOH with Cl_2 d. NaCl with O_2 e. NaCl with O_3

8. The acid present in vinegar is:

a. H_2CO_3 b. HNO_3 c. HCl d. $HC_2H_3O_2$ e. H_2S

9. Which Soluble Group anion is detected by a test that uses $La(NO_3)_3$ as a reagent?

a. NO_3^- b. NO_2^- c. ClO_3^- d. $C_2H_3O_2^-$

10. Which one of the following anions must be removed before testing for NO_3^-?

a. NO_2^- b. NH_4^+ c. Cl^- d. Ag^+ e. Na^+

<u>Test Problems</u>

1. Write a balanced net ionic equation for the reaction of

a. the nitrite ion with water

b. the acetate ion with hydrochloric acid

c. the Fe^{2+} ion with nitric oxide

d. copper with nitric acid

2. An unknown may contain any of the following anions, but no others:

Cl^-, SO_4^{2-}, ClO_3^-, NO_3^-

When barium chloride solution is added to a portion of the unknown, a white precipitate forms. When silver nitrate is added to another portion, no precipitate forms; however, the solution produced gives a brown ring test. Which ions are present? absent? in doubt? Explain your reasoning.

3. How would you determine whether the anion in a sodium salt was:

a. Cl^- or ClO_3^-? b. NO_2^- or NO_3^-? c. $C_2H_3O_2^-$ or SO_4^{2-}

4. K_a of HNO_2 is 4.5×10^{-4}. At what pH will the equilibrium concentration of HNO_2 be half that of NO_2^-?

5. K_a of $HC_2H_3O_2$ is 1.8×10^{-5}; K_a of HNO_2 is 4.5×10^{-4}.
Consider the reaction:

$$HC_2H_3O_2(aq) + NO_2^-(aq) \rightarrow C_2H_3O_2^-(aq) + HNO_2(aq)$$

a. Calculate K for this reaction.

b. If one starts with 0.10 M solutions of acetic acid and sodium nitrite, what will be the equilibrium concentration of nitrous acid?

Answers to Problems

36.1 a. $2H^+(aq) + CO_3^{2-}(aq) \longrightarrow H_2CO_3(aq) \longrightarrow CO_2(g) + H_2O$

b. $BaCO_3(s) + 2H^+(aq) \longrightarrow Ba^{2+}(aq) + CO_2(g) + H_2O$

c. $ZnS(s) + 2H^+(aq) \longrightarrow Zn^{2+}(aq) + H_2S(g)$

36.2 a. $NiCO_3(s) + 2H^+(aq) \longrightarrow Ni^{2+}(aq) + H_2CO_3(aq)$

b. $NiCO_3(s) + 2HC_2H_3O_2(aq) \longrightarrow Ni^{2+}(aq) + H_2CO_3(aq) + 2C_2H_3O_2^-(aq)$

c. $NiCO_3(s) + 6NH_3(aq) \longrightarrow Ni(NH_3)_6^{2+}(aq) + CO_3^{2-}(aq)$

36.3 a. $Ba^{2+}(aq) + SO_3^{2-}(aq) \longrightarrow BaSO_3(s)$

b. $H_2O_2(aq) + SO_3^{2-}(aq) \longrightarrow SO_4^{2-}(aq) + H_2O$

c. $SO_2(aq) + SO_3^{2-}(aq) + H_2O \longrightarrow 2HSO_3^-(aq)$

36.4 a. $H_2CO_3(aq) + 2\ OH^-(aq) \longrightarrow CO_3^{2-}(aq) + 2H_2O$

b. $SO_2(g) + OH^-(aq) \longrightarrow HSO_3^-(aq)$

c. $SO_3^{2-}(aq) + \frac{1}{2}\ O_2(g) \longrightarrow SO_4^{2-}(aq)$

d. $Pb^{2+}(aq) + H_2S(aq) \longrightarrow PbS(s) + 2H^+(aq)$

36.5 a. $2H^+(aq) + SO_3^{2-}(aq) \longrightarrow SO_2(g) + H_2O$

b. $SO_3^{2-}(aq) + H_2O \rightleftharpoons HSO_3^-(aq) + OH^-(aq)$

c. $S^{2-}(aq) + H_2O \rightleftharpoons HS^-(aq) + OH^-(aq)$

36.6 a. $H_2S(aq) + 2\ OH^-(aq) \longrightarrow S^{2-}(aq) + 2H_2O$

b. $SO_3^{2-}(aq) + 2H^+(aq) \longrightarrow SO_2(g) + H_2O$

c. $HSO_3^-(aq) \rightleftharpoons H^+(aq) + SO_3^{2-}(aq)$

36.7 a. $Cu^{2+}(aq) + S^{2-}(aq) \longrightarrow CuS(s)$

b. $S^{2-}(aq) + ClO^-(aq) + H_2O \longrightarrow S(s) + Cl^-(aq) + 2\ OH^-(aq)$

c. $S^{2-}(aq) + H_2O \longrightarrow HS^-(aq) + OH^-(aq)$

d. $S^{2-}(aq) + 2H^+(aq) \longrightarrow H_2S(g)$

36.8 a. Cl^-

b. $Cl_2(g) + 2e^- \longrightarrow 2Cl^-(aq)$

$SO_3^{2-}(aq) + 2\,OH^-(aq) \longrightarrow SO_4^{2-}(aq) + H_2O + 2e^-$

$Cl_2(g) + SO_3^{2-}(aq) + 2\,OH^-(aq) \longrightarrow 2Cl^-(aq) + SO_4^{2-}(aq) + H_2O$

c. moles $Cl_2 = 0.0500\text{ L} \times 0.100\,\frac{\text{mol}}{\text{L}} = 5.00 \times 10^{-3}$ mol

$V = \frac{nRT}{P} = \frac{(5.00 \times 10^{-3})(0.0821)(273)}{1.00}\text{ L} = 0.112\text{ L}$

36.9 a. $3[HSO_3^-(aq) + H_2O \longrightarrow SO_4^{2-}(aq) + 3H^+(aq) + 2e^-]$

$2[NO_3^-(aq) + 4H^+(aq) + 3e^- \longrightarrow NO(g) + 2H_2O]$

$3HSO_3^-(aq) + 2NO_3^-(aq) \longrightarrow 3SO_4^{2-}(aq) + H^+(aq) + 2NO(g) + H_2O$

b. $H_2S(aq) \longrightarrow S(s) + 2H^+(aq) + 2e^-$

$2[NO_2^-(aq) + 2H^+(aq) + e^- \longrightarrow NO(g) + H_2O]$

$H_2S(aq) + 2NO_2^-(aq) + 2H^+(aq) \longrightarrow S(s) + 2NO(g) + 2H_2O$

c. $3[SO_3^{2-}(aq) + 2\,OH^-(aq) \longrightarrow SO_4^{2-}(aq) + H_2O + 2e^-]$

$2[CrO_4^{2-}(aq) + 4H_2O + 3e^- \longrightarrow Cr(OH)_3(s) + 5\,OH^-(aq)]$

$3SO_3^{2-}(aq) + 2CrO_4^{2-}(aq) + 5H_2O \longrightarrow 3SO_4^{2-}(aq) + 2Cr(OH)_3(s) + 4\,OH^-(aq)$

36.10 a. $E^o_{tot} = E^o_{red}Fe^{3+} + E^o_{ox}H_2S = +0.63\text{ V}$

b. $E^o_{tot} = E^o_{red}Fe(OH)_3 + E^o_{ox}S^{2-} = -0.13\text{ V}$

c. $E^o_{tot} = E^o_{ox}S^{2-} + E^o_{red}O_2 = +0.83\text{ V}$

36.11 a. $E_{ox} = -0.14\text{ V} - \frac{0.0591}{2}\log_{10}\frac{(1 \times 10^{-3})^2}{0.10} = +0.01\text{ V}$

b. $0.14 = -\frac{0.0591}{2}\log_{10}\frac{(\text{conc. }H^+)^2}{0.10}$; $\log_{10}\frac{(\text{conc. }H^+)^2}{0.10} = -4.7$

$\frac{(\text{conc. }H^+)^2}{0.10} = 1.8 \times 10^{-5}$; conc. $H^+ = 1.3 \times 10^{-3}$ M;

pH = 2.9

36.12 a. $K = K_{sp}Ag_2S/(1 \times 10^{-20}) = 1 \times 10^{-29}$

b. $\frac{(2s)^2(s)}{(0.10)^2} = 1 \times 10^{-29}$; $4s^3 = 1 \times 10^{-31}$; $s = 3 \times 10^{-11}$ M

36.13 a. $[Cu^{2+}] = K_{sp}CuCO_3/0.10 = 2 \times 10^{-9}$ M

b. $[Cu^{2+}] = K_{sp}CuS/0.10 = 1 \times 10^{-34}$ M

c. $[S^{2-}] = \frac{(1 \times 10^{-20})(0.10)}{(1 \times 10^{-6})^2} = 1 \times 10^{-9}$ M

$[Cu^{2+}] = 1 \times 10^{-26}$ M

36.14 Approximately 20 volumes $NaHCO_3$ to 1 volume Na_2CO_3

36.15 a. 0.2, 2 b. 2, 0.2

36.16 a. $\frac{[HSO_3^-]}{[H_2SO_3]} = \frac{1.7 \times 10^{-2}}{[H^+]}$; 0.17, 1.7

b. $\frac{[HSO_3^-]}{[SO_3^{2-}]} = \frac{[H^+]}{5.6 \times 10^{-8}}$; 1.8, 0.18

36.17 a. $\frac{1.0 \times 10^{-14}}{4.8 \times 10^{-11}} = 2.1 \times 10^{-4}$

b. $\frac{[OH^-]^2}{0.10} \approx 2.1 \times 10^{-4}$; $[OH^-] = 0.0046$ M

$[H^+] = 2.2 \times 10^{-12}$ M; pH = 11.66

36.18 $[H_2S] = \frac{1 \times 10^{-26}}{1 \times 10^{-20}} \times [S^{2-}] = (1 \times 10^{-6}) \times [S^{2-}]$

a. 5×10^{-8} M b. 2×10^{-8} M c. 1×10^{-8} M

36.19 a. $K = \frac{K_a HSO_3^-}{K_a HCO_3^-} = \frac{5.6 \times 10^{-8}}{4.8 \times 10^{-11}} = 1.2 \times 10^3$

b. $K = K_a\ H_2CO_3 \times K_a\ HCO_3^- = 2.0 \times 10^{-17}$

c. $K = K_{sp}CuS/(1 \times 10^{-20}) = 1 \times 10^{-15}$

36.20 a. $2H^+(aq) + S^{2-}(aq) \longrightarrow H_2S(g)$

$Pb^{2+}(aq) + H_2S(g) \longrightarrow PbS(s) + 2H^+(aq)$

b. $2H^+(aq) + SO_3^{2-}(aq) \longrightarrow SO_2(g) + H_2O$

36.21 a. A few S^{2-} ions are converted to H_2S, which has a powerful odor.

b. Oxidized to SO_4^{2-} by H_2O_2

c. SO_3^{2-} gives precipitate with Ba^{2+}; SO_4^{2-} does not give SO_2

36.22 Step 1: Add HNO_3; observe effervescence of CO_2 if CO_3^{2-} is present

Step 2: To part of HNO_3 solution, add excess Ag^+; precipitate of AgBr indicates Br^-.

Step 3: To solution remaining after Br^- precipitation
- add $KMnO_4$; if decolorized, $C_2O_4^{2-}$ is present
- add NO_2^-; precipitate indicates ClO_3^- is present

Step 4: To acidified solution, add Ba^{2+}; precipitate indicates SO_4^{2-}.

36.23 S^{2-} absent; would give odor of H_2S

SO_3^{2-} present; sharp odor with acid, precipitate with Ba^{2+}

CO_3^{2-} in doubt; precipitate could be mixture of $BaSO_3$, $BaSO_4$

36.24 CrO_4^{2-} absent; yellow color

I^-, Cl^- absent; AgI, AgCl would not dissolve in HNO_3

SO_3^{2-} present; Ag_2SO_3 dissolves in acid

36.25 $\Delta H = \Delta H_f CO_3^{2-}(aq) + \Delta H_f H_2O(l) - \Delta H_f CO_2(g) - 2\,\Delta H_f OH^-(aq)$

$= -676.3\ kJ - 285.8\ kJ + 393.5\ kJ + 459.8\ kJ = -108.8\ kJ$

$\Delta S^o = S^o CO_3^{2-}(aq) + S^o H_2O(l) - S^o CO_2(g) - 2S^o\ OH^-(aq)$

$= (-53.1 + 69.9 - 213.6 + 21.0) J/K = -175.8\ J/K$

$\Delta G^o = -108.8\ kJ + 298(0.1758) kJ = -56.4\ kJ$

$\log_{10}K = 9.91$; $K = 8.1 \times 10^9$

36.26 H_2CO_3, HCO_3^-; HSO_3^-, SO_3^{2-}; H_2S, HS^-

36.27 a, b, c, e

36.28 a. precipitates S b. oxidized by air

c. CO_3^{2-} in equilibrium with HCO_3^-, OH^-

36.29 a. Test pH; Na_2SO_3 is basic

b. Smell; Na_2S has strong odor of H_2S

c. Add Ca^{2+}; CO_3^{2-} gives precipitate

36.30 a. $H - \overset{\cdot\cdot}{\underset{\cdot\cdot}{S}} :$ b. $H - \overset{\cdot\cdot}{\underset{\cdot\cdot}{O}} - \overset{}{\underset{\underset{:\overset{\cdot\cdot}{O}:}{|}}{S}} - \overset{\cdot\cdot}{\underset{\cdot\cdot}{O}} :$ c. $H - \overset{\cdot\cdot}{\underset{\cdot\cdot}{O}} - \overset{}{\underset{\underset{:O:}{||}}{C}} - \overset{\cdot\cdot}{\underset{\cdot\cdot}{O}} :$

36.31 a. $2H^+(aq) + CaCO_3(s) \longrightarrow Ca^{2+}(aq) + H_2CO_3(aq)$

$$\downarrow$$

$$CO_2(g) + H_2O$$

b. $MnS(s) + 2H^+(aq) \longrightarrow H_2S(aq) + Mn^{2+}(aq)$

c. $BaSO_3(s) + 2H^+(aq) \longrightarrow Ba^{2+}(aq) + SO_2(g) + H_2O$

36.32 a. $Ag_2SO_3(s) + 2H^+(aq) \longrightarrow 2Ag^+(aq) + SO_2(g) + H_2O$

b. $Ag_2SO_3(s) + 2HC_2H_3O_2(aq) \longrightarrow 2Ag^+(aq) + SO_2(g) + H_2O$

$$+ 2C_2H_3O_2^-(aq)$$

c. $Ag_2SO_3(s) + 4NH_3(aq) \longrightarrow 2Ag(NH_3)_2^+(aq) + SO_3^{2-}(aq)$

36.33 a. $Cu^{2+}(aq) + S^{2-}(aq) \longrightarrow CuS(s)$

b. $2H^+(aq) + S^{2-}(aq) \longrightarrow H_2S(g)$

c. $S^{2-}(aq) + H_2O + \frac{1}{2} O_2(g) \longrightarrow S(s) + 2 OH^-(aq)$

36.34 a. $H_2CO_3(aq) + OH^-(aq) \longrightarrow HCO_3^-(aq) + H_2O$

b. $SO_2(g) + 2 OH^-(aq) \longrightarrow SO_3^{2-}(aq) + H_2O$

c. $BaSO_3(s) + 2H^+(aq) \longrightarrow Ba^{2+}(aq) + SO_2(g) + H_2O$

d. $ZnS(s) + 2H^+(aq) \longrightarrow Zn^{2+}(aq) + H_2S(aq)$

36.35 a. $H^+(aq) + CO_3^{2-}(aq) \longrightarrow HCO_3^-(aq)$

b. $CO_3^{2-}(aq) + H_2O \rightleftharpoons HCO_3^-(aq) + OH^-(aq)$

c. $H_2S(aq) + OH^-(aq) \longrightarrow HS^-(aq) + H_2O$

36.36 a. $Ba^{2+}(aq) + SO_3^{2-}(aq) \longrightarrow BaSO_3(s)$

b. $Pb^{2+}(aq) + H_2S(aq) \longrightarrow PbS(s) + 2H^+(aq)$

c. $Ca^{2+}(aq) + CO_3^{2-}(aq) \longrightarrow CaCO_3(s)$

36.37 a. redox   b. redox, precipitation   c. acid-base

d. acid-base, precipitation

36.38 a. O_2; OH^- ions

b. $S^{2-}(aq) + H_2O + \frac{1}{2} O_2(g) \longrightarrow S(s) + 2\ OH^-(aq)$

c. $5.0 \times 10^{-3}\ L \times 0.10\ \frac{mol}{L} \times \frac{32.06\ g}{1\ mol} = 0.016\ g$

36.39 a. $3[SO_3^{2-}(aq) + H_2O \longrightarrow SO_4^{2-}(aq) + 2H^+(aq) + 2e^-]$

$Cr_2O_7^{2-}(aq) + 14H^+(aq) + 6e^- \longrightarrow 2Cr^{3+}(aq) + 7H_2O$

$3SO_3^{2-}(aq) + Cr_2O_7^{2-}(aq) + 8H^+(aq) \longrightarrow 3SO_4^{2-}(aq) + 4H_2O + 2Cr^{3+}(aq)$

b. $HSO_3^-(aq) + H_2O \longrightarrow SO_4^{2-}(aq) + 3H^+(aq) + 2e^-$

$2[NO_3^-(aq) + 2H^+(aq) + e^- \longrightarrow NO_2(g) + H_2O]$

$HSO_3^-(aq) + 2NO_3^-(aq) + H^+(aq) \longrightarrow SO_4^{2-}(aq) + 2NO_2(g) + H_2O$

c. $S^{2-}(aq) \longrightarrow S(s) + 2e^-$

$2[Fe^{3+}(aq) + 2\ OH^-(aq) + e^- \longrightarrow Fe(OH)_2(s)]$

$S^{2-}(aq) + 2Fe^{3+}(aq) + 4\ OH^-(aq) \longrightarrow S(s) + 2Fe(OH)_2(s)$

36.40 a. $E^o_{tot} = -0.57\ V + E^o_{ox}S^{2-} = -0.14\ V$

b. $E^o_{tot} = +0.57\ V + E^o_{red}O_2 = +0.97\ V$

c. $E^o_{tot} = -0.57\ V + E^o_{ox}NO_2^- = -0.58\ V$

36.41 a. $E_{ox} = +0.89\ V - \frac{0.0591}{2} \log_{10} \frac{(\text{conc. } SO_4^{2-})}{(\text{conc. } SO_3^{2-})(\text{conc. } OH^-)^2}$

$= +0.89\ V - \frac{0.0591}{2} \log_{10} \frac{0.10}{(1.0 \times 10^{-4})^2} = +0.68\ V$

b. $-0.14 = -\frac{0.0591}{2} \log_{10} \frac{1}{(\text{conc. } OH^-)^2} = 0.0591 \log_{10}(\text{conc } OH^-)$

$\log_{10}(\text{conc. } OH^-) = -2.4$; conc. $OH^- = 4 \times 10^{-3}\ M$

conc. $H^+ = 2.5 \times 10^{-12}\ M$; pH = 11.6

36.42 a. $K = K_{sp}CdS/(1 \times 10^{-20}) = 1 \times 10^{-6}$

b. $s^2 = (0.10)^2(1 \times 10^{-6})$; $s = 1 \times 10^{-4}\ M$

36.43 a. $[Pb^{2+}] = K_{sp}PbCO_3/0.010 = 1 \times 10^{-11}\ M$

b. $[Pb^{2+}] = K_{sp}PbS/0.010 = 1 \times 10^{-25}\ M$

36.43 c. $[S^{2-}] = \frac{(1 \times 10^{-20}) \times 0.010}{(1 \times 10^{-4})^2} = 1 \times 10^{-14}$ M

$[Pb^{2+}] = 1 \times 10^{-13}$ M

36.44 Mix about 1 part $NaHSO_3$ to 5 parts Na_2SO_3

36.45 a. 0.4, 4 b. 2, 0.2

36.46 a. $\frac{[HCO_3^-]}{[H_2CO_3]} = \frac{4.2 \times 10^{-7}}{[H^+]}$; 0.42, 4.2

b. $\frac{[HCO_3^-]}{[CO_3^{2-}]} = \frac{[H^+]}{4.8 \times 10^{-11}}$; 2.1, 0.21

36.47 a. $K_b = (1.0 \times 10^{-14})/(5.6 \times 10^{-8}) = 1.8 \times 10^{-7}$

b. $[OH^-]^2 \approx 1.8 \times 10^{-8}$; $[OH^-] = 1.3 \times 10^{-4}$ M

$[H^+] = 7.7 \times 10^{-11}$ M; pH = 10.11

36.48 $[H_2S] = \frac{[H^+]^2 \times [S^{2-}]}{1 \times 10^{-20}} = 1 \times 10^{19} \times [H^+]^2$

a. 1×10^{-5} M b. 2×10^{-6} M c. 1×10^{-6} M

36.49 a. $K = \frac{K_a\ H_2S}{K_a\ H_2CO_3} = \frac{1 \times 10^{-7}}{4.2 \times 10^{-7}} = 0.2$

b. $K = K_a\ H_2SO_3 \times K_a\ HSO_3^- = (1.7 \times 10^{-2})(5.6 \times 10^{-8})$

$= 9.5 \times 10^{-10}$

c. $K = K_{sp}\ ZnS/(1 \times 10^{-20}) = 1$

36.50 a. $SO_3^{2-}(aq) + H_2O_2(aq) \longrightarrow SO_4^{2-}(aq) + H_2O$

$Ba^{2+}(aq) + SO_4^{2-}(aq) \longrightarrow BaSO_4(s)$

b. $2H^+(aq) + CO_3^{2-}(aq) \longrightarrow CO_2(g) + H_2O$

$CO_2(g) + Ba^{2+}(aq) + 2\ OH^-(aq) \longrightarrow BaCO_3(s) + H_2O$

36.51 a. Forms gas with acid; gives precipitate with $Ba(OH)_2$

b. Oxidized to SO_4^{2-} by H_2O_2

c. Precipitates and hence removes S^{2-} and SCN^-

36.52 Step 1: Add $AgNO_3$ to precipitate AgSCN (white) and Ag_2S (black)

Step 2: Treat precipitate in (1) with NH_3; black residue indicates Ag_2S

Step 3: Treat NH_3 solution with H^+ and then Fe^{3+}; blood-red color indicates SCN^-

Step 4: Treat portion of solution from (1) with acid, heat; odor of vinegar indicates $C_2H_3O_2^-$

Step 5: To another portion of solution from (1), add KNO_2, $AgNO_3$; white precipitate indicates ClO_3^-

36.53 S^{2-} present (odor)

CO_3^{2-} present (gives precipitate with $Ba(OH)_2$)

SO_3^{2-} in doubt (would be oxidized by H_2O_2)

36.54 CrO_4^{2-} present (yellow, orange colors)

CO_3^{2-} present (effervescence)

SO_4^{2-}, NO_3^- in doubt

36.55 $\Delta H = \Delta H_f\, Cu^{2+}(aq) + 2\,\Delta H_f NO_2(g) + 2\,\Delta H_f H_2O(l) - \Delta H_f CuS(s) - 2\,\Delta H_f NO_3^-(aq)$

$= (64.4 + 67.8 - 571.6 + 48.5 + 413.2)\text{kJ} = +22.3\text{ kJ}$

$\Delta S^\circ = S^\circ Cu^{2+}(aq) + S^\circ S(s) + 2S^\circ NO_2(g) + 2S^\circ H_2O(l) - S^\circ CuS(s) - 2S^\circ NO_3^-(aq)$

$= (-98.7 + 31.9 + 481.0 + 139.8 - 66.5 - 292.8)\text{J/K}$
$= -194.7\text{ J/K}$

$\Delta G^\circ = \Delta H - T\Delta S^\circ = 22.3\text{ kJ} - 298(0.1947)\text{kJ} = -35.7\text{ kJ}$

$\log_{10}K = \dfrac{35.7}{(0.0191)(298)} = 6.27;\ K = 1.9 \times 10^6$

36.56 a. H_2SO_3-HSO_3^- b. HCO_3^--CO_3^{2-} c. HS^--S^{2-}

36.57 b, c, e

36.58 a. oxidized to sulfate b. CuS very insoluble

c. solubility $CO_2 < SO_2$

36.59 a. Acidify and heat; strong odor indicates SO_3^{2-}

b. Smell; H_2S indicates S^{2-}

c. Treat with acid; note effervescence with CO_3^{2-}

36.60 a. $H - \overset{..}{\underset{..}{O}} - C - \overset{..}{\underset{..}{O}}:$ b. $H - \overset{..}{\underset{..}{O}} - \overset{..}{\underset{..}{S}} - \overset{..}{\underset{..}{O}} - H$ c. $H - \overset{..}{\underset{..}{S}} - H$
$\quad\quad\quad\quad \underset{:O:}{\overset{\|}{}}$ $\quad\quad\quad\quad\quad \underset{:O:}{\overset{|}{}}$

*36.61 n CO_2 = 5.0 x 10^{-3} L x 0.10 $\dfrac{mol}{L}$ = 5.0 x 10^{-4} mol

CO_2 in solution = 0.035 $\dfrac{mol}{L}$ x 0.010 L = 3.5 x 10^{-4} mol

V CO_2 = $\dfrac{(1.5 \times 10^{-4} \text{ mol})(0.0821 \text{ L·atm/mol·K})(298 \text{ K})}{1.00 \text{ atm}}$

$\quad\quad$ = 0.0037 L = 3.7 mL

*36.62 $\dfrac{x^2}{0.10 - x}$ = 0.10; $x^2 + 0.10x - 0.010 = 0$

$\quad\quad$ x = $\dfrac{-0.10 \pm \sqrt{0.010 + 0.040}}{2}$; x = 0.06 = $\left[OH^-\right]$

$\quad\quad$ $\left[H^+\right] \approx 1.7 \times 10^{-13}$; pH = 12.8

*36.63 36.19, 36.20, 36.30, 36.31, 36.32, 36.33, 36.34; 7

*36.64 K = $\dfrac{(0.01)(0.01)(0.02)^2}{1}$ = 4 x 10^{-8}

$\quad\quad$ = $\dfrac{K_{sp} \times (K_a HC_2H_3O_2)^2}{K_a H_2CO_3 \times K_a HCO_3^-}$

$\quad\quad$ K_{sp} = $\dfrac{(4 \times 10^{-8})(4.2 \times 10^{-7})(4.8 \times 10^{-11})}{(1.8 \times 10^{-5})^2}$ = 2 x 10^{-15}

*36.65 $\dfrac{\left[H_2CO_3\right]}{\left[CO_2\right] + \left[H_2CO_3\right]}$ = 0.0017

$\quad\quad$ K_a(true) = $\dfrac{\left[H^+\right] \times \left[HCO_3^-\right]}{\left[H_2CO_3\right]}$ = 4.2 x 10^{-7} x $\dfrac{\left[H_2CO_3\right] + \left[CO_2\right]}{\left[H_2CO_3\right]}$

$\quad\quad\quad\quad\quad\quad\quad\quad$ = $\dfrac{4.2 \times 10^{-7}}{1.7 \times 10^{-3}}$ = 2.5 x 10^{-4}

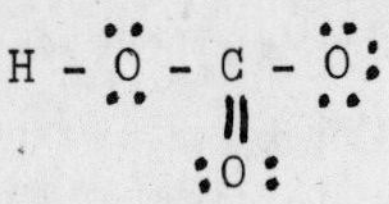

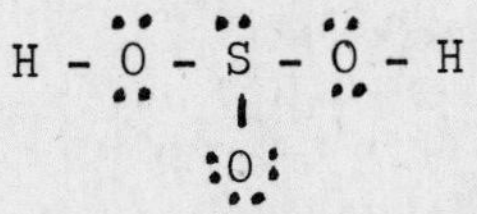

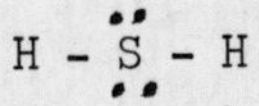

Test Questions, Multiple Choice

1. Which one of the following is the strongest acid?

   a. H_2CO_3 b. HCO_3^- c. CO_3^{2-} d. none are acids

2. K_a for H_2CO_3 is 4.2×10^{-7}. K_b for the CO_3^{2-} ion is:

   a. 2.4×10^{6} b. 4.2×10^{-7} c. 2.4×10^{-8}

   d. cannot tell without further information

3. Two reagents used to test for the presence of CO_3^{2-} are

   a. HCl and NaOH   b. $CaCl_2$ and NaOH   c. HCl and $Ba(OH)_2$

   d. HCl and $Pb(NO_3)_2$ e. none of these

4. Solutions containing the SO_3^{2-} ion are usually contaminated with

   a. H^+ b. S^{2-} c. SO_4^{2-} d. H_2S

5. The sulfite ion can be prepared in good yield by bubbling SO_2 through a solution of

   a. H_2SO_4 b. $AgNO_3$ c. $Fe(OH)_3$ d. NaOH   e. $NaNO_3$

6. Which one of the following gives off a gas most readily upon addition of acid?

   a. Cl^- b. SO_4^{2-} c. NO_3^- d. SO_3^{2-} e. SCN^-

7. K_a for H_2S is 1×10^{-7}; K_a for the HS^- ion is 1×10^{-13}. K for the reaction: $2H^+(aq) + S^{2-}(aq) \longrightarrow H_2S(aq)$ is:

   a. 1×10^{-20} b. 1×10^{-6} c. 1×10^{6} d. 1×10^{14}

   e. 1×10^{20}

8. Hydrogen sulfide can be prepared by adding acid to a solution containing:

   a. SO_3^{2-} b. SO_4^{2-} c. CH_3CSNH_2 d. SO_2 e. CH_3NH_2

9. Which one of the following anions is detected by a test that uses $Pb(NO_3)_2$ solution?

   a. Cl^- b. NO_3^- c. CrO_4^{2-} d. CO_3^{2-} e. S^{2-}

10. Which one of the following reagents is not effective in dissolving metal sulfides?

    a. HNO_3 b. aqua regia   c. NH_3 d. NaOH   e. Zn

Test Problems

1. Write a balanced net ionic equation to explain why

a. a foul odor is detected when a solution of sodium sulfide is acidified.

b. a water solution of $NaHCO_3$ is basic.

c. a precipitate forms when carbon dioxide is bubbled through a solution of barium hydroxide.

2. Describe how the following anions are detected in qualitative analysis.

a. S^{2-} b. CO_3^{2-}

3. Using appropriate tables in your text, calculate ΔG° at 25°C for the reaction:

$$Mg(OH)_2(s) + CO_2(g) \longrightarrow MgCO_3(s) + H_2O(l)$$

4. Taking K_a H_2CO_3 = 4.2×10^{-7}, K_a HCO_3^- = 4.8×10^{-11}, calculate the ratios $[HCO_3^-]/[H_2CO_3]$ and $[HCO_3^-]/[CO_3^{2-}]$ at pH 5.52.

5. Taking K_a HSO_3^- = 5.6×10^{-8}, calculate

a. K_b of the SO_3^{2-} ion

b. the pH of a 0.10 M Na_2SO_3 solution

Answers to Problems

37.1 a. $Al(OH)_3(s) + 3H^+(aq) \longrightarrow Al^{3+}(aq) + 3H_2O$

b. $CaC_2O_4(s) + 2H^+(aq) \longrightarrow Ca^{2+}(aq) + H_2C_2O_4(s)$

c. $NiCO_3(s) + 2H^+(aq) \longrightarrow Ni^{2+}(aq) + H_2CO_3(aq)$

d. $ZnS(s) + 2H^+(aq) \longrightarrow Zn^{2+}(aq) + H_2S(aq)$

37.2 a. $Al^{3+}(aq) + 4\ OH^-(aq) \longrightarrow Al(OH)_4^-(aq)$

b. $Ca^{2+}(aq) + H_2C_2O_4(aq) + 2OH^-(aq) \longrightarrow CaC_2O_4(s) + 2H_2O$

c. $Ni^{2+}(aq) + 2\ OH^-(aq) \longrightarrow Ni(OH)_2(s)$

d. $Zn^{2+}(aq) + 4\ OH^-(aq) \longrightarrow Zn(OH)_4^{2-}(aq)$

37.3 $Pb(OH)_2(s) + 2H^+(aq) \longrightarrow Pb^{2+}(aq) + 2H_2O$

$Pb^{2+}(aq) + 2\ OH^-(aq) \longrightarrow Pb(OH)_2(s)$

$Pb(OH)_2(s) + OH^-(aq) \longrightarrow Pb(OH)_3^-(aq)$

37.4 a. $Bi_2S_3(s) + 3Zn(s) + 6H^+(aq) \longrightarrow 3H_2S(aq) + 2Bi(s) + 3Zn^{2+}(aq)$

b. $CuI_2(s) + 2NO_3^-(aq) + 4H^+(aq) \longrightarrow Cu^{2+}(aq) + I_2(g) + 2H_2O + 2NO_2(g)$

c. $PbCl_2(s) + CO_3^{2-}(aq) \longrightarrow PbCO_3(s) + 2Cl^-(aq)$

37.5 $SCN^-(aq) + 4H_2O \longrightarrow SO_4^{2-}(aq) + HCN(aq) + 7H^+(aq) + 6e^-$

$6[NO_3^-(aq) + 2H^+(aq) + e^- \longrightarrow NO_2(g) + H_2O]$

$SCN^-(aq) + 6NO_3^-(aq) + 5H^+(aq) \longrightarrow SO_4^{2-}(aq) + HCN(aq) + 2H_2O + 6NO_2(g)$

37.6 a. $Cr(s) + 3H^+(aq) \longrightarrow Cr^{3+}(aq) + 3/2\ H_2(g)$

b. $3Ag(s) + NO_3^-(aq) + 4H^+(aq) \longrightarrow 3Ag^+(aq) + NO(g) + 2H_2O$

c. $Hg(l) + 2NO_3^-(aq) + 4Cl^-(aq) + 4H^+(aq) \longrightarrow HgCl_4^{2-}(aq) + 2NO_2(g) + 2H_2O$

37.7 a. $Al(OH)_3(s) + 3H^+(aq) \longrightarrow Al^{3+}(aq) + 3H_2O$

b. $Fe(OH)_2(s) + 2HC_2H_3O_2(aq) \longrightarrow Fe^{2+}(aq) + 2C_2H_3O_2^-(aq) + 2H_2O$

37.7 c. $Ni(OH)_2(s) + 6NH_3(aq) \rightarrow Ni(NH_3)_6^{2+}(aq) + 2\ OH^-(aq)$

37.8 a. $Na_2SO_3(s) + 2H^+(aq) \rightarrow 2Na^+(aq) + SO_2(g) + H_2O$

b. $Ag_2CO_3(s) + 2H^+(aq) \rightarrow 2Ag^+(aq) + CO_2(g) + H_2O$

c. $NaC_2H_3O_2(s) + H^+(aq) \rightarrow Na^+(aq) + HC_2H_3O_2(aq)$

37.9 a. $AgBr(s) + 2NH_3(aq) \rightarrow Ag(NH_3)_2^+(aq) + Br^-(aq)$

b. $PbSO_4(s) + 3\ OH^-(aq) \rightarrow Pb(OH)_3^-(aq) + SO_4^{2-}(aq)$

c. $BaSO_4(s) + CO_3^{2-}(aq) \rightarrow BaCO_3(s) + SO_4^{2-}(aq)$

d. $PbO_2(s) + 4H^+(aq) + 4Cl^-(aq) \rightarrow PbCl_2(s) + Cl_2(g) + 2H_2O$

37.10 a. $HC_2H_3O_2$, NH_3, NaOH   b. no   c. no; $CaCrO_4$ soluble

37.11 a. $BaSO_4$, $Bi_2(SO_4)_3$, $SnSO_4$, $Sb_2(SO_4)_3$, . . .

b. $SnSO_4$, $PbSO_4$ c. NaOH, Na_2CO_3, . .

37.12 Ag^+; anion could be NO_3^-, ClO_3^-, $C_2H_3O_2^-$, NO_2^- or SO_4^{2-}

37.13 A = Mg^{2+}, Fe^{3+}, possibly Cr^{3+} or Co^{2+}, Mn^{2+}, Hg^{2+}

B = Ni^{2+}, Cu^{2+}, or Cd^{2+}

C = Sn^{2+}, Sn^{4+}, or Sb^{3+}

37.14 a. Precipitate as $PbCl_2$, dissolve in hot water, precipitate $PbCrO_4$ from hot water solution

b. Precipitate as CuS at pH = 0.5, dissolve in HNO_3, add NH_3 to give deep blue color

c. Precipitate as ZnS at pH 9, dissolve in HCl, convert to $Zn(NH_3)_4^{2+}$, and finally precipitate as $K_2Zn_3[Fe(CN)_6]_2$

d. Heat with OH^- to form NH_3

37.15 a. $Ag(NH_3)_2^+$ b. $Cu(NH_3)_4^{2+}$ c. $Fe(OH)_3$ e. $Mg(OH)_2$

37.16 a. Ca^{2+}, K^+ b. Pb^{2+}, Al^{3+}, Ca^{2+}, K^+

c. Ca^{2+}, K^+ d. Ag^+, Cu^{2+}, Zn^{2+}, Cr^{3+}

37.17 a. Ba^{2+}, Na^+ b. Hg^{2+}, Sn^{4+} c. Na^+

d. Hg^{2+}, Cd^{2+}, Fe^{2+}, Ni^{2+}, Co^{2+}

37.18 a. ~ 0   b. 0.5   c. 9   d. 10

37.19 a. Precipitate as AgSCN, dissolve with Zn, convert to $Fe(SCN)^{2+}$

b. Precipitate as $BaCrO_4$, dissolve in acid, convert to CrO_5

c. Brown-ring test

d. Acidify to form H_2S, which reacts with Pb^{2+} to form PbS

37.20 a. NO_3^-, NO_2^- b. CrO_4^{2-} c. SO_4^{2-}, CrO_4^{2-}

d. SO_4^{2-}, NO_3^-

37.21 a. SCN^-, Br^-, ClO_3^- b. SCN^-, Br^-, PO_4^{3-}, $C_2O_4^{2-}$

c. SCN^-, Br^-, $C_2O_4^{2-}$ d. ClO_3^-

37.22 Cl^-, I^-, SO_4^{2-}, CrO_4^{2-}, SCN^-, Br^-, PO_4^{3-}, $C_2O_4^{2-}$

37.23 Fe (Fe^{3+} forms $Fe(OH)_3$); Co, Ni (Group III sulfides insoluble in HCl)

37.24 $MnCO_3$ (Mn^{2+} is light pink, oxidized to purple MnO_4^- by BiO_3^-; CO_3^{2-} converted to CO_2 by acid)

37.25 $(NH_4)_2C_2O_4$ (NH_4^+ gives NH_3 when treated with NaOH; $C_2O_4^{2-}$ gives ppt. of $CaC_2O_4 \cdot H_2O$ when treated with Ca^{2+})

37.26 Bi_2S_3 (Foul-smelling gas is H_2S. Precipitate with NaOH or NH_3 is $Bi(OH)_3$; precipitate with HCl and water is BiOCl)

37.27 CrO_4^{2-} and SCN^-; Cr^{3+}, SO_4^{2-}, HCN

37.28 a. Will give redox reaction in acid (Table 37.6)

b. Both give brown-ring test

c. Precipitates BaC_2O_4

d. Gives gas with H^+ which gives precipitate with $Ba(OH)_2$

37.29 Solid soluble in water and all acids, giving a blue solution. Water solution gives blue precipitate with 6M NaOH, 6M NH_3, and 2M Na_2CO_3. With excess 6M NH_3, deep blue solution is formed. This identifies Cu^{2+} as cation; solubility tests would rule out anions of Volatile Acid Group. Add Ba^{2+} in acid to show presence of SO_4^{2-}.

37.30 Solid is black, partially soluble in water and cold acids. Treatment with hot HNO_3 produces a green solution, which implies Ni^{2+}, and a white precipitate ($PbSO_4$). Green solution gives a green ppt. with NaOH, NH_3, or Na_2CO_3; with excess NH_3, deep blue solution is formed, confirming Ni^{2+}. Odor above H_2SO_4 solution would be that of vinegar, indicating $C_2H_3O_2^-$. Behavior of solid in hot HNO_3 would imply presence of S^{2-}.

Further tests: residue from treatment with hot HNO_3 could be shown to be $PbSO_4$. Reduction with Zn in acid would produce H_2S, confirming S^{2-}. Ca^{2+} would be most difficult to identify. Could add excess SO_4^{2-} to water solution, removing all Pb^{2+} and leaving Ca^{2+} as only cation. Presence of Ca^{2+} could then be shown by precipitating CaC_2O_4.

37.31 a. $FePO_4(s) + 3H^+(aq) \rightarrow Fe^{3+}(aq) + H_3PO_4(aq)$

b. $2BaCrO_4(s) + 2H^+(aq) \rightarrow 2Ba^{2+}(aq) + Cr_2O_7^{2-}(aq) + H_2O$

c. $PbCO_3(s) + 2H^+(aq) \rightarrow Pb^{2+}(aq) + H_2CO_3(aq)$

d. $Cu(OH)_2(s) + 2H^+(aq) \rightarrow Cu^{2+}(aq) + 2H_2O$

37.32 a. $Fe^{3+}(aq) + 3\ OH^-(aq) \rightarrow Fe(OH)_3(s)$

b. $2Ba^{2+}(aq) + Cr_2O_7^{2-}(aq) + 2\ OH^-(aq) \rightarrow 2BaCrO_4(s) + H_2O$

c. $Pb^{2+}(aq) + 3\ OH^-(aq) \rightarrow Pb(OH)_3^-(aq)$

d. $Cu^{2+}(aq) + 2\ OH^-(aq) \rightarrow Cu(OH)_2(s)$

37.33 $CuCO_3(s) + 2H^+(aq) \rightarrow Cu^{2+}(aq) + H_2CO_3(aq)$

$Cu^{2+}(aq) + 2NH_3(aq) + 2H_2O \rightarrow Cu(OH)_2(s) + 2NH_4^+(aq)$

$Cu(OH)_2(s) + 4NH_3(aq) \rightarrow Cu(NH_3)_4^{2+}(aq) + 2\ OH^-(aq)$

37.34 a. $AgSCN(s) + 2NH_3(aq) \rightarrow Ag(NH_3)_2^+(aq) + SCN^-(aq)$

b. $Ba^{2+}(aq) + C_2O_4^{2-}(aq) \rightarrow BaC_2O_4(s)$

c. $CO_3^{2-}(aq) + 2H^+(aq) \rightarrow CO_2(g) + H_2O$

37.35 $3SO_3^{2-}(aq) + 2CrO_4^{2-}(aq) + 10H^+(aq) \rightarrow 3SO_4^{2-}(aq) + 2Cr^{3+}(aq) + 5H_2O$

37.36 a. $Zn(s) + 2H^+(aq) \rightarrow Zn^{2+}(aq) + H_2(g)$

b. $3Cu(s) + 2NO_3^-(aq) + 8H^+(aq) \rightarrow 3Cu^{2+}(aq) + 2NO(g) + 4H_2O$

c. $Bi(s) + 3NO_3^-(aq) + 4Cl^-(aq) + 6H^+(aq) \rightarrow BiCl_4^-(aq) + 3NO_2(g) + 3H_2O$

37.37 a. $Fe(OH)_3(s) + 3H^+(aq) \rightarrow Fe^{3+}(aq) + 3H_2O$

b. $Mg(OH)_2(s) + 2HC_2H_3O_2(aq) \rightarrow Mg^{2+}(aq) + 2C_2H_3O_2^-(aq) + 2H_2O$

c. $Cu(OH)_2(s) + 4NH_3(aq) \rightarrow Cu(NH_3)_4^{2+}(aq) + 2\,OH^-(aq)$

37.38 a. $K_2SO_3(s) + 2H^+(aq) \rightarrow 2K^+(aq) + SO_2(g) + H_2O$

b. $CaCO_3(s) + 2H^+(aq) \rightarrow Ca^{2+}(aq) + CO_2(g) + H_2O$

c. $ZnS(s) + 2H^+(aq) \rightarrow Zn^{2+}(aq) + H_2S(g)$

37.39 a. $AgSCN(s) + 2S_2O_3^{2-}(aq) \rightarrow Ag(S_2O_3)_2^{3-}(aq) + SCN^-(aq)$

b. $PbCrO_4(s) + 3\,OH^-(aq) \rightarrow Pb(OH)_3^-(aq) + CrO_4^{2-}(aq)$

c. $Hg_2Cl_2(s) + 2NH_3(aq) \rightarrow HgNH_2Cl(s) + Hg(l) + NH_4^+(aq) + Cl^-(aq)$

d. $MnO_2(s) + 4H^+(aq) + 2Cl^-(aq) \rightarrow Mn^{2+}(aq) + Cl_2(g) + 2H_2O$

37.40 a. no   b. HCl, H_2SO_4, NaOH   c. no

37.41 a. $CuCrO_4$, $Cu_3(PO_4)_2$, CuC_2O_4, $CuCO_3$, . .

b. $HC_2H_3O_2$, HCl, H_2SO_4, NH_3 c. HNO_3

37.42 $CuCO_3$ or $CuSO_3$

37.43 A = OH^-; B = SO_4^{2-}; C = S^{2-}

37.44 a. Precipitated as Bi_2S_3 at pH = 0.5; Bi_2S_3 dissolved in HNO_3, Bi^{3+} precipitated as $Bi(OH)_3$ or BiOCl

b. Precipitated as CdS at pH = 0.5, dissolved in HNO_3 and later reprecipitated as CdS

c. Precipitated as NiS at pH 9; NiS dissolved in aqua regia; Ni^{2+} precipitated with dimethylglyoxime

d. Flame test (yellow)

37.45 a. $Pb(OH)_2$ b. $Bi(OH)_3$ c. $Cd(NH_3)_4^{2+}$ d. $Ni(NH_3)_6^{2+}$

e. NH_4^+

37.46 a. Ag^+, Pb^{2+}, Zn^{2+}, Al^{3+}, Ca^{2+}, K^+ b. Pb^{2+}, Zn^{2+}, Al^{3+}

c. Al^{3+}, Ca^{2+}, K^+ d. Ag^+, Cu^{2+}

37.47 a. Fe^{2+}, Co^{2+} b. Cd^{2+}, Ni^{2+}, Co^{2+} c. Hg^{2+}, Sn^{4+}

d. none

37.48 a. Ag^{+}, Pb^{2+}, Cu^{2+}, Hg^{2+}, Cd^{2+}, Sn^{4+} b. none

c. Zn^{2+}, Al^{3+}, Cr^{3+}, Ca^{2+}, Fe^{2+}, Ni^{2+}, Co^{2+}, Ba^{2+}

37.49 a. Precipitated as $FePO_4$ which is dissolved in acid and treated with $(NH_4)_2MoO_4$ to form precipitate.

b. Treated with $La(NO_3)_3$ and KI_3 to form deep blue color

c. Precipitated as AgCl, dissolved in NH_3, reprecipitated with HNO_3

d. Precipitated as $BaSO_4$

37.50 a. Cl^-, I^-, SO_4^{2-}, CrO_4^{2-}, NO_3^- b. SO_4^{2-}, CrO_4^{2-}

c. NO_3^-, NO_2^- d. Cl^-, I^-

37.51 a. PO_4^{3-}, $C_2O_4^{2-}$ b. PO_4^{3-}, $C_2O_4^{2-}$ c. SCN^-, $C_2O_4^{2-}$

d. PO_4^{3-}

37.52 I^-, NO_2^-, SCN^-, Br^-

37.53 Hg and either Sn or Sb; HgS only Group II sulfide insoluble in HNO_3

37.54 $NiSO_4$; green ppt. is $Ni(OH)_2$, white ppt. is $BaSO_4$

37.55 CdI_2 or possibly $CdBr_2$; white ppt. is $Cd(OH)_2$, yellow ppt. is either AgI or AgBr

37.56 PbO; white solids are $PbCl_2$, $PbSO_4$, $Pb(OH)_2$. No anions in qual scheme are present

37.57 SO_3^{2-}, I^-

37.58 a. Oxidizes I^- to I_2 b. Oxidized to SO_4^{2-} by H_2O_2

c. Could be OH^- or O^{2-} d. SO_3^{2-} oxidized by HNO_3

37.59 Will dissolve in water and all acids to give yellow solution. Addition of OH^-, NH_3, or CO_3^{2-} gives red ppt. insoluble in excess reagent. Might be able to identify Fe^{3+} on this basis; could confirm with SCN^-. To identify Cl^-, add Ag^+ to give AgCl (Chapter 33).

37.60 Soluble in water and all acids to give blue solution. Addition of OH^- gives blue ppt. Addition of excess NH_3 gives deep blue solution. Addition of CO_3^{2-} gives blue ppt.

Color changes would identify Cu^{2+}. To identify Zn^{2+}, work with water solution after addition of NaOH. Adjust pH to 9, saturate with H_2S, forming white ppt. of ZnS. Identify SO_4^{2-} by adding Ba^{2+} to water solution, NO_3^- by brown-ring test.

*37.61 Ca^{2+}, K^+, CrO_4^{2-}, $C_2O_4^{2-}$

*37.62 Contains Pb, Bi, Cd, Sn. Gives chloride precipitate with HCl, soluble in hot water; addition of K_2CrO_4 to the hot solution gives a yellow precipitate. In Group II, a dark precipitate of Bi_2S_3, CdS, and SnS_2 forms. This precipitate is partially soluble in NaOH to give $Sn(OH)_6^{2-}$. Neutralization with HCl gives a yellow precipitate of SnS_2. The residue from the NaOH treatment dissolves in HNO_3 to give Bi^{3+} and Cd^{2+}. Addition of NH_3 to the HNO_3 solution gives a white precipitate of $Bi(OH)_3$ and a colorless solution. Addition of H_2S to this solution gives a yellow precipitate of CdS.

*37.63 a. $K = \dfrac{K_{sp}Zn(OH)_2}{(K_w)^2} = 5 \times 10^{11}$

b. $K = \dfrac{K_{sp}Zn(OH)_2 \times (K_aHC_2H_3O_2)^2}{(K_w)^2} = 2 \times 10^2$

K will always be smaller for reaction with $HC_2H_3O_2$

*37.64 $5[2I^-(aq) \longrightarrow I_2(s) + 2e^-]$

$2ClO_3^-(aq) + 12H^+(aq) + 10\ e^- \longrightarrow Cl_2(g) + 6H_2O$

$2ClO_3^-(aq) + 10\ I^-(aq) + 12H^+(aq) \longrightarrow 5I_2(s) + Cl_2(g) + 6H_2O$

ClO_3^- is in excess; I^- is limiting. I^- consumed, some ClO_3^- remains

*37.65 $CaC_2O_4(s) + 2H^+(aq) \longrightarrow Ca^{2+}(aq) + H_2C_2O_4(aq)$

$CO_3^{2-}(aq) + 2H^+(aq) \longrightarrow CO_2(g) + H_2O$

$Ca^{2+}(aq) + CO_3^{2-}(aq) \longrightarrow CaCO_3(s)$

$CaCO_3(s) + 2H^+(aq) \longrightarrow Ca^{2+}(aq) + CO_2(g) + H_2O$

$CO_3^{2-}(aq) + 2H^+(aq) \longrightarrow CO_2(g) + H_2O$

Test Questions, Multiple Choice

1. Which one of the following cations forms a stable complex with both NH_3 and OH^-?

   a. Ag^+ b. Hg^{2+} c. Cu^{2+} d. Zn^{2+} e. Ca^{2+}

2. Which one of the following cations forms a water-soluble hydroxide but an insoluble sulfate?

   a. Pb^{2+} b. Bi^{3+} c. Ni^{2+} d. Ba^{2+} e. Na^+

3. Which one of the following cations forms a water-insoluble chloride?

   a. Hg_2^{2+} b. Hg^{2+} c. Al^{3+} d. Fe^{3+} e. NH_4^+

4. The compound $Fe(OH)_3$ is marked A^- in the Solubility Table. In how many of the following solvents does $Fe(OH)_3$ dissolve?

   H_2O, $HC_2H_3O_2$, HCl, HNO_3, NH_3

   a. 1 b. 2 c. 3 d. 4 e. 5

5. How many of the following cations are neutral (pH ~ 7) in water?

   Pb^{2+}, Al^{3+}, Zn^{2+}, Ba^{2+}, NH_4^+

   a. 1 b. 2 c. 3 d. 4 e. 5

6. How many of the following compounds are water-soluble?

   Na_2S, $MgSO_4$, $(NH_4)_2CO_3$, $ZnCl_2$, $Ni(NO_3)_2$

   a. 1 b. 2 c. 3 d. 4 e. 5

7. Which one of the following anions is colored?

   a. SCN^- b. I^- c. $C_2H_3O_2^-$ d. CrO_4^{2-} e. NO_2^-

8. Which one of the following anions gives a precipitate with $AgNO_3$ in acidic solution?

   a. SCN^- b. $C_2H_3O_2^-$ c. CrO_4^{2-} d. $C_2O_4^{2-}$ e. NO_3^-

9. Which one of the following anions, in the qual scheme, is precipitated as the calcium salt?

   a. Cl^- b. SO_4^{2-} c. CrO_4^{2-} d. $C_2O_4^{2-}$ e. S^{2-}

10. Which one of the following solvents brings $PbSO_4$ into solution?

   a. H_2O b. HCl c. HNO_3 d. NH_3 e. $NaOH$

Test Problems

1. Write a balanced net ionic equation to show why

   a. $ZnCl_2$ dissolves in water

   b. $NiCO_3$ dissolves in ammonia

   c. $Al(OH)_3$ dissolves in sodium hydroxide

   d. $Fe(OH)_3$ dissolves in hydrochloric acid

2. Describe, as best you can, what would be observed when stainless steel, an alloy containing Fe, Cr and Ni, is subjected to the qualitative analysis scheme for cations.

3. A pure white ionic compound is is insoluble in water, but soluble in HCl, NaOH, or NH_3 to give a colorless solution. With HCl, an odorless gas is given off. Identify the compound as best you can; explain your reasoning.

4. For each of the following ions, list two other ions which fall in the same group in qualitative analysis.

   a. Ag^+, ____, ____ d. SO_3^{2-}, ____, ____

   b. Ni^{2+}, ____, ____ e. SCN^-, ____, ____

   c. Na^+, ____, ____ f. SO_4^{2-}, ____, ____

5. What is the pH of a buffer prepared by adding 25 mL of 0.20 M NH_4Cl to 75 mL of 0.10 M NH_3? (K_a NH_4^+ = 5.6×10^{-10}).

ANSWERS TO TEST QUESTIONS

Chapter 29

1. b 2. b 3. d 4. c 5. d 6. d 7. c 8. d 9. b 10. e

1. $3Pb(s) + 2NO_3^-(aq) + 6Cl^-(aq) + 8H^+(aq) \rightarrow 3PbCl_2(s) + 2NO(g) + 4H_2O$

2. a. $AgCl(s) + 2NH_3(aq) \rightarrow Ag(NH_3)_2^+(aq) + Cl^-(aq)$

   b. $PbSO_4(s) + 3\ OH^-(aq) \rightarrow Pb(OH)_3^-(aq) + SO_4^{2-}(aq)$

   c. $Pb^{2+}(aq) + CrO_4^{2-}(aq) \rightarrow PbCrO_4(s)$

   d. $Ag(NH_3)_2^+(aq) + Cl^-(aq) + 2H^+(aq) \rightarrow AgCl(s) + 2NH_4^+(aq)$

3. a white ppt. b. no ppt. c. grey ppt. d. no ppt.

4. 0.003 M

5. yes

Chapter 30

1. d 2. d 3. c 4. c 5. c 6. d 7. d 8. a 9. d 10. b

1. Sb^{3+}, Cd^{2+} present; Cu^{2+}, Bi^{3+}, Hg^{2+}, Sn^{4+} absent

2. a. $Cd(NH_3)_4^{2+}(aq) + 4H^+(aq) \rightarrow Cd^{2+}(aq) + 4NH_4^+(aq)$

   b. $Bi^{3+}(aq) + 3NH_3(aq) + 3H_2O \rightarrow Bi(OH)_3(s) + 3NH_4^+(aq)$

   c. $Sn(OH)_6^{2-}(aq) + 6H^+(aq) \rightarrow Sn^{4+}(aq) + 6H_2O$

   d. $Cu^{2+}(aq) + 4NH_3(aq) \rightarrow Cu(NH_3)_4^{2+}(aq)$

3. $3HgS(s) + 2NO_3^-(aq) + 12Cl^-(aq) + 8H^+(aq) \rightarrow 3HgCl_4^{2-}(aq) + 2NO(g) + 4H_2O$

4. a. 1×10^{-6} b. 0.006 M

5. a. 1×10^{-7} b. 0.02 M

Chapter 31

1. d 2. b 3. c 4. d 5. a 6. e 7. a 8. d 9. d 10. c

1. a. $2Fe^{3+}(aq) + 3S^{2-}(aq) \rightarrow 2FeS(s) + S(s)$

   b. $Fe^{3+}(aq) + 3NH_3(aq) + 3H_2O \rightarrow Fe(OH)_3(s) + 3NH_4^+(aq)$

1. c. $Al^{3+}(aq) + 4\ OH^-(aq) \longrightarrow Al(OH)_4^-(aq)$

d. $Cr^{3+}(aq) + 3S^{2-}(aq) + 3H_2O \longrightarrow Cr(OH)_3(s) + 3HS^-(aq)$

2. 0.63 V

3. a. 5 b. 1 M

4. a. Adjust pH to 9, precipitate $Fe(OH)_3$ and $Al(OH)_3$, form $Ni(NH_3)_6^{2+}$ and $Zn(NH_3)_4^{2+}$ complexes

b. Test for Co^{2+} and Fe^{3+} by forming colored complexes

c. Oxidize Mn^{2+} to MnO_2, Cr^{3+} to CrO_4^{2-}

d. Dissolve $Fe(OH)_3$, $Ni(OH)_2$

5. All ions are absent except Co^{2+} and/or Ni^{2+}. Dissolve sulfides in aqua regia, test portions of solution with SCN^- and DMG.

Chapter 32

1. c 2. c 3. d 4. b 5. e 6. d 7. b 8. c 9. d 10. c

1. a. $Mg^{2+}(aq) + NH_3(aq) + HPO_4^{2-}(aq) \longrightarrow MgNH_4PO_4(s)$

b. $CaC_2O_4(s) + 2H^+(aq) \longrightarrow Ca^{2+}(aq) + H_2C_2O_4(aq)$

c. $NH_4^+(aq) + OH^-(aq) \longrightarrow NH_3(aq) + H_2O$

d. $Ba^{2+}(aq) + CrO_4^{2-}(aq) \longrightarrow BaCrO_4(s)$

2. a. Mg^{2+} b. Na^+, K^+ c. Ba^{2+} d. Na^+, NH_4^+

3. Na^+ present, Ba^{2+} absent, others in doubt

4. 45 mL

5. a. 6×10^{-4} b. 0.003 M

Chapter 33

1. d 2. d 3. b 4. a 5. c 6. c 7. c 8. e 9. c 10. d

1. a. $Ag(NH_3)_2^+(aq) + 2H^+(aq) \longrightarrow Ag^+(aq) + 2NH_4^+(aq)$

b. $Zn(s) + 2AgSCN(s) \longrightarrow Zn^{2+}(aq) + 2Ag(s) + 2SCN^-(aq)$

c. $Cl_2(g) + 2I^-(aq) \longrightarrow 2Cl^-(aq) + I_2(s)$

d. $Ag^+(aq) + Cl^-(aq) \longrightarrow AgCl(s)$

2. a. Brings silver salts into solution

b. Oxidizes I^-

c. Oxidizes Br^- to Br_2

3. SCN^- present (color); Br^-, I^- absent (would be oxidized by Cl_2) Cl^- in doubt

4. a. $2MnO_4^-(aq) + 16H^+(aq) + 10\ Br^-(aq) \longrightarrow 5Br_2(l) + 2Mn^{2+}(aq) + 8H_2O$

b. 0.20 mL

5. 0.11 M

Chapter 34

1. a 2. c 3. b 4. e 5. a 6. c 7. c 8. d 9. c 10. b

1. a. $Ba^{2+}(aq) + SO_4^{2-}(aq) \longrightarrow BaSO_4(s)$

b. $2BaCrO_4(s) + 2H^+(aq) \longrightarrow 2Ba^{2+}(aq) + Cr_2O_7^{2-}(aq) + H_2O$

c. $Fe^{3+}(aq) + PO_4^{3-}(aq) \longrightarrow FePO_4(s)$

d. $PO_4^{3-}(aq) + H_2O \rightleftharpoons HPO_4^{2-}(aq) + OH^-(aq)$

2. 12.8

3. CrO_4^{2-} is absent (color); $C_2O_4^{2-}$, PO_4^{3-} absent (basic)

4. $HC_2O_4^-$

5. 8×10^2

Chapter 35

1. c 2. a 3. d 4. d 5. c 6. c 7. c 8. d 9. d 10. a

1. a. $NO_2^-(aq) + H_2O \rightleftharpoons HNO_2(aq) + OH^-(aq)$

b. $C_2H_3O_2^-(aq) + H^+(aq) \longrightarrow HC_2H_3O_2(aq)$

c. $Fe^{2+}(aq) + NO(g) \longrightarrow Fe(NO)^{2+}(aq)$

d. $Cu(s) + 2NO_3^-(aq) + 4H^+(aq) \longrightarrow Cu^{2+}(aq) + 2NO_2(g) + 2H_2O$

2. SO_4^{2-} present (white ppt. is $BaSO_4$); Cl^- absent (would precipitate AgCl); other ions in doubt

3. a. Add Ag^+; Cl^- gives precipitate of AgCl

b. Test pH; NO_2^- is basic

c. Add Ba^{2+}; SO_4^{2-} gives precipitate of $BaSO_4$

4. 3.65

5. a. 0.040 b. 0.017 M

Chapter 36

1. a 2. d 3. c 4. c 5. d 6. d 7. e 8. c 9. e 10. c

1. a. $2H^+(aq) + S^{2-}(aq) \longrightarrow H_2S(g)$

b. $HCO_3^-(aq) + H_2O \rightleftharpoons H_2CO_3(aq) + OH^-(aq)$

c. $Ba^{2+}(aq) + 2\ OH^-(aq) + CO_2(g) \longrightarrow BaCO_3(s) + H_2O$

2. a. Add H^+; test for H_2S by odor and with Pb^{2+}

b. Add H^+; test for CO_2 with $Ba(OH)_2$

3. -39 kJ

4. 0.14, 6.2×10^4

5. a. 1.8×10^{-7} b. 10.13

Chapter 37

1. d 2. d 3. a 4. c 5. a 6. e 7. d 8. a 9. d 10. e

1. a. $ZnCl_2(s) \longrightarrow Zn^{2+}(aq) + 2Cl^-(aq)$

b. $NiCO_3(s) + 6NH_3(aq) \longrightarrow Ni(NH_3)_6^{2+}(aq) + CO_3^{2-}(aq)$

c. $Al(OH)_3(s) + OH^-(aq) \longrightarrow Al(OH)_4^-(aq)$

d. $Fe(OH)_3(s) + 3H^+(aq) \longrightarrow Fe^{3+}(aq) + 3H_2O$

2. No precipitate in Groups 1 or 2; black ppt. in Group 3. Precipitate is partially soluble in HCl; residue dissolves in aqua regia, gives test for Ni^{2+} with DMG. HCl solution when treated with OCl^- in base gives a red ppt. which can be identified as $Fe(OH)_3$ and a yellow solution (CrO_4^{2-}). No ion present in Group 4.

3. $ZnCO_3$

4. a. Pb^{2+}, Hg_2^{2+} b. Co^{2+}, Fe^{3+}, Al^{3+}, Cr^{3+}, Zn^{2+}, Mn^{2+}

4. c. K^+, Ba^{2+}, Ca^{2+}, Mg^{2+}, NH_4^+ d. CO_3^{2-}, S^{2-}

e. Cl^-, Br^-, I^- f. $C_2O_4^{2-}$, CrO_4^{2-}, PO_4^{3-}

5. 9.43

NOTES

NOTES

NOTES

NOTES

NOTES